South East Asian Development:
Geographical Perspectives

Edited by Denis Dwyer

Longman
Scientific &
Technical

Copublished in the United States with
John Wiley & Sons, Inc., New York

Longman Scientific & Technical,
Longman Group UK Ltd,
Longman House, Burnt Mill, Harlow,
Essex CM20 2JE, England
and Associated Companies throughout the world.

Copublished in the United States with
John Wiley & Sons, Inc., 605 Third Avenue, New York, NY 10518

HC
441
S67
1990

First published 1990

British Library Cataloguing in Publication Data
South East Asian development: geographical perspectives.
  1. Latin America. Human geographical features to 1986
  I. Dwyer, Denis
  304.20959

ISBN 0-582-30149-1

Library of Congress Cataloging-in-Publication Data
South East Asian development: geographical perspectives/edited
  by Denis Dwyer.
    p. cm.
  Includes bibliographical references.
  ISBN 0-470-21665-4 (Wiley)
  1. Asia, Southeastern – Economic conditions. 2. Asia.
Southeastern – History.   I. Dwyer, D. J. (Denis John)
HC441.S67 1990
338.959 – dc20

89-13809
CIP

Set in 10/11 pt Palatino, Linotron 202

Produced by Longman Group (FE) Limited
Printed in Hong Kong

In memory of
Professor William Kirk
eminent geographer
1921–87

Contents

List of Figures vii
List of Plates viii
List of Contributors viii
Introduction x
Acknowledgements xi

I Geographical perspectives on development 1

CHAPTER 1 **South East Asia in the world today** 1
Denis Dwyer

CHAPTER 2 **South East Asia in the colonial period: cores** 15
and peripheries in development processes
William Kirk

CHAPTER 3 **Concepts of development** 48
David Drakakis-Smith

II Resources for development 78

CHAPTER 4 **Environmental resources** 78
Chris Barrow

CHAPTER 5 **Human resources** 110
Chris Dixon

CHAPTER 6 **Energy resources** 140
John Soussan

III Economic activities and development 168

CHAPTER 7 **Agriculture and fisheries** 168
Donald Fryer

CHAPTER 8 **Mining and manufacturing** 193
Peter Dicken

CHAPTER 9   **Trade, aid and regional integration**   225
              George Cho and Stephen Wyn Williams

IV Adjustments to development   ·   256

CHAPTER 10   **Population mobility**   ·   256
              Paul Lightfoot

CHAPTER 11   **Urbanization**   278
              Denis Dwyer

Index   309

List of Figures

Fig. 1.1	Population and gross national product per capita	5
Fig. 1.2	Strategic straits and US military bases	7
Fig. 2.1	Core areas before 1500AD	19
Fig. 2.2	Cores of the European colonial period	25
Fig. 2.3	Political boundaries and capital cities	45
Fig. 3.1	Relationships between development theory and planning	51
Fig. 3.2	A framework for the study of development theories	53
Fig. 3.3	Rostow's stages of growth model	55
Fig. 3.4	Modernization surfaces in Malaysia	57
Fig. 3.5	Diagrammatic representation of cores and peripheries in the world-system	64
Fig. 3.6	Mode of production components	69
Fig. 3.7	Relationships within peripheral capitalism	70
Fig. 4.1	Generalized landforms	79
Fig. 4.2	Generalized structures	80
Fig. 4.3	Principal mineral deposits	82
Fig. 5.1	Age–sex pyramids	113
Fig. 5.2	Generalized population density	114
Fig. 5.3	Languages of indigenous peoples	119
Fig. 5.4	Religions of indigenous peoples	121
Fig. 6.1	The biomass fuel situation in South East Asia	152
Fig. 6.2	Indonesia: fuel resources	159
Fig. 7.1	The Pahang Tenggara and Johore Tenggara integrated development areas	184
Fig. 8.1	Level and composition of manufacturing value added in South East Asia 1982	200
Fig. 8.2	Distribution of mining, oil and petroleum production in South East Asia 1982	203
Fig. 8.3	Estimated employment in export processing zones in South East Asia 1980	215
Fig. 11.1	World urban population, 1950–2000	278
Fig. 11.2	Manila in 1671, after the map of Ignacio Muñoz	284
Fig. 11.3	Proposed racial groupings in Singapore 1828	287
Fig. 11.4	The municipal boundaries of Kuala Lumpur 1895–1974	288
Fig. 11.5	Population and employment in Malaysia 1980 and 1990	296
Fig. 11.6	Location of squatter settlements and resettlement sites in Kuala Lumpur and the Federal Territory 1976	303

List of Plates

Plate I   Malay wedding, Kuala Lumpur
Plate II   Indian festival of Thaipusam, Penang
Plate III   Chinatown, Penang
Plate IV   A typical rice landscape of lowland South East Asia
Plate V   Experimental high-yielding rice, Thailand
Plate VI   Harvesting oil palm, Malaysia
Plate VII   Oil palm processing mill, Malaysia
Plate VIII   A Thai village house
Plate IX   Rain-making ceremony, North East Thailand
Plate X   Village industry: weaving, Thailand
Plate XI   Village industry: charcoal making, Thailand
Plate XII   Malay-style coastal fish trap, Singapore
Plate XIII   Commercial sea-fishing, southern Thailand
Plate XIV   Jakarta: the central core
Plate XV   Jakarta: suburban sprawl
Plate XVI   Jakarta: inner-city infilling: squatter huts along a storm drain
Plate XVII   Food hawker, Bangkok
Plate XVIII   Suburban shops along the *klongs* of Bangkok
Plate XIX   Small-scale industry, Bangkok: making lamps from old car-springs
Plate XX   Small sawmill , Kuala Lumpur
Plate XXI   The industrial estate and the multinational company, Petaling Jaya, Malaysia

List of Contributors

Dr Chris Barrow, Lecturer, Centre for Development Studies, University College Swansea

Dr George Cho, Senior Lecturer, Canberra College of Advanced Education

Professor Peter Dicken, Professor of Geography, University of Manchester

Professor Chris Dixon, Professor of Geography, City of London Polytechnic

Professor David Drakakis-Smith, Professor of Development Geography, University of Keele

Professor Denis Dwyer, Research Professor of Geography, University of Keele

Professor Donald Fryer, Professor of Geography, University of Hawaii

Professor William Kirk,[*] Professor of Geography, Queen's University, Belfast

Dr Paul Lightfoot, Bank for Agriculture and Agricultural Co-Operatives, Bangkok

Dr John Soussan, Lecturer in Geography, University of Reading
Dr Stephen Wyn Williams, Lecturer, Department of Geography and
   Recreation Studies, Staffordshire Polytechnic

* Deceased.

Introduction

This book analyses the major aspects of development problems and policies in contemporary South East Asia that are the subject of geographical research. It does not, of course, present a complete analysis of development problems and policies in the region; rather, it demonstrates the significant contribution geographers have made to the study of the problems of the developing countries in recent decades. In Britain, an active group of such geographers works within the Institute of British Geographers as the Developing Areas Research Group (DARG). The Group welcomes additional members, geographers and non-geographers. It is under the auspices of DARG that this volume appears, as one of a series of such volumes written by members of the Group and close associates. The first, *Latin American Development: Geographical Perspectives*, edited by David Preston, was published by Longman in 1987. As with that book, the chapters which follow reflect not only ideas but also to a certain extent differences in interpretation that have arisen in the regular series of conferences and seminars held by the Group. As editor of this volume, it is a pleasure to acknowledge the willing cooperation of my fellow authors, even at times when repeated editorial requests must have been very trying.

Denis Dwyer
January 1990

Acknowledgements

We are grateful to the following for permission to reproduce copyright figures and tables:

Association of American Geographers for fig. 11.4 from pp. 546–63 of "Malaysia's Emerging Conurbation" by Aiken & Leigh in *Annals of the Association of American Geographers* Vol. 65:4, 1975 (Hill, 1986); Australian National University (Depart. of Human Geography) for fig. 3.6 (Peet, 1980); Food & Agriculture Organization of the United Nations for tables 5.3, 5.4 (F.A.O., 1986), 6.4 (F.A.O., 1981b) & 7.4 (F.A.O., 1986b); International Development Research Centre for table 11.4 (McGee & Yeung, 1977) © 1977 International Development Research Centre; International Labour Organisation for table 3.4 (Sethuraman, 1976); Journal of Contemporary Asia for table 3.3 (Clairmonte & Cavanagh, 1983); the author, Prof. T. R. Leinbach for fig. 3.4 (Leinbach, 1972); Longman Group UK Ltd. for fig. 7.1 (Ooi, 1976); Methuen & Co. for table 11.3 (Santos, 1979); Oxford University Press (Malaysia) for figs. 4.1, 4.2 & table 4.1 (Swan, 1979), figs. 4.3 (Courtenay, 1979) & 5.2 (Neville, 1979); Singapore Journal of Tropical Geography (National University of Singapore) for fig. 11.3 (Hodder, 1953); The Unesco Press for table 4.3 (Tran Van Nao, 1974) © 1974 Unesco; United Nations for table 6.2 (United Nations, 1982a); University of California Press Journals and the respective authors for an extract in chapter 1 (Khong, 1987) & fig. 1.2 (Simon, 1985) © 1985 & 1987 by The Regents of the University of California; The World Bank for fig. 1.1 & tables 1.1, 3.1, 3.2, 5.9 (World Bank, 1987), fig. 8.1 (World Bank, 1985) and tables 9.6 & 9.7 (World Bank, 1987b).

Whilst every effort has been made to trace the owners of copyright material, in a few cases this has proved impossible and we take this opportunity to offer our apologies to any copyright holders whose rights we may have unwittingly infringed.

CHAPTER 1

South East Asia in the world today

Denis Dwyer

It is something of a paradox that although the very concept of South East Asia as a geographical region is relatively recent, in terms of international relationships its significance in the world today is profound. As Fryer (1970, p. 1) has pointed out, 'Before World War II South East Asia was scarcely even a geographical expression. For the West it was little more than an undifferentiated part of Monsoon Asia, the teeming eastern and southern margins of the great Asian continent; for Asians themselves, it had no significance at all.' The powers of the day commanded the disposition of the resources of South East Asia completely, even though one state, Thailand (then Siam), remained independent of colonial political control; but except for the Dutch in relation to their colony of what is now Indonesia, in the eyes of the colonial powers South East Asian possessions were of relatively minor significance. It was the Japanese invasion of South East Asia that changed a colonial perception perhaps best symbolized by the fact that when the Japanese invaded it was discovered that the heavy guns of Singapore were pointing out to sea, emphasizing the accepted view of South East Asia as a mere crossing place between the greater realms of India and China, and the fact that the countries of the region themselves were held to be of relatively little account. The Japanese invaded overland revealing *en route* the fragility of the existing regimes and initiating a series of political changes that were to culminate with the wholesale withdrawal of the colonial powers from political control in the region at the conclusion of hostilities.

It was during the Second World War that the South East Asia Command was established by the Western powers and the region began to be recognized internationally as an entity. Comprising only 3 per cent of the land area of the globe, but with about 8 per cent of its population, from some points of view South East Asia can be viewed as something of a residual area, with significant negative characteristics in terms of its regional character: that is, as the collection of peninsulas and archipelagos which lies between the major continental and sub-continental land masses of Australia, India and China. Nevertheless, its individuality compared with those areas is well marked. In the first place the region straddles the Equator and is almost wholly within the humid tropics. Its agricultural potentialities are thus distinctive, and during and since the colonial period it has become a major world supplier of a range of tropical crops: rubber, copra, abaca (Manila hemp), and more recently palm oil, especially. It is a significant fact in terms of development possibilities that because of the accessibility provided by deep interpenetration by arms and gulfs of the sea within the region, and because of its exceptionally elaborate network of rivers (many with great deltas), and also because of volcanic enrichment of some of its soils, its potential agricultural productivity is far above the average for the humid tropical zone. Moreover, apart from the island of Java, the cultivated areas of North Vietnam in general and some highly localized but relatively restricted rural areas elsewhere, South East Asia does not suffer from the problems of excessive population pressure so prevalent in India and China.

Secondly, South East Asia is distinctive within the greater Asian-Australian area in being a major global crossing place. In both pre-historic and historical times, its predominantly north-south topographic alignments have influenced successive flows of peoples into the region from the north. The ethnic diversity resulting from these migrations has been augmented in more recent historical periods by further population movements associated with the region's position as a great maritime crossroads. Seaborne invasions, of peoples and of cultures, have been characteristic of South East Asia and have added to its remarkable diversity of land and sea, of island and of mainland, and of rugged forest-covered mountain and intensely cultivated valley floors and deltas. Cultural influences from India and China, and more recently the West via the colonial powers, have added to the mix. Religious diversity is also striking.

It was against this background of geographic difference that Charles Fisher introduced the concept of South East Asia as the Balkans of the Orient in an important paper more than twenty-five years ago (Fisher 1962). Fisher saw South East Asia as a great marchland like the Balkans, an area of transition and instability, a cultural and political fault zone. He referred to the similarity of the geopolitical role of the Balkans, forming a bridge from Asia Minor into the heart of Europe, with that of South East Asia forming the strategic route between Asia and Australasia. He also drew a striking parallel between Constantinople and Singapore, both significant in international trade and both, during certain periods of their history, symbolizing ambitions to hold the key to vital international crossroads, the Straits of Marmora and Malacca respectively. Most of all, however, he drew attention to the political tremors that have stressed both regions as alien overlords have withdrawn and new national states have come into being in their place. After centuries of subjugation, Fisher pointed out, the Balkans successfully united to expel the Turk shortly before the First World War, only quickly to fall victim to internecine conflict within the region. It was an open question, he wrote, whether the ending of European rule over South East Asia would bring lasting stability or merely prepare the way for a renewed Balkanization there.

Fisher noted that virtually all the states in the region contained a much greater diversity of population than would usually be associated with the concept of a nation state in the West and that, in particular, in addition to the various groups in each state that were officially considered indigenous, there was a very large group of Chinese, considered as immigrant and to be found in every state, with particularly large concentrations in Singapore and Malaysia. In no case did a single language or one particular religion provide a common denominator for a whole national population; indeed, wrote Fisher, what appeared to be holding the states in the region together was a combination of three forces. Firstly, a kind of political-geographical inertia, associated with patterns of administration, transport and communication – and the political boundaries – taken over from the former colonial regimes. Secondly, there was the sense of recent history, associated particularly with the various struggles for national independence; and thirdly the desire of small educated elites to control their respective states and demonstrate capabilities of national government. These, as he saw it, were possibly fragile assets to set against considerable external pressures – in particular the Communist thrust from the north through Vietnam – and fractioning forces within most states caused by regional ethnic differences among the indigenous peoples and the unsolved problem of the Chinese, who were almost universally looked upon by national governments as aliens if not actual subversives.

South East Asia in the 1980s

As Fisher rightly pointed out, in determining the political future of South East Asia the progress of economic development would be critical. Shortly after Fisher wrote, Donald Fryer was in process of distinguishing two South East Asias in terms of development. He suggested a division into two major economic groupings: Malaysia, Singapore, Thailand and the Philippines on the one hand, and Indonesia, Burma and the Indochinese states on the other. This division was also basic in the political field, he observed. Fryer saw the first group of countries as outward-looking in terms of their development goals. 'They recognize that they are part of a wider world to which they must make adjustments; they have been active in promoting schemes for regional cooperation and integration; and on all major world issues involving a confrontation of the Western and the Communist powers, they have tended to support the West' (Fryer 1970, p. 20). Fryer's second grouping consisted of the inward-looking countries in which economic growth had not always been accorded a high degree of priority, and which tended to follow doctrinaire rather than pragmatic policies and sometimes to concern themselves more with the preservation of political or cultural values than with accommodation with the modern world. Generally, these states adopted anti-Western positions but at least one, Indonesia, following the fall of the Sukarno regime in 1965, had come to realize that problems of stagnating production had to be faced and was beginning to seek increased international economic integration. There was little in their respective physical backgrounds, Fryer observed, that helped to account for the superior economic performance of the outward-looking group of countries. Thailand resembled more its neighbours in mainland South East Asia than the Malaysia-Singapore-Philippines group. Indonesia was more like Malaysia than the impoverished states of Indochina. 'It is difficult to resist the conclusion,' Fryer wrote (p. 20), 'that it is largely to the preservation of a mainly free and open economy that Malaysia, Thailand and the Philippines owe their higher degree of well-being.'

During the intervening period, at least until the onset of world economic downturn and the collapse of the oil price of the mid-1980s, the countries of South East Asia that Fryer identified in his outward-looking grouping enjoyed relatively high rates of economic growth, though with one important exception. This was the Philippines, where the imposition of a dictatorship by the Marcos regime in 1972 produced accelerating economic decline and general internal chaos. By 1984, the total GNP was actually declining. Since the overthrow of Marcos in February 1986 and the inauguration of the government of President Aquino, the Philippines has been desperately searching both for internal stability and for economic recovery. On the other hand, since Fryer wrote, Indonesia has, broadly, taken an opposite trajectory. The years of economic disorganization under Sukarno have been succeeded by a centralized, autocratic 'New Order' political system under President (former General) Suharto and a series of pragmatic economic measures introduced that have transformed an economy with negative growth rates in the late 1960s and with inflation running at 600 per cent a year into one with a record of growth which has equalled that of Thailand and exceeded that of the Philippines. Nevertheless, as Figure 1.1 shows, Indonesia's economic problems remain grave in part because of collapsing oil prices but more fundamentally because of a population which is huge for the region and implies only very difficult prospects in terms of substantially raising GNP per capita.

The other countries within Fryer's second grouping remain beset with serious economic problems. Although the available data is not very reliable, Burma, with an estimated per capita GNP of only US $190, appears to be among the twenty most impoverished nations in the world. So most certainly are Kampuchea and Laos – one (1985) estimate for the per capita GNP of Kampuchea being as little

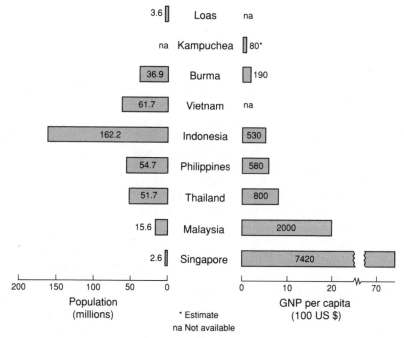

Fig. 1.1 Population and gross national product per capita. (Source: World Bank, *World Development Report 1987*)

as US $80 – while the per capita GNP of Vietnam is certainly less than that of China.

After 25 years of highly individual variety of socialism/ communism following the military coup of 1962, under General Ne Win, Burma almost reached the point of a general agrarian rebellion by its peasant rice growers to add to those which had already been in existence for several decades involving its important ethnic minorities, the Kachin, Shan and Karens, as well as the Burmese Communist Party. When Ne Win travels, according to one report (*The Independent*, London, 17 June 1987), 'Officials shield him from the real world, like the king in Mandalay a hundred years before him. The road leading to his bungalow in a Rangoon suburb is swept daily, and three service stations nearby always have petrol. When he travels through the capital, a route is chosen where decay is not so obvious. When he travels to his home village near Prome, some 180 miles from Rangoon, roadside stall holders are primed in advance in case he stops to ask them the cost of vegetables. They are told what price to give. In every case it is a figure which bears no relation to reality . . . . Even Vietnam is beginning to look prosperous by comparison. It seems that only the General does not know his country is coming to the end of the

5

road.' In 1988, after a brief period of instability following what was announced as the retirement of Ne Win, the army took control. It appears that the combination of internal repression and international isolation which has characterized Burma since 1962 is to be continued.

Vietnam is in an equally desperate situation, with a siege economy constantly near the point of collapse. In 1987, its foreign exchange reserves fell to US $15 million, barely enough to pay for two weeks of imports, and if life in Hanoi has officially been described as 'miserable' (Esterline 1988, p. 89) then conditions in the countryside, and especially in the rice-deficient north, are certainly very much worse. In nearby Kampuchea, since the invasion by Vietnam very much a client state of that country as well as of the Soviet Union, exports currently cover only 20 per cent of imports, international aid, very largely from Russia and its satellites, is both relatively small and not used effectively. An estimated 216 out of every thousand children die before the age of five in Kampuchea, an infantile mortality rate that is one of the highest in the world.

The geopolitical situation

Today, as in the earlier phases of the Second World War, South East Asia is a region of prime significance in terms of the search for positions of geopolitical leverage by the Great Powers. Much of the region's importance derives from the geographical juxtaposition of two ideologically opposed groupings which now embrace all of the countries of the region except for isolated Burma: the states of the Association of South East Asian Nations (ASEAN) – Singapore, Malaysia, Brunei, Indonesia, Thailand and the Philippines (all of the higher-growth countries of the region, it is worth noting); and the Communist countries of Indochina, dominated by Vietnam and supported by Russia. In the calculations of the Great Powers, the major international sea lanes which pass through the region, and indeed the region's position as a whole *vis à vis* the strategically vital Indian Ocean and Western Pacific, also figure prominently.

For the United States, it appears that in recent years 'An emphasis on close relations with market-oriented economies in Asia congenial to American trade and investment and the development of military capabilities to protect the sea-lanes of international commerce appear to be the primary components of Washington's East Asia policy' (Simon 1985, p. 920). One important lesson of the Vietnam War for the United States has been the necessity for reliance upon air and naval power rather than on ground forces. Thus American naval and air facilities at Subic Bay and Clark Field have gained enhanced significance and were undoubtedly a major factor in determining the United States' attitudes to the Marcos regime at a time when that government had become completely discredited internationally. ASEAN as a group is the United States' fifth largest trading partner and support for its further economic development is clearly regarded as of importance by the United States not only for this reason but also as a successful example of market-oriented economic philosophy. As Simon (1985) points out, the ASEAN states are also an integral component of a United States strategy that now rests heavily upon rapid deployment of naval forces from the Indian Ocean to the western Pacific and vice versa. In this context, the relatively narrow straits which separate the islands of the region are vital in terms of both monitoring and, possibly, interdicting naval movements. Only a few of the channels through the Indonesian archipelago are wide enough and deep enough to permit the safe passage of submerged submarines. These are shown on Fig. 1.2 and it is to be assumed that US ballistic missile nuclear submarines from Guam transit them on their way to the Indian Ocean. For the Soviet Union, these straits could be important choke points where the Soviets would have some chance of locating US submarines through acoustic devices.

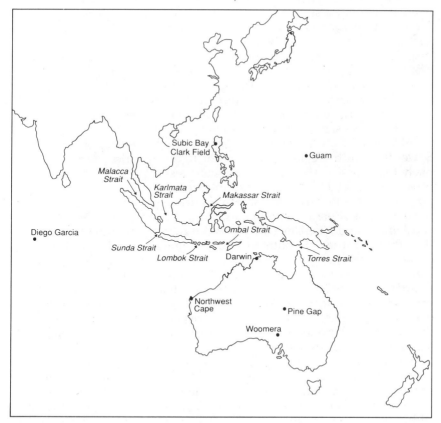

Fig. 1.2 Strategic straits and US Military Bases. (Source: Simon, 1985)

Of more local significance to the United States is the geopolitical problem of Vietnam, and significant military assistance has been provided to Thailand to resist Vietnamese encroachments along the Thai frontier. Until their withdrawal in 1989, a medium-term threat to Thailand was seen as being posed by the continuing Vietnamese occupation of Kampuchea but there was still the possibly destabilizing impact of 275 000 refugees from Kampuchea in camps inside the Thai border. It is perceived as possible that the traditional buffer provided by Kampuchea between the historic rivals of Vietnam and Thailand could eventually be removed, and that in the future Thailand itself could become a buffer between an expansionist Vietnam and Malaysia. However, as yet there is little evidence of this; indeed, it was in large part because of its own economic difficulties, that Vietnam responded to pressure for withdrawal from Kampuchea.

Another aspect of the geopolitical influence of Vietnam within the region is the Soviet use of the naval facilities at Cam Ranh Bay, for this base not only facilitates a Russian naval presence in the region, it also flanks the major trade artery running between Japan and its main source of oil supply, the Persian (Arab) Gulf. At least before the recent international initiatives of Mr Gorbachev, 'Perceived as Vietnam's patron, a state which has violated the independence of its neighbour, the Soviets seem prepared to write off relations with the rest of the region for the time being. Soviet influence in Indochina and the facilitation of naval passage are the gains Moscow sees in South East Asia toward the enhancement of its total Asian presence.' (Simon 1985, p. 931.) In addition, possession of the Cam Ranh Bay base significantly contributes towards the Soviet encirclement of China. Both the size and the fire power of the Soviet fleet in the region is far inferior to the American, and it is likely that Russian strategic thinking at present is not in terms of offensive capability but rather of more general influence, including possible claims to participate in any future negotiations on a regional basis. Russia continues to add to Vietnam's military capabilities (and Vietnam now has more men under arms than all of the other countries of South East Asia together); Russia also supplies Vietnam with vital grain, all of its oil and most of its other imports. Hanoi's options *vis à vis* Russia are therefore very few but so far there is little evidence that Russia wishes to abandon its most important client in the region, despite the damage this relationship does to Sino-Soviet relations. Not surprisingly in the circumstances, the Soviet presence in South East Asia is viewed with considerable alarm by the ASEAN nations.

The other Great Power actively involved in the region is, of course, China, its position being considerably complicated by the presence of a substantial minority of some 19 million ethnic Chinese in the various South East Asian countries, for an abiding concern of all the region's governments is that Beijing might seek to politicise their Chinese communities. Unlike the other two Great Powers,

China currently has a direct and individual adversary in the region in Vietnam and a major objective of Chinese policy, now achieved, has been to coerce Vietnam, if necessary by force, into withdrawing completely from Kampuchea. Chinese troops invaded northern Vietnam for a period in early 1979 but this succeeded only in driving Vietnam closer to Russia. In addition to continuing its active support of the remnants of the previous barbarous regime in Kampuchea, the Khmer Rouge, China has therefore turned much more towards cooperation with the United States and ASEAN in the region as a means of countering Soviet influence. The justification for this appears to be that within South East Asia 'Soviet hegemony directly threatens China's security, while US hegemony only offends its moral sensibilities.' (Simon 1985, p. 938.) Despite the apparent congruence of US-China security relations in the region, however, the ASEAN members, and particularly Indonesia and Malaysia, remain suspicious of China's long-term objectives. They fear both a conflagration between China and Vietnam – a conflagration which would send destabilizing shock waves throughout the region – and also the fact that the present coincidence of interests between China and the United States in the region might provide a basis for Chinese hegemony in the long term.

A fourth power in the region is undoubtedly Japan, though through economic rather than military might. As with China, attitudes to Japan in the non-Communist states in the region can best be described as both apprehensive and equivocal. As far as Malaysia is concerned for example, to quote Khong (1987, pp. 1095, 1102),

'Among older people in Malaysia the memory of Japan is one of brutality and atrocities. To the postwar generation, Japan is the source for many of their needs – cars, radios, televisions, videos, and all kinds of electronic and electrical gadgets. The first was created by the activities of the Japanese troops that occupied the country during World War II, and the latter by the success of Japanese penetration, if not domination. Today, it looks as if Japan has been able to establish its "Co-prosperity Sphere" without force of arms. Japan is the single most dominant economic power in Malaysia. . . . However, it should be noted that the Malaysian experience with Japanese investment has not been entirely satisfactory. Several criticisms have been levelled: (1) that the focus of investment is mostly in resource-oriented industries or in those that involve no significant transmission of technology and with low value added; (2) that some Japanese investments in industries are established solely to enjoy the preferential privileges offered by the developed countries to the manufactured exports of less developed countries, an approach seen as an attempt by Japanese investors to reap the benefits of access to markets of other developed countries, which is inconsistent with the original purpose of such preferential treatment – that is, to promote the industrial development of less developed countries; (3) that Japanese private investment (like foreign private investment generally) is producing a new

9

form of colonializm in which the host country is exploited; (4) that package arrangements whereby local firms agree to purchase material from the parent firm increase dependency on Japan; (5) that profits go disproportionately to nonlocals; (6) that management and control, in fact, remain in Japanese hands; (7) and that Japanese investors have no regard for the promotion of national policy in the country.'

Similar feelings have been expressed in other ASEAN countries, and there have been sporadic popular demonstrations against the Japanese presence. In Washington, however, there is undoubtedly a growing feeling of resentment over what is perceived as Japan's 'free ride' in defence in respect of eastern Asia. Should the growth of such feelings eventually result in an even larger presence of Japan in South East Asia in the future, with some form of military as well as economic involvement, local suspicions of Japan would be even further intensified.

Finally, the disputes over the Paracel and Spratly Islands between certain of the countries of South East Asia, and which also involve China, remain a serious issue. China and Vietnam have clashed militarily over both the Paracel Islands and the Spratly Islands. China is now in occupation of the Paracels but the ownership of the Spratly Islands remains a matter of dispute, China, Taiwan, Malaysia, Vietnam and the Philippines all having staked claims over one or more of the thirty-three small islands. Their importance is both strategic and as bases for off-shore oil exploration. With six islands occupied by Vietnam, seven by the Philippines and one by Taiwan, and control over the whole archipelago having been formally claimed by Vietnam and the Philippines in addition to China, conflict could be a future consequence. These disputes could possibly involve all the Great Powers, for in addition to Russian support for Vietnam's claims, United States oil interests have been drilling in the area with permission from the Philippines (Lee 1982, p. 57).

The future outlook

In terms of the future, to these geopolitical concerns must be added additional ones, though it must be stressed that this is one part of the Third World in which, at least for the non-Communist countries, the general long-term development prospects are reasonably encouraging. As outlined above, a potential for instability is present both in superpower rivalries and in the role of Vietnam within the region. In addition, the influence of Islam is becoming ever more apparent, sometimes in constructive ways but also in destabilizing ways. Indonesia is the country with the largest Muslim population in the world. Malaysia, Singapore, Brunei, the Philippines and Thailand all have significant Muslim populations. Powerful Muslim voices, both from within the region and from the wealthy Gulf states, are now claiming that Islam should be the determining factor in the lives of citizens and that political authority over Muslims should reflect Islamic principles. This view is not universally accepted by Muslims within the region and neither do some of the governments with Muslim minorities pay much attention to it. As Federspiel (1985, p. 806) has observed,

'Significantly, the political leaders of South East Asian nations have had their own responses to the role of Islam in national life. Confronted with gigantic problems of structuring workable political systems, developing functional economic systems, and surviving in a complicated international community, they have placed their attention on those areas. In the minds of these state leaders, Islamic values have not appeared to offer immediate and practical solutions to the paramount issues the nations must face in contemporary times. Rather, other values taken from abroad, such as economic modernisation, or those adopted from traditional South East Asian cultures, such as the Javanese bureaucratic system or simple nationalism, have assumed great importance. Islam has had to compete with those other value systems and has fared differently in the different nations, sometimes assuming considerable importance, as in Malaysia and Brunei, sometimes winning secondary consideration, as in Singapore and Indonesia, or being largely ignored, as in Thailand and the Philippines.'

Nevertheless, Islamic religious fundamentalism is steadily becoming more apparent in some of these nations, particularly Malaysia, whilst in the Philippines the Moro National Liberation Front has been in conflict with the government in the Muslim south of the country for more than a decade. Numbering only about 8 per cent of the total Philippines population, the Muslims of Mindanao, Sulu and Palawan have only a tenuous contact with the concept of Philippine nationality and their own ethnic divisions have been cemented by Islam into a unified opposition with a goal of secession from the Philippines to form an independent Islamic state.

There are other existing or potential internal conflicts within the region. One major group arises out of ethnic conflicts, a reflection of the difficulty of relatively new and fragile states accommodating South East Asia's multiplicity of ethnic groups, religions, cultures and languages. Burma's dissident minority groups have already been mentioned, their potential for instability being enhanced because of their location adjacent to and even astride sections of the country's political frontiers. There are similar problems, though fortunately not so serious, in the south of Thailand with Muslim minority groups but currently ethnic problems are probably at their most acute in Malaysia, here concerning the position of the Chinese, and to a lesser extent that of the second minority group associated with immigration in colonial times, the Indians. As a consequence of very serious riots in 1969, Malaysia instituted a New Economic Policy with a view to increasing rapidly the share of the Malays in the national economy, the Malays being held officially to be indigenous to the country, though in fact a large proportion are also relatively recent immigrants, from adjacent parts of what is now Indonesia. A government politically dominated by Malays has added discriminatory cultural and educational policies to the economic ones. The result today is a country even further than in the colonial years from the state of ethnic harmony that is officially desired. Despite its economic success of recent decades, therefore, the potentiality for serious ethnic conflagration in Malaysia has only been increased.

Throughout the region, the Chinese communities have frequently been subject to the wrath of national governments. Because of the relative economic success of some Chinese, conflicts have arisen based not only upon ethnic difference but also differences in standards of living. Additionally, the majority of rich immigrants have become urbanized, whereas national governments have depended for their legitimacy upon the consent of a majority of indigenous people, who are poor and living in rural areas. Anti-Chinese economic nationalism has been and continues to be a significant phenomenon within the region. In the Philippines periodically, but more especially in Indonesia over a long period of time, the Chinese minority has suffered from anti-Chinese official attitudes. In Indonesia, as in Malaysia, Islamic attitudes have in some instances recently added a new element to the mix, highlighting religious difference as an additional discriminatory factor. In September 1984, for example, there were serious anti-Chinese riots in Jakarta in which 28 people were killed. Muslim extremist agitation was clearly the major factor in their occurrence. The riots followed the massacres of Communist-oriented and non-Communist Chinese alike which followed the overthrow of Sukarno in 1965, violence in Bandung in 1973, in Jakarta in early 1974 and in South Sulawesi and Central Java in 1980. Once again, however, the most Draconian measures of the recent period have been taken by Vietnam, which during 1978 started to expel Chinese from the towns of the south, precipitating the tragic exodus of tens of thousands of 'boat people' and an even larger number of 'land' refugees who crossed neighbouring borders. Some of the refugees were Vietnamese but at least two-thirds were ethnic Chinese.

Another major group of internal problems revolves around the prospects for economic growth, for such growth in relation to population growth, and also for prospects for individual welfare, particularly of the poorest members of the national communities. The mid-1980s saw a severe blow given to the confidence in the economic future of the previously buoyant non-Communist countries of South East Asia. The general depression in world trade, the decline in commodity prices and the collapse of the oil price combined to reduce severely, and even in some cases (notably that of Singapore, the most prosperous state in the region and that with the highest growth) to reverse at least temporarily, the progress of economic growth. Only in the later 1980s, were prospects once again beginning to look better. Meanwhile, at least two long-standing problems of major dimensions remain to be resolved. The first concerns population. Although the region as a whole is not overcrowded, Java has one of the most severe problems in the world in terms of population density. Schemes to send settlers to Indonesia's Outer Islands dating back to before the Second World War having proved of only symbolic significance in addressing Java's population problem, it is clear that the future can best be approached in terms of relief *in situ*, on the basis of comprehensive birth-control policies over a lengthy period, and with the acceptance that Java must be supported economically, as it is now, by the remainder of the nation.

Cognisant of the problem of Java, and faced with the reality of an overcrowded planet, it might be thought that birth control would be a primary objective of all the region's governments. However, this is not the case at present, and in particular two of the most pragmatic and outward-looking states in terms of economic policy have recently adopted policies of population increase. In 1982 the Prime Minister of Malaysia, Dr Mahathir Mohamad, announced a policy of increasing Malaysia's population to no less than 70 millions (Dwyer 1987). This startling departure from economic and demographic reality was followed the next year by Mr Lee Kuan Yew, the Prime Minister of Singapore, expressing concern over the growing number of unmarried graduate women in the state and the possible resulting deterioration in the intellectual quality of its population. By mid-1986, these concerns had been overtaken by official attention directed towards the fact that the average number of children had fallen to a level just below the replacement rate and an inter-ministerial committee had been established to examine ways in which to raise the birthrate. The official population slogan had become, 'At least two. Better three. Four if you can afford it.', its bizarre nature possibly being exceeded only by that of Dr Mahathir's 'Go for five.'

Finally, it needs to be observed that all the perturbations outlined above are taking place within a region in which, even in the

Table 1.1 Percentage share of household income by percentile groups of households

Country and year of data	Lowest 20%	Second quartile	Third quartile	Fourth quartile	Fifth quartile	Highest 10%
Philippines (1985)	5.2	8.9	13.2	20.2	52.5	37.0
Thailand (1975–6)	5.6	9.6	13.9	21.1	49.8	34.1
Malaysia (1973)	3.5	7.7	12.4	20.3	56.1	39.8
UK (1979)	7.0	11.5	17.0	24.8	39.7	23.4
USA (1980)	5.3	11.9	17.9	25.0	39.9	23.3
Japan (1979)	8.7	13.2	17.5	23.1	37.5	22.4

Source: World Bank, *World Development Report 1987*.

relatively rich countries, the basic problem of ensuring a fair share of economic growth for the poorest has yet to be solved. As Table 1.1 indicates, in the three countries for which information is available, and they are typical of the region, an unjustifiable share of national income rests in the hands of small, privileged groups. About 40 per cent of total household income accrues to only 10 per cent of families in Malaysia, for example. This concentration of wealth contrasts very markedly with the situation in the advanced industrial countries. At the other end of the scale, the poorest 20 per cent of families receive only a minute share of national household income. To be fair, the proportions for this category of families are not startlingly better in the industrialized countries but in those countries the number of the very poor is proportionately much smaller, the actual income of poor families higher and social services for the poor very much better developed. If the problem of income distribution is not successfully addressed within South East Asia during the years ahead, it could prove to be the most fundamental of all in terms of future stability in the region.

Further reading

Dwyer D J (1987) 'New population policies in Malaysia and Singapore', *Geography* **72**, 248–50.

Esterline J H (1988) 'Vietnam in 1987: Steps towards rejuvenation', *Asian Survey* **28**, 86–94.

Federspiel H M (1985) 'Islam and development in the nations of ASEAN', *Asian Survey* **25**, 805–21.

Fisher C A (1962) 'South East Asia: the Balkans of the Orient?', *Geography* **47**, 347–67.

Fryer D W (1970) *Emerging South East Asia: A Study in Growth and Stagnation*, Philip, London.

Khong K H (1987) 'Malaysia-Japan relations in the 1980s', *Asian Survey* **27**, 1095–1108.

Lee Y L (1982) *South East Asia: Essays in Political Geography*, Singapore University Press, Singapore.

Simon S W (1985) 'The Great Powers and South East Asia: Cautious minuet or dangerous tango?', *Asian Survey* **25**, 918–42.

World Bank (1987) *World Development Report 1987*, Oxford University Press, New York.

CHAPTER 2

South East Asia in the colonial period: cores and peripheries in development processes

William Kirk

If one accepts that 'development' necessarily involves change and structural transformation and that processes of transformation are almost invariably initiated by some innovative agent or by tensions between interactive structural systems, then it follows that development will never occur simultaneously in all places. Differentiation of one kind or another will result. Some places or communities, by reason of special natural or human endowments, or the stimulus of perceived problems and opportunities, or by being at the right place at the right time to take advantage of historical trends, find themselves at the centre of development processes and for a time reap the benefits and bear the responsibilities that such structural centrality confers. This can find expression in a variety of forms. It might take the form of political leadership and hegemony, whereby the location becomes the core of a new political system and territory – a power centre where major decisions are taken, policies formulated and the institutions and symbols of political primacy are concentrated, usually in a capital city. It might take the form of economic centrality, whereby the location becomes the core of an economic system – a focus of wealth and accumulation of capital; a nursery of commercial enterprise; a centre of a transportation network; a market; an attraction for those industrial activities and financial institutions which stand to gain from the economies of agglomeration at a central location. Alternatively, centrality may be expressed in cultural terms, whereby the ideas, values, beliefs and modes of social behaviour of a core community are diffused to other communities by processes of migration, culture contact or active proselytization. In many cases, however, a place which has achieved core status in one field acquires similar status in others through the association of behavioural systems. Even though the systems can operate independently, and often do at times of structural stress, in periods of vigorous growth of cores, political hegemony becomes identified with economic hegemony and the cultural *mores* of the core comes to be regarded as the means to achieve progress and as a standard against which 'modernization' can be measured.

For every core, however, there must be a periphery. Indeed the concept of periphery is inseparable from the concept of centrality. Peripheral areas are those areas external to the core but within the sphere of influence of the core, in contrast to extra-peripheral areas that lie outside the systems and structure of which the core is the centre, and many of the central issues in the study of development can be expressed in terms of core-periphery relationships. To some students of development (see Ch. 3), cores are the powerhouses of development, transmitting pulses of energy outward through peripheral systems, or acting as 'growth poles' within regional structures, and as such have been created as development planning devices on the understanding that over time they would raise the level of activity of peripheral areas. Advocates of this mode of development planning recognize that initially the planting of a core will lead to inequality between core and periphery but argue that in time, as the stimulus of the core takes effect, such inequalities will be reduced. Initial divergence will be replaced by convergence.

Other students of development, however, argue that this is not the invariable outcome. They contend that inequalities continue to increase and that instead of radiating growth-energy cores sap the energy of peripheral areas in order to maintain their own growth and superiority. The periphery is seen to be tributary to the core, contributing its resources of men and materials to the aggrandisement of the centre but receiving little in return except perhaps political 'protection'. The periphery continues to labour under the handicap of distance and experiences the deprivations arising from increasing political, economic and social distance from the centre, both real and perceived. In comparison with the sophistication, urbane and complex civilization of the core it appears to be culturally backward, with primary social groups engaged in primary production and following old, traditional codes of behaviour rooted in the countryside. It loses its potential political and economic leaders by migration to the core, attracted by the greater opportunities for advancement afforded by the core environment, where politics is an urban game and political power provides economic rewards. To participate in core systems peripheral communities must learn the language of the core, sacrifice some measure of local independence, substitute core value and belief systems for their own traditional systems and convert productive activities to meet the demands of the core market. In the process the periphery, it is argued, becomes increasingly dependent on the core and as the demands of the core increase this has a depressive effect on the life, ecology and economic growth of the periphery. According to such analysis, 'underdevelopment' is considered to be a process rather than a state.

Core-periphery relationships, however, are complex and much influenced by the scale and historical context of the structures within which they are operative. It is possible to identify a series of such structures, ranging from small, highly localized and self-sufficient entities at one extreme to world-wide systems at the other. Primary, metropolitan cores of one period, and at one scale of operation, can become secondary, peripheral cores at a later date or in a wider structure. Indeed it is not unknown for important cores of one period to disintegrate in succeeding periods and sink back into an undifferentiated periphery. Spatial structures have geographical momentum but are not eternal, and regional inequalities characteristic of one developmental phase can be reversed by a change in technology or the discovery of a new resource.

Modern development theorists are concerned with the impact of European (Western) colonialism on former imperial territories, with the continuation of economic imperialism in world capitalistic systems even after the end of political empire, and with the growing differentiation between the 'developed' and the 'underdeveloped' regions of the world, but tend to neglect the fact that for most of human history what are now some of the cores of world development were peripheral territories to tropical and sub-tropical cores. In emphasizing world systems, it is also possible to overlook the continued expression of core-periphery contrasts in localized terms in both 'developed' and 'underdeveloped' countries. To understand development processes in a world divided into states and other political theatres of action, problems of internal colonialism and regional inequality within states merit investigation as well as the problems of large scale, multi-national or inter-state forms of external colonialism.

For the study of such issues South East Asia comprises classic ground, not only because it was the provenance of some of the earlier contributions to development theory such as Furnivall's (1948) studies of 'plural societies' and Boeke's (1942) consideration of economic dualism' in Indonesia, but because of the richness and variety of its development history (see Evers 1980). As a large, tropical forest-and-sea region, including the catchment basins of the Irrawaddy, Salween, Menam, Mekong and Sang-koi rivers and the peninsulas and archipelagos of the Malay world, it has experienced many forms of colonization and colonialism. Many of its peoples came from the north, from the uplands of Yunnan and the forests of southern China and moved southward along the valleys of the great rivers or across the inland plateaux in a series of waves of colonization that carried them to the tropical seas and island arcs extending eastward to the Melanesian world of New Guinea. Occupation of new ground and the need to adjust to a wide variety of environmental resources resulted in cultural change and much local differentiation of ecological structures. Some localities were able to support large numbers of people by intensive wet-rice agriculture, in others lower densities of population were sustained by various forms of shifting (swidden) agriculture, while in coastal locations it was possible to widen the resource base by different combinations of farming and fishing (Gourou 1953). In general cultural terms, the region received transforming inputs from its two large neighbours – India and China – and some of its characteristics are derived from its position as an Indo-Chinese convergence zone. India certainly played an important role in the early cultural and political development of the region, through its Buddhist and Hindu missionary activities, its settlements of political adventurers and the extension of its trading system across the Bay of Bengal (Wales 1951). It also provided the base from which Islam was transplanted from the deserts of South West Asia to the forests of South East Asia in the later centuries of the medieval period. Chinese impact on the other hand was initially constrained largely to the Song-koi delta area and the Vietnamese coast, and to intermittent bursts of maritime activity in the South East Asian seas and the Indian Ocean, and it was not until Chinese migrants began to enter South East Asia in large numbers in the modern era that its presence was fully realized in developmental processes. By that time, the initiative had passed to European powers and South East Asia became the scene of Portuguese, Spanish, Dutch, English and French rivalry for commercial supremacy and, later, political empire. The immediate context of the study of development problems in South East Asia is provided by the respective colonial structures created by these European powers and the difficulties inherited by their successor states, but it should not be forgotten that some of the problems of core and periphery relationships and regional inequalities were deeply entrenched in the landscape long before the Portuguese fleet appeared off Malacca in 1511.

Cores: pre 1500 AD

In some cases, the core areas pre-date the coming of the Europeans (Fisher 1964, Hall 1964, Coedes 1966, Smith and Watson 1979). In Burma, for example, Mon people who lived on the margins of the Irrawaddy and Sittang deltas came under Indian (particularly Buddhist) influence in the early centuries AD and Thaton and Pegu constituted urban centres of an important political, cultural and commercial core of considerable influence and durability (Fig. 2.1). It was linked *via* the Three Pagodas Pass across the Martaban Range with a similar Mon core in the deltaic lowlands of the Menam Chao Phraya, centred on the ancient port city of Nakhon Pathom, some 40 miles west of present-day Bangkok. Both cores had lasting cultural impact on migrants from the interior highlands who moved southward along their respective river valleys in succeeding centuries. In the Irrawaddy Valley the Pyus, who reached the lower valley in the sixth century, established their capital at Hmawza (Prome) at the head of the delta, but later Burman migrants occupied the dry zone of the middle Irrawaddy valley and built

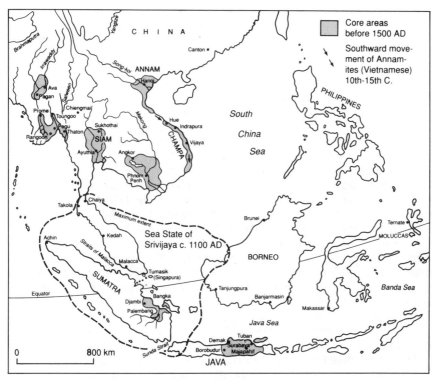

Fig. 2.1 Core areas before 1500 AD

their capital city at Pagan in *c*.850 AD. Ultimately the dry zone, with its successive capital cities such as Ava, Sagaing, Amarpura and finally Mandalay, constituted the central political core of the kingdom of Burma but the deltaic south preserved much of its regional identity and this duality played an important role in the British occupation of Burma during the 19th century. In the Menam lowlands similarly the ancient Dvaravati core attracted the Thai people who entered the head of the valley in the 12th century. From small kingdoms such as Chiengmai in the north they pressed southward, occupying Sukhothai in 1238 – sometimes described as the cradle of Siamese civilization – and founding the city and kingdom of Ayuthia in 1350. It was not until 1767, following the earlier destruction of Ayuthia by invading Burmese forces and the transfer of the capital to Thonburi, that the capital of Siam (Thailand) was moved to Bangkok.

The lower valley and delta of the Mekong also sustained an early core. In the early centuries AD this area constituted the centre of the powerful Indianized kingdom of Funan with its capital at Vyadhapura at the head of the delta, but by the sixth century pressure from the Khmer people of the middle valley led to the disintegration of Funan and its replacement by a number of Khmer states, including Water Chenla in the delta and Land Chenla in the middle valley. In 802 the two Chenlas were unified by the Angkor Dynasty and the new kingdom of Kambujadesa (Cambodia: Kampuchea) created. From the late ninth century until 1432, the capital of the Khmer Empire was located near the northern shores of the great lake of the Tonle Sap in a remarkable series of urban and temple sites of which Angkor Thom and Angkor Vat are the most famous (Groslier 1966). The size and grandeur of the ruins of Angkor bear witness to the food and population potential of the expanded core of the Tonle Sap basin and the Mekong alluvial lowlands as well as the wealth-creating power and commerce of the medieval Khmer Empire which at its peak in the 12th century extended from the Indian Ocean to the seaboard of the South China Sea. During the following century, however, the Cambodian core came under pressure from Thai/Lao movements, from Siam and the upper Mekong where Lao tribes established the kingdom of Lan Chang (Luang Prabang). After the sacking of Angkor by the Siamese in 1432 the city and its irrigation system was left to the jungle and the capital moved in 1434 to Phnom Penh at the confluence of the Tonle Sap and Mekong drainage (Osborne 1985). Although never recapturing its former architectural glory the essential Cambodian core survived as an important unit into colonial times and beyond.

The only other early cores in mainland South East Asia were located on the seaboard of the South China Sea. In the early centuries AD small rice-growing communities between the rugged Annamite Range and the sea were organized into the Indianized Kingdom of Champa, with capital cities of Indrapura and Vijaya indicating the core area of the state. A more populous and powerful core, however, lay to the north in the Song-koi Delta and the narrow zone of coastal lowlands extending southward to the Port d'Annam. This was the original hearthland of the Annamite (Vietnamese) people, but from the earliest recorded evidence their culture was profoundly influenced by China and from c.200 BC the area was subject to Chinese colonization. The intensive forms of land-use and irrigation systems introduced by the Chinese ensured substantial food supplies and produced population growth and high population densities. For centuries Annam was caught between the frontier forces of Imperial China and the northward expansionary ambitions of Champa and it was not until 939 that a national dynasty established itself at the capital Hanoi, asserted its independence of China and began to press south against Champa. It took half a millenium of inter-core warfare, however, before the Chams capitulated in 1471 and opened the way south for the colonial surges of the Vietnamese that carried them into the Mekong Delta by the 18th century. It was not an easy task to maintain political unity in this narrow, attenuated coastal zone where Chinese-derived culture of the north met Indian-derived culture and state forms of the south and there was no dominant central location on which a political core could be developed. The violent history of Vietnam owes not a little to its geographical characteristics.

It will be noted that, with the exception of Champa, all the early cores of mainland South East Asia were based in major riverine environments, particularly in the lower valleys where extensive alluvial areas could be reclaimed from tropical forest and freshwater swamp for wet-rice cultivation. It was in such terrains that maximum returns could be achieved by communal effort and ecological and hydrological management. In contrast to the various forms of shifting agriculture practised in the forested highlands settlement was fixed and systems of land use evolved in relation to the regime of the rivers, seasonal flooding in the deltas and inland basins such as the Tonle Sap, and communal modes of land-tenure. Wealth resided in land and revenue derived from rural taxation supported the managerial structures represented by various state organizations. With the exception of Annam, the basic state model was derived from India, with its concept of kingship, the court, the army, the civil and religious bureaucracy and royal overlordship of land. Administration was centred on a primate, royal capital city where the wealth from land taxes was concentrated and used to support the royal retinue, to finance military operations for the defence or expansion of the state, to engage in trade, to foster craft-industry for the immediate market of the court and the capital city, and, particularly in Buddhist kingdoms, to carry out 'good works', either in the form of great religious edifices or public works such as new irrigation canals. The construction of these great monuments absorbed considerable resources but provided some of the centralizing symbols of statehood, as well as personal memorials. Their remains, such as at Pagan and Angkor, mark the sites of former primate central places within core areas which have functioned as nurseries of capital cities.

Of course at this stage of development there were many areas and communities that had virtually no links with the core areas and thus can, in a strict sense, be regarded as 'extra-periphery'. Indeed as one extends the analysis into the island world or archipelago of South East Asia, the extra-peripheral category of territory increases. In fact, it is possible to identify only two significant core areas in the archipelago during this period. The first of these emerged in the seventh century in South East Sumatra as a result of the union of the two small Indianized port city-states of Djambi (Malaya) and Drivijaya, with the core capital at Palembang. The wealth of the Palembang core was derived from its position in relation to the seaways between India and China *via* the Straits of Malacca and the South China Sea. It provided an important entrepôt for the growing long-distance sea-trade between the two great civilizations and for the produce of the Indonesian archipelago which was fed into that trade. As its power grew it extended its commercial monopoly by occupying West Java, thereby gaining control of the Sunda Straits, and by consolidating its command of the Straits of Malacca through the establishment of outposts on both shores. At the height of its power under Sailendra dynasty *c.*1100 AD the sea-state of Srivijaya had achieved an almost complete monopoly of trade in this strategic region and had connections with both the Coromandel coast of India and Canton in China. Palembang became a wealthy city and a centre of Mahayana Buddhist learning. In 1025, however, the Cholas of Southern India, increasingly frustrated in their oriental trade by the Srivijayan middlemen, raided the Malayan empire. Srivijaya never quite recovered from this assault on its monopoly although its identity and some of its functions continued into the 13th century. By the arrival of the Portuguese, however, these functions had been inherited by Malacca, a Sumatran colony on the west coast of Malaya.

The other core of the archipelago was located in Java. Here a number of small Indianized states had been established in the early centuries AD in the valleys of central and eastern Java and much of the early history of the island consists of the struggle for dominance among the ruling families of these states. The first to achieve more than local significance was Mataram in central Java and the Sailendras, responsible for great architectural works such as the magnificent Borobudur eighth-century Buddhist stupa and for the beginnings of trade to the eastern islands. The centre of power shifted eastward at the beginning of the 10th century to the Brantas valley where major reclamation of the deltaic swamps was carried out for intensive cultivation by the rulers of the state of Kediri. This state and its successor in the 13th century – Singhasari – developed overseas trade with the Moluccas, noted for their nutmegs and cloves, and made great inroads into the former sea-space of Srivijaya. The appearance of Arab merchants stimulated the trade as well as introducing Islam into many of the ports of the area, and the ports of Surabaya, Tuban and Demak developed extensive overseas connections as well as mixed sea-faring and merchant populations. By 1500, however, the structure was disintegrating and most of the coastal areas were in the hands of Islamic rulers.

At least eight core areas of considerable antiquity can thus be identified prior to the impact of European imperialism. Each constituted the nucleus of a system with dependent peripheral terrains and communities. Each represented an organized, dynamic, unit of clearance and development in the great forest of South East Asia. With the exception of Annam, each duplicated, with local variations arising from varying environmental and cultural contexts, a model of land colonization first developed in the forests of India. Although it is now generally accepted that few people of Indian origin actually participated in this colonial surge, in a cultural sense the inspiration is clear and the old name for South East Asia – 'Further India' – has much to commend it. It was a momentous period in the development of South East Asia and many of the structural transformations initiated at that time have yet to run their full course.

European impact on the early cores

During the period 1500–1800, European seapowers competed for strategic control of the Indian Ocean and the seaways to China, and their merchants vied with each other to gain trading privileges from suppliers of profitable commodities for the European market (Tate, 1970, 1979). During the 16th century the Portuguese had a veritable monopoly of long-distance sea trade around the Cape of Good Hope, but in the succeeding centuries the Dutch East India Company, the English East India Company, the French, the Danes and other European traders broke the Portuguese monopoly and established their own trading networks in the oriental seas. Each of the trading networks varied in spatial dimensions and commercial rhythms but had certain features in common. For each of them India occupied a key position. Indian textiles and other manufactured products provided much needed trade goods for European merchant adventurers trading to the East and the impact of Europe on South East Asia during the mercantilist era in many respects was a continuation of earlier Indian impact on the area. For each of the European networks the first priority was the commerce of existing cores rather than territorial acquisition, thereby strengthening these and confirming indigenous core-periphery relationships.

By the beginning of the 19th century, however, important structural changes had occurred within the European maritime networks, often emanating from distant European cores. The British victories over France and Holland in the Napoleonic wars left Britain master of the Indian Ocean and the English East India Company in a powerful position in India. The Dutch fell back on Java and the Indonesian Archipelago, while the French, thwarted in their Indian ambitions, turned to construct a second empire in Indo-China (Cady 1954). The nature of trade was also changing as a result of economic developments in Europe. For the first time in their dealings with the East, European traders found they had European manufactured goods to market and raw materials to seek for European industry. 'Trade followed the flag' and led to a new approach to the political and economic potentials of South East Asia. Each of the European powers sought to establish colonial control over land as well as sea networks and territorial acquisition became important.

In analysing the development processes in South East Asia and the contribution of European colonialism to those processes it is thus important to remember that in most of the area the basic spatial structures of cores and peripheries had already been established in earlier colonial periods. In the main the European colonial powers, first as traders and later as rulers, were attracted to existing cores and with only two exceptions half a millenium of European control simply enhanced those cores and further entrenched the existing spatial patterns. The human and capital resources they were able to apply to this huge and diverse area did little to reduce regional inequalities, indeed by concentrating on core development they exacerbated such inequalities.

Cores of the European colonial period: 1500–1950

During the five centuries of European imperialism in South East Asia only two completely new cores were created, namely the Manila core within the Philippines and the West Coast core of Malaya, but as will be seen considerable transformations were effected in the ancient core areas and their structural relationships with their respective peripheries (Fig. 2.2).

The Manila core

Although Magellan on his circumnavigation of the world claimed the Philippines for Spain in 1521, the first permanent Spanish settlement within the islands did not occur until 1565. Initially the island of Cebu, within the central Visayan Sea, was selected as the Spanish base, but in 1572 the small Moslem stronghold of Manila on the island of Luzon was captured and two years later central headquarters was transferred to that site by Legaspi. Manila Bay provided a large sheltered harbour commanding the northern strait into the Visayan Sea on the one hand and access to the central plain of Luzon on the other. Here a new port city was built in the Spanish colonial manner, characteristic of the conquistador cities of Latin America, and within a few decades the port had established itself as an important entrepôt in the western Pacific. Regular sailings of the so-called Manila galleons between the port and Acapulco in Mexico brought Mexican silver in exchange for the silks and porcelain of China and spices from the island trade, and commercial development attracted Chinese and Japanese merchants to the city.

It was not until 1834, however, that the port was opened by Spain to world trade. British and US traders moved in and profits from trade particularly in tobacco and hemp provided finance for substantial improvements to the harbour and port facilities, for the building of feeder roads and the development of inter-island shipping, and for the establishment of an efficient postal and telegraph service. Banking and commercial agencies facilitated the inflow of new capital and provided opportunities for a new educated Filipino middle class, often with Chinese or Spanish ethnic connections, to participate in expanding urban functions.

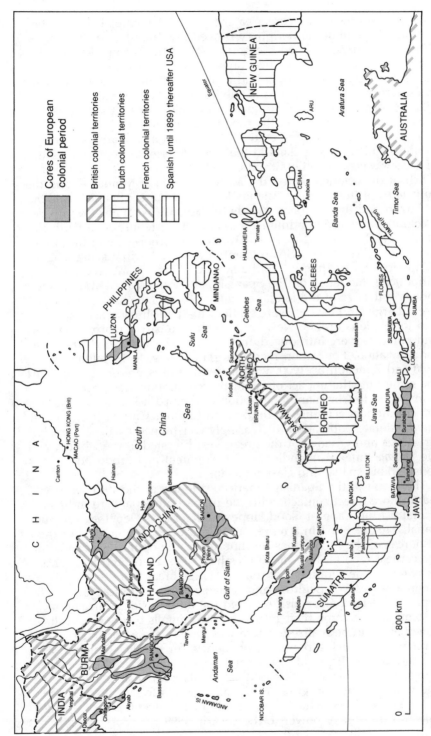

Fig. 2.2 Cores of the European colonial period

Other features of the core area and the relationships between the core and its Philippine periphery proved less tractable to change. When the Spanish arrived the population of the islands was small (about 0.5 million) and political organization almost non-existent above the local level. They found communities of peasant-fisherman in estuarine and coastal sites in the Visayan Sea area, wet-rice villages in dispersed lowland areas and valleys or by the side of great flights of terraces on the mountain sides of north Luzon, groups of dry-rice shifting cultivators on the forest margins, and hunting and collecting clans in the forest interiors – representative of earlier waves of settlers from elsewhere in South East Asia. For much of the colonial period the Moros (Moors) of Mindanao and the sultanates of the southern islands lay outside direct Spanish rule but in Luzon and the northern islands the Spanish were able to impose a colonial structure of administration and land tenure on a veritable *tabula rasa*. The Philippines constituted a governor-generalship under the viceregality of New Spain (Central America) – in a sense then a periphery of a periphery – and a system of land grants was introduced, modelled on the Spanish-American colonial system and indeed reflective of the feudal estates of the home core in Spain itself. Large land estates were granted to Spanish conquistadores and to religious orders. New crops such as tobacco, maize, peanuts and potatoes were introduced and labour levies used to clear land for cultivation. Population increased remarkably, to $1\frac{1}{2}$ million by 1800 and 7 million by 1900, with increasing pressure on the land. At the same time the impact of Spanish acculturation was profound. Conversion to Christianity, adoption of the Spanish language, and the educational activities of the Roman Catholic Church produced a highly literate population, increasingly constrained by the colonial structures and envious of the great wealth vested in the Church and the *'cacique'* (meztizo) landlords. Development had produced considerable spatial and class inequality.

The revolt of the Spanish American colonies in the 19th century led to more direct contacts with the primary imperial core in Spain, at first *via* the Cape of Good Hope and later (post 1869) *via* the Suez Canal. This resulted in a more varied and open commerce, but the resources of the Philippines did not justify the optimism of some of the earliest reports. Few metal ores had been discovered to offer a basis for the development of mining. Large areas were too mountainous for agriculture and remained under tropical forest. The separate lowland areas also required the introduction of irrigation systems to overcome their relatively low rainfalls before their full potential could be realized. There was little to attract Spanish investment in such a remote dependency, for the development of the core or the dispersal or economic activity into peripheral areas. Congestion in the core increased as population densities heightened and political agitation grew. Nationalist groups fought the Spanish in the war of 1898, which saw the end of three centuries of Spanish rule and transfer of power to the USA in 1899.

The USA was a reluctant colonial power. In order to achieve a naval base in the western Pacific, as part of a grand oceanic strategy, it found itself as ruler of the island group administered from the Manila core. From the outset the declared intention was to facilitate Filipino movement to full independence and this was ultimately realized in the inauguration of the Republic of the Philippines in 1946. In a spirit of paternal benevolence the American administrators also planned to develop the economy and social welfare of the islands. To relieve congestion of the core planned-agriculture colonies were established in Mindanao, in Bohol, and in the Cagayan valley of North East Luzon, and substantial numbers were financially assisted in such migrations (Pelzer 1945). Road building, especially in the military-administered terrains of Mindanao, was encouraged; railway systems were constructed in Central Luzon, in Cebu and in Panay; but the major effort was put into the development of inter-island shipping which had constituted the most important element in the transport infrastructure. Agricultural development was also fostered. Extension of the irrigation system in the central plain of Luzon increased the production of wet-rice; the acreage under maize, the other important food of the Filipinos since Spanish colonial times, was expanded; fruit and vegetable production for the urban markets was encouraged; sugar growing in central Luzon and north-western Negros was improved and linked to 'centrals' for refining; and other crops such as coconuts, cigar tobacco, and abaca (Manila hemp) were provided with improved processing plant and outlets. Some of the processing was located at the point of production or at minor ports such as Cebu, Iloilo, Davao and Aparri, but most factory development occurred near and in Manila. Here were the large rice-mills, the cigar and cigarette factories, hat factories, embroidery manufacturers, boot and shoe factories, etc., as well as the main warehousing and retailing organizations. United States ambitions to reinstate Manila to its former commercial pre-eminence were frustrated by the role of British Hong Kong, but the city still was of central significance to the economy of the Philippines. By 1935 some 120 000 were employed in crop processing industries using modern factories, and the majority of these were situated in the core area.

The US period thus pointed the way forward for economic development and modernization, but left many problems unanswered. The anti-colonialism of the US administration had protected the Filipinos from some of the worst effects of the intrusion of foreign capital and the alienation of land to overseas corporations, but this did not prevent the concentration of wealth in the hands of Chinese and Japanese merchants, the *cacique* families, and some Spanish and US firms. Nor did it prevent increasing dependence on the USA. Some 80% of exports and imports were to and from the USA, largely comprising of primary produce from the Philippines and manufactured goods from the USA. The cultivated area expanded from about 7 million acres in 1903 to 16.5 million acres in 1939, but a good deal of the expansion was in cash cropping rather than food production, which left a growing population increasingly dependent on imports. Under US administration health and living conditions improved, the death rate fell and the population increased from 7.6 million in 1903 to 16.5 million in 1939. Although the average per capita income rose, there was still widespread poverty in the rural areas and substantial inequality in the provision of services between the Manila core area and the communities of the periphery. Further development in answer to such problems was, however, halted by the occupation of the Philippines by Japanese forces in December 1941.

The Malayan west coast core

In 1511 the Portuguese captured Malacca, heralding the arrival of European seafarers in South East Asia and indicating their intention to wrest control of the eastern spice trade from Arab/Islamic merchants (Mehmet 1977). Malacca was the western terminus of a network of inter-island seaways that reached out eastward to the Moluccas and brought cloves, nutmegs, pepper, ginger and other produce of the islands to the gateway to the Indian Ocean. It was also a calling place on inter-ocean voyages between India and China, with power to exact tolls on vessels using the Malacca Strait, and a vital link in the chain of sea stations the Portuguese built in the 16th century leading from Goa in India to Macao (1557) in southern China. It had historic connections with Sumatra and the ports of eastern Java and had played an important part in the spread of Islam to the Indonesian archipelago. Undoubtedly it was a centre of considerable commercial, political and cultural status and at the time of the Portuguese occupation was thronged with Javanese merchant princes, shipwrights and soldiers. During the 130 years of Portuguese rule that status was confirmed and Malacca was able to resist attacks from the Islamic powers of Achin at the northern extremity of Sumatra and Johore on the southern coast of the Malay peninsula. The Dutch, from their base in Java, took over the port in 1641 and used it as means of exerting influence on the Malayan sultanates and the growing trade in tin they were generating, but apart from architectural alterations to the town and its river frontage did not develop the location. It was perceived as an out-port of Batavia. Similarly when, in turn, the English East India Company captured Malacca in 1795 they regarded it as an out-station of their newly acquired base in Penang (1786) and for a time the Directors considered the possibility of closing down Malacca entirely as a potential rival. It is ironic that the most persuasive intervention to preserve Malacca came from Stamford Raffles, at that time an agent in Penang who was aware of the historic status of the town from his readings of Malay history, but who in 1819 by the founding of Singapore established a rival port with which neither Penang or Malacca could compete.

Under the British, the island of Penang (with its naval base of Georgetown and a small strip of adjacent mainland territory named Province Wellesley), the Dindings (a promontory and group of islands off the Perak coast), Malacca, and Singapore constituted the Straits Settlements. In origin the pattern was strategic, designed as part of the outer oceanic defences of India and as stations securing the route to the South China Sea. As with the Portuguese before them there was originally no intention of becoming involved in landward expansion, and yet within a century these coastal settlements had become nuclei of a developed core embracing most of the western lowlands of the Malay Peninsula (Swettenham, 1929).

This transformation was brought about by a number of factors. Initially the Straits Settlements had encouraged the immigration of Indians and Chinese to provide a workforce to sustain the growing commercial activities of the port cities (Purcell 1951). Indian settlement was a marked feature in the early development of Penang and, with the encouragement of Raffles, Singapore island, which was virtually unpopulated at the time of the foundation of British Singapore on the southern coast of the island, received large numbers of Chinese. By the middle of the 19th century over half of the 60 000 population of Singapore were Chinese and from there Chinese groups had spread, under British protection, to other parts of the Straits Settlements. Once established there, however, it was difficult to contain their pioneering energy. They pressed in to the hinterlands of the ports exploiting opportunities in a landscape the British authorities had dismissed as of low commercial potential (Jackson 1968). In the hinterland of Malacca for example, thousands were involved in forest clearance for the commercial cultivation of gambier but from the middle of the 19th century tin was the greatest force of attraction into the interior.

Tin ore (cassiterite) originated along with other minerals in the contact zone between the central granitic mass that comprises the backbone of the Malay Peninsula and Triassic limestones and shales, but large quantities have been washed out of the veins, pipes and lodes of the ore bodies and deposited by the rivers at breaks of slope on the margins of the highlands.

Following the discovery of the rich Larut field in 1848, Chinese miners swarmed into the interior and within a few decades their mining camps and settlements led to the growth of townships such as Ipoh, Taiping and Kuala Lumpur, and sizeable Chinese populations in the sultanates of Perak, Selangor and Negri Sembilan (Gullick 1983). Disputes with the local Malay population ensued and led to the intervention of British authority from the Straits Settlements. To restore order, not to mention the realization of economic opportunities, the tin-bearing states were declared British Protectorates in 1874 and Residents appointed to their courts (Andaya and Andaya 1982). By 1896 the states of Perak, Selangor, Negri Sembilan and Pahang were federated into the Federated Malay States with Kuala Lumpur as the capital, but in order to preserve the free trade polities of the Straits Settlements as opposed to the 'protected' economy of the federation the two administrations were kept separate. Later British suzerainty was extended to the remaining unfederated states. In 1909 the states of Kedah, Kelantan, Trengganu and Perlis, formerly under Siamese jurisdiction, were transferred to the British sphere and in 1914 the old sultanate of Johore also came to a political agreement with the British. Thus in forty years a loose political structure embracing the entire Peninsula came into being and provided the context for economic development; but within this the western, tin-bearing, Federated Malay States played a central, core role (Allen and Donnithorne 1957).

External capital, largely British, flowed into the area, with associated commercial and banking concerns, although the labour force throughout remained essentially Chinese. Since the tin ore was smelted in Penang and Singapore there was also need for improved transport arrangements. Initially the ore was carried to the coast by Chinese coolies, or by river boats – which put a premium on those mining areas within easy access of water transport – and small ports such as Port Weld, Klang and Port Dickson came into being as trans-shipment points. The first railways in Malaya were built to link the mining areas with such ports and by 1906 a rail network had been constructed which extended from opposite Penang in the north to Malacca in the south. In the years that followed the network was extended beyond the Federated Malay States. By 1909 the Johore State Railway linked the system to Singapore, and by 1918 the Kedah Railway provided a through route to the Siamese border.

The rail network thus constructed, with its feeder roads and links with the Straits Settlements and west coast ports, provided the infrastructure for further economic development. Devised originally to meet the needs of the tin traffic and movement of food supplies and general merchandise to the mining centres, it facilitated the opening up of the interior for commercial agriculture. Chinese farmers had for some time been cultivating pepper and gambier for export in the vicinity of the ports and European planters from Ceylon had experimented with coffee in Selangor since 1885 but it was not until 1890 that rubber planting proved to be the key to commercial success. First introduced in Malaya in 1877 the *Hevea brasiliensis* was shown to be well suited to the climate and soil conditions of the better drained lands of western Malaya, and world demand for rubber increased dramatically in the early part of the 20th century. By 1920 $2\frac{1}{2}$ million acres of former tropical forest were under rubber and by the outbreak of World War II rubber plantations constituted two-thirds of the 5 million areas of cultivated land in Malaya. Much of the forest clearance was carried out by Chinese contractors and there was also considerable Chinese capital invested in the plantation economy, but ownership increasingly passed to large British-financed joint stock companies such as

Dunlop and the London Asiatic Rubber company; the managerial staff was European but the labour force on the plantations was drawn from southern India in the form of indentured Tamil workers. Some of the larger estates exceeded 100 000 acres. They provided a leadership which was followed by hundreds of small estates of 100 acres or less and many Chinese owned plantations of 100–500 acres. By 1939 Malayan rubber production stood at about 500 000 tonnes per annum. This development of commercial agriculture, based on the plantation system, reinforced the dominance of the western core. Development to the east of the mountains was restricted to small, isolated, coastal pockets. Large areas of the east remained under forest and were occupied by primitive hunters and collectors, shifting cultivators, traditional rice-growing villages along the main rivers, and communities of peasant fishermen along the surf-beaten coasts of the South China Sea. There were schemes for economic development of these terrains but depression of the world economy and world prices for tin and rubber in the inter-war years reduced the impetus of Malayan expansion from the western core into the eastern periphery and perpetuated a regional inequality that has persisted to the present day.

Population, wealth, transport facilities, accessibility to medical and educational establishments, employment opportunities in private and public undertakings, prospects of political advancement, higher per capita income, and the benefits of urbanization processes were phenomena of the western coastal zone. Towns such as Kuala Lumpur and Ipoh grew apace and the port cities of Penang and Singapore benefited hugely. The latter had become the capital of the Straits Settlements in 1837, handled some two-thirds of Malaya's considerable foreign trade, functioned as one of the world's great immigrant ports, became a trans-shipment port of world status and a naval base, developed a major dock system at Keppel Harbour to supplement the Singapore river docks and roadstead of Raffles' port-city, possessed a number of processing industries, preserved its initial 'free port' status, provided banking and other commercial facilities, and attracted a cosmopolitan population. By World War II the population of Singapore island had reached about 1 million, of which about 80% were Chinese. It was a bustling, energetic city, reflective of the economic progress that had made western Malaya one of the most advanced cores of the tropical world.

British Burma

British political involvement with the ancient kingdom of Burma arose in the early decades of the 19th century from troubles on the eastern frontiers of British Bengal, and from then until the government of Burma was separated from the Indian administration in 1935, Burma constituted an eastern frontier zone of the British Indian Empire (Tate 1979). The annexation of the country took place in three stages. In 1826 by the Treaty of Yandabo at the end of the First Burma War Burma ceded the coastal territories of Arakan and Tenasserim as well as relinquishing suzerainty over Assam and Manipur. In 1852, following the Second Burma War, the British took possession of Lower Burma comprising the deltaic lands of the Irrawaddy and Sittang, south of a Prome-Toungoo line. In 1886, as the result of a third war, Upper Burma was annexed, the Burmese king deposed and in the following years British authority extended into the Shan States and into the far north of the Irrawaddy Valley towards the frontiers of China. The conquest thus took some seventy years, with various regions experiencing different durations of British rule.

During this period of frontier expansion the main nucleus of the Burma core area lay in the dry zone of the central Irrawaddy valley. This was the most populous region of the kingdom, with irrigation systems in the Kyaukse, Myitnge and Mu valleys and a diversified agriculture including the cultivation of millets, peas, oilseeds, pulses and cotton, as well as rice. Successive royal capital cities, such as Ava and Amarapura, had been built on the banks of the great river in the eastern part of the lowlands and in 1857 Mandalay was constructed – its antique morphology belying the recency of its creation. In contrast, the southern (Pegu) core had lost much of its ancient status and many of its people as a result of successive military devastations, and when the British occupied Lower Burma Rangoon was a small river port on the margin of the delta. Its selection as the capital city, initially of Lower Burma and after 1886 of all Burma, is characteristic of European colonial impact whereby the urban nuclei of cores tend to be ports and power is drawn down from inland locations to tide water. During the period of British rule in Burma Mandalay hardly grew at all and its population was only 163 000 at the outbreak of World War II, while Rangoon grew from a city of 92 000 in 1871 to 500 000 in 1941. River, rail and later road transport and oil pipe-lines converged on Rangoon and its industrial satellite at Syriam. The city became the primate centre of the Burma core – the colonial capital, the main commercial hub, the chief industrial centre with rice-mills, saw-mills (teak), engineering workshops and an oil refinery. It became a university town, and a religious centre associated with the great Shwe-Dagon pagoda. A new city, on a rectilineal pattern, grew up behind the port with a cosmopolitan population including a considerable Indian element.

Rangoon acted as a growth pole for other changes within the core. It was the base from which was organized the remarkable transformation of the delta lands from a low level of utilization to one of the largest rice surplus regions of southern Asia (Cady 1958, Adas 1974). The process began in the 1860s with the building of embankments and other protective works against the summer floods and the ingress of salt water impelled by the monsoonal surges from the Gulf of Martaban, the institution of surveys to assess land potential, the establishment of markets and rice milling centres such as Bassein, and the provision of a transport infrastructure to serve incoming colonists. As the pioneer fringe gained momentum it drew in Burmans from central Burma, Karens from the eastern hills, and large numbers of immigrants from India. In the 1920s Rangoon was one of the greatest immigrant ports in the world, receiving large numbers of Indians and Chinese. The Indians provided a labour force and a peasantry, but as time went on became increasingly involved with marketing, money lending and landlordism. The initial colonizing unit was about 15 acres under private, peasant ownership but eventually Indian landlords held large tracts of land worked by Burmese tenants, and by 1941, before the Japanese invasion and local hostility caused the flight of the Indian population, the investment of Indian capital in Burma, chiefly in agriculture and agro-industries, was as large as the entire British investment in the country. The consequence in economic terms was a steep rise in exports of cleaned rice, from 0.5 million tons in 1886 to 3.5 million tons in 1941, as well as substantial shipments to the Dry Zone and to the towns. The consequence in social terms was the creation of minority problems, increasing social dualism and conflict.

British investment from the Rangoon base was more widely disposed. In addition to commercial rice agriculture capital was invested in transport, trade and manufactures, teak and mining. The teak-yielding monsoonal forests of the Pegu Yoma, the Salween valley and the eastern margins of the Indo-Burmese mountains were worked by elephant power and the timber rafted down to saw mills at river mouths. Forestry and conservation were highly developed by the excellent Burma Forestry Service – an arm of government – which provided an overall ecological supervision within which the timber companies operated. After 1886 British capital also moved into the Dry Zone with the encouragement of commercial production of cotton and groundnuts and the growth of agro-industries at Mandalay and Myingyan. The most successful venture, however, was the development of the Dry Zone oil fields. The formation of the Burmah Oil Co. in 1886 transformed the perception of the resource. Fields were developed at Yenangyoung and Chauk and the petroleum transported initially by river barges and later by pipeline to the refinery and bulk-storage depots at Syriam. By modern measures production was never large (275 million gallons in 1939) but it represented a fifth of the total oil production of the British Empire during the 1930s and found a ready market in India.

Most of this economic growth led to structural changes within the lowland core and increased the regional inequality between the core and the highland periphery. There were growth points in the latter but these were small pockets in a large area known officially as the 'Backward Tracts', and were usually associated with extractive industries. Economically, socially and politically the hill tribes were remote from the peripheral lowland communities, let alone the commercialized and urbanized population of the core. It can be argued that the colonial power neglected such peripheral communities because little economic benefit could accrue from the large expenditure required to 'open up' the hills to forces of modernization. It should not be forgotten, however, that liberal schools of thought within the policy-making institutions of the colonial power tended to be paternalistic and protective in their attitudes towards 'primitive societies'. In Burma, as in India, regulations were formulated to prevent invasion of the tribal areas by commercial or other agents who might destroy the distinctive traditional cultures of the hills and forests.

Dutch Indonesia

The Dutch East India Company, or Vereenigde Oost-Indische Campagnie (VOC), under its charter awarded by the States-General of the Netherlands in 1602, was granted a monopoly of Dutch trade east of the Cape of Good Hope with the object of breaking the monopoly the Portuguese had established in the Indian Ocean during the 16th century (Ricklefs 1981). Initially there was no intention to acquire territorial possessions or to assume the responsibilities and expense of political administration in areas distant from the homeland, and the directors of the VOC at Amsterdam attempted to maintain that position right up to the dissolution of the Company in 1798. Representatives of the Company in the East Indies, however, caught up in the intense rivalry for trade in distant waters, found themselves increasingly drawn in to local political and military activities with native states as well as European commercial competitors in order to establish and maintain Dutch control over the sources and channels of sought-after commodities, especially spices. The Company seized Malacca from the Portuguese in 1641, drew up treaties with harbour sultans fixing quotas and prices, and weakened the authority of the Indonesian states. From the Company headquarters at Batavia (now Djakarta), established in 1621, it controlled the Sunda and Malacca Straits, and extended its power in Java, exacting tribute from petty rulers, enforcing levies of rice and timber from the states of Central Java, and introducing commercial crops such as coffee and sugar on a system of enforced deliveries.

The Netherlands Government who took over from the VOC in 1798 faced similar problems and it was not until 1830 that the 'take-off' stage of economic growth began. The assumption of political control over the entire area that was ultimately to comprise the Netherlands East Indies (and Indonesia) was a long, protracted affair extending in the case of some inland territories well into the 20th century, but for most of the 19th century Java was the focus of Dutch development and the scene of a remarkable transformation of the earlier core. Some measure of that change can be derived from the facts that between 1830 and 1930 the population of Java increased from 7 million to 42 million; that two-thirds of its area was brought under cultivation in spite of the problems of the volcanic, mountainous interior of the island; that 8 million acres of land were under intense wet rice (*sawah*) agriculture supported by irrigation systems built by Dutch hydrological engineers; that a further 11 million acres of dry field (*tegalan*) cultivation were devoted to a rich diversity of crops; that 2.5 million acres were leased to estates growing rubber, cinchona, tea and coffee in the uplands and sugar and tobacco in the lowlands; that the island had achieved a transport network that was the envy of many a 'developed' country, with no location more than 50 miles from rail service; and that its health, educational, banking, agricultural research, and communication facilities had been developed to a level superior to any area in the tropical world. Such transformation brought its own developmental problems but the rapidity and dimensions of the change cannot be gainsaid.

The key to the economic 'take-off' was the introduction in 1830 of the so-called Culture System (*cultuur stelsel*), and during the formative years 1830–70 it constituted the central policy of the Dutch for land development in their eastern empire. In essence it was an extension of the corvée system, not unknown to South East Asian or European landlords, to substitute the product of local labour for the capital investment the Dutch of the home core in Holland were unable to provide. Local farmers were given remission of land taxes on condition that they undertook to cultivate government-nominated cash crops on one-fifth of their land-holdings or alternatively worked for 66 days of the year on government estates or projects. Annual crops such as tobacco, indigo and particularly sugar cane were cultivated on *sawahs* in rotation with rice, and perennials such as coffee, tea, pepper, cinchona and cinnamon were grown on estates in enclaves separate from indigenous systems. Sugar did exceptionally well in the wet rice area of the North East Java coastlands and became the dominant export commodity. The building of dams and irrigation systems by corvée labour under the Culture System legislation was aimed at both a sugar-rice cropping combination and the need to overcome water shortages during the dry season in the eastern parts of the island. Coffee on the other hand was grown on plantations under European or Chinese management, and this system of land-use spread into the hill areas of East, Central and then West Java.

Under the Culture System, Java was made to pay its way for the first time, but at a price. It accentuated the dual nature of the economy, whereby a capital-intensive Western sector began to differentiate from a labour-intensive Eastern sector. The sugar cane and wet rice, growing side-by-side in the fields of the *sawah* villages, belonged to two distinct systems – one an element in a 'production for gain' system, the other belonging to a 'production for use' system. For the Dutch, the operation of the former system required the latter system to remain static, and as Geertz (1963) has pointed out the policy shielded indigenous agriculture from the forces of change at a time when reform would have been possible and resulted in an involution not only of peasant agriculture but of peasant culture as a whole. As population grew, as a result of famine-relief services and improved water supplies as well as the institution of *pax Nederlandica*, the additional numbers were absorbed into the *sawah* system. There were few alternative outlets, since the Dutch did not encourage industrialization, and urbanization was a process confined to other ethnic groups. The native peasant society turned in upon itself, producing more and more complex social and role relationships rather than developmental structures open to modernization. The Culture System had other effects on developmental processes. Apart from some localities in Sumatra it was applied to the Java core only, since by the time it was abandoned in 1870, few areas in the peripheral Outer Isles had been brought under full Dutch colonial administration. As a consequence regional inequality increased.

In 1870 an Agrarian Land Law was enacted, which while prohibiting the outright alienation of settled land to foreigners also confirmed that all uncultivated 'waste' was the inalienable property of the state. Henceforth such land could be leased to private plantation companies on leases of up to 75 years. Mechanization of sugar milling also encouraged the construction of 'centrals' to which the produce of a number of villages could be sent for processing, and this made the arrangements of the Culture System increasingly obsolete. At the same time, several large corporations which had made money out of the trade in commodities generated by the Culture System began to invest in the sources of the trade. This infusion of capital led to substantial structural changes on the Dutch side of the economy without major impact on the village economy of Java. There was a further diversification of cropping patterns in the *sawah* districts, with the introduction of maize, soyabeans and groundnuts, and villages leased land to the state or corporations for the cultivation of sugar for the centrals, but the essential, involuted structures of the wet-rice communities did not change. As the tempo of capital-engendered economic growth in the Dutch sector increased, they withdrew still further into the 'little tradition' of the early core culture.

In the Outer Isles, however, the dualist character of development took a different form, much more akin to the segregative structure of Malaya than the symbiotic relationships of Java. In the periphery, economic development followed an enclave pattern of clearances in a great forest of shifting cultivation (*ladang*). In the main, growth was highly localized and capital-intensive, and it focussed on specific agricultural and mineral products for the world market. In 1863, for example, tobacco-planter Jacob Nienhuys moved from East Java to the Deli area of North East Sumatra and initiated there a remarkable commercial agricultural enclave with a highly organized plantation economy. Known as the *Cultuurgebied*, it ultimately comprised a tobacco area in the Deli valley around Medan, a rubber zone to the north and south of this location, coconut estates along the coast, tea estates on the eastern slopes of the Batak highlands and oil-palms and sisal estates as later accretions on the periphery. By 1930 the 10 000 km^2 of the *Cultuurgebied* contributed about one-third of the total export earnings of the Netherlands East Indies. Sumatra also made a substantial contribution from the oilfields of Palembang-Djambi and the tin deposits of the islands of Bangka, Belitung and Singkep off its south east coast. Elsewhere, there were vast areas in the interior of Borneo, New Guinea and in the Lesser Sunda islands where little change occurred.

As population pressure built up in the Java core the government attempted to relieve it in various ways. Under an 'Ethical Policy' measures were taken to improve health and welfare provisions in some of the heavily populated and distressed parts of the core. There was also a belated attempt in the 1930s to broaden the range of industrial employment. Increasing urbanization also provided tertiary, service industry and part-time employment in the 'informal sector' of the larger towns. But rural–urban migration was not large and many of the towns remained alien to the predominantly village population. Batavia, for example, had grown to house a population of half a million by 1940 and had established a new artificial harbour at Tandjong Priok six miles to the north east of the old, canalized river port, but the old town – the chief business and industrial centre – had become largely Chinese while the Dutch Europeans and Eurasians had moved out to the higher class residential suburbs of Weltevreden and Meester Cornelis. The Indonesian population of the town lived in *kampongs* in the interstices of the urban fabric, carrying on *rus in urbe* activities, but never fully integrated into the urban economy and way of life. Urban-rural dualism did not provide the structure colonial Java required to relieve the pressures building up in rural deprived areas.

In such circumstances great store was set on plans to move population from Java to the Outer Isles. As early as 1905 attempts were made to resettle Javanese on pioneer agricultural tracts in South Sumatra but in spite of considerable government support the outcome was disappointing. Later schemes, which involved more careful selection of pioneers, greater reliance on individual effort, more preparation of the ground and the use of earlier settlers to employ later settlers until these were established, proved to be more successful, and in the late 1930s about 100 000 people per year were migrating from the overcrowded parts of Java to the Outer Isles – and particularly to Sumatra. Since at this time the population of Java was increasing by 600 000 per year this did not solve the problem, but at least was a contribution to redressing the great imbalance between the population resources of the core and the periphery. Many migrants moved to the agricultural plantation enclaves as well as the *sawah* pioneer fringe, e.g. in the 1930s the population of the *Cultuurgebied* comprised 35% Javanese, 20% indigenous Malays, 20%

Batak, 11% Chinese as well as other groups such as British Indians.

Thus the operation of core-peripheral processes in a colonial setting is well-exemplified by the Dutch experience in Indonesia. For various historical and environment reasons, the development of Java took a different course from that of the Outer Islands, and within Java itself dualism achieved classical dimensions. The 'modern' elements of the economy of the Netherlands East Indies were an extension of the Dutch economy into what was significantly called 'Tropical Holland' and in discussions on 'dependency' it should not be overlooked that more than any other Western power in Asia the Dutch economy was dependent on its Eastern colony. It was this commitment and identification that underlay the great reluctance of the Dutch to grant independence to Indonesia after World War II. Equally it can be argued that the indigenous economy and culture was immobilized by the Dutch presence, and was dependent on the primary core in Holland. Unlike Britain, Holland was not a great industrial power seeking markets in its tropical possessions and was unwilling to encourage the raising of expectations and demands that it could not satisfy. By restricting indigenous change and by blocking the intrusion of other forces of change such as international capitalism it can be argued that the Dutch 'underdeveloped' Indonesia – but when one examines the intensity of land-use and magnificently engineered landscapes of Java one wonders whether geographically this is the right word.

French Indo-China

Although French traders and missionaries had previously been active in mainland South East Asia, France did not enter the colonial field there until the middle of the 19th century. By that time its dreams of an Indian empire and a French-dominated Indian Ocean had been forsaken and imperial ambitions had turned to Africa, but it could not resist competing with its old adversary, Britain, for the potential rewards of opening up a back door into southern China. Thus while Britain was extending its influence in Burma and raising possibilities of connections with Yunnan via the Shan States, the French turned to Annam and the potential routeways of the Mekong and Song-koi into the same territory. In 1859 a French force captured Saigon, established the colony of Cochin China astride the Mekong delta and within five years proclaimed a protectorate over Cambodia and began to explore the Mekong valley as a possible route north. In 1882, following Dupuis' exploration and recommendation of the Song-koi route, a French force occupied Hanoi and a French fleet bombarded Hue the imperial capital until the emperor conceded Tonkin and Annam as French protectorates. In 1887 these territories were combined in the French Indo-Chinese Union. During the 1890s various Laotian states on plateaux overlooking the Mekong were brought together under the protectorate of Laos and added to the Union.

Of all the European colonial powers in South East Asia, the approach of France was the most hierarchical administratively, exploitive economically, and elitist culturally towards its colonial dependency. The colony was regarded as an extension of France – *'la France d'outremer'* – in which the needs of metropolitan France took priority and in which French culture was to be the civilizing force. Indo-China was to provide a market for French industry, an outlet for French investment, a supplier of raw materials and food for metropolitan France, a route to China and a sphere of operation of the French Roman Catholic missions. In the event, however, the achievement of these aims was modified by developmental processes already operative within the area before the French arrived – particularly in relation to the early cores already identified.

By the time the French captured Saigon and converted it into a replica of a French provincial town, Annamite colonists from the northern coastlands had already established pioneer settlements in the largely depopulated Mekong Delta as far as the Bessac River. By the construction of major canals, for drainage and water transport, French planners promoted a further advance of this frontier into the delta and resurrected its ancient status as one of the rice bowls of South East Asia (Robequair 1944). Land-holdings granted by the French, however, were large – up to 3000 acres – with French and Vietnamese landlords, and tenant farmers who paid 40% of their crop as rent. Most of the transport, milling and marketing of the rice in the delta was undertaken by Chinese, and the largely Chinese town of Cholon, at the meeting of the canalized waterways, was the main centre of the rice trade. Saigon and Cholon constituted the urban nucleus of this rejuvenated southern core and when the commercial system was extended to Cambodia, with the cultivation of maize as well as rice, Phnom Penh, the old capital, was also reactivated as a central place. The Chinese became an important element in all the major urban places.

The development of the infrastructure of the southern core also encouraged the introduction of plantation agriculture. Coffee and tea estates were located on the hillocky red earths on the margins of the delta, but after World War I there was a major expansion of rubber planting on old alluvial deposits of former deltaic margins. Rubber planting and processing was almost exclusively a French concern, with labour drawn from Tonkin on three-year contracts, and by 1941 some 300 000 acres were planted. In total the southern area contained some 80% of Indo-China's plantation acreage, and by 1938 74% of the value of exports from French Indo-China were from the southern core.

Although the French took great pride in the development of the southern coast of Cochin China and its appendage in Cambodia, the superior demographic, economic and political 'weight' of the northern core in Tonkin eventually began to attract greater attention. In 1902 the capital of the Indo-Chinese Union was moved to Hanoi and the French invested substantial capital and energy in rebuilding and extending the town and its outport Haiphong. In 1910 the Kunming-Hanoi-Haiphong railway was completed. A very dense road network was built which centred on the city, encouraging its growth as a major market and commercial centre. It and its associated urban nuclei were also industrial centres. Hanoi specialized in rubber, glass, pottery and leather manufactures; Haiphong in ship-building, cotton spinning, cement making and rice mills; and Nam Dinh in cotton spinning and weaving, silk, glass making and distilling. At first sight, the development of cotton textile manufacture would appear to conflict with French colonial policy but production was aimed at the lower end of the market and competed with Indian rather than up-market French textiles.

Around the urban-industrial nuclei stretched the densely populated terrains of the Song-koi delta. Here pressure on the land was already a problem before the French assumed control and measures taken by them to reduce natural hazards, such as the great floods of 1926, and to reduce the death rate from diseases such as malaria, smallpox and cholera, increased population growth. The increase of food production by improvements in the irrigation system and expansion of the area double-cropped with rice did something to counterbalance this but the Tonkin core became increasingly dependent on rice imports from the south. Subdivision of holdings went on apace and in 1940 92% of the Tonkin peasantry farmed holdings of less than one acre. Poverty and rural indebtedness were widespread and a class of landless agricultural labourers and vagrants began to emerge.

As the social transformation proceeded the centre of Vietnamese nationalism became increasingly associated with the northern core. It was the social and economic inequities of the north, allied to an educational system that produced a local intelligensia, which provided the base for the rise of Ho Chi Minh's Revolutionary Association in the 1930s and ultimately the power struggle between North and South Vietnam in the post-war period. During the French colonial period Cambodia and Laos stagnated, and Central Annam with its royal court at Hue was allowed to persist in its traditional structures; but of the two areas of development, the Mekong Delta began to assume more peripheral characteristics as power of the northern core increased.

Thailand

Of all the early cores, that of Thailand (Siam) was the only one to avoid colonial rule by a Western power, and consequently the development history of the country provides an important control against which the role of colonialism elsewhere in South East Asia can be assessed. It raises the important question of the extent to which structural transformation could occur in a tropical country without the experience of Western political imperialism. In the case of Thailand there was no primary European core to which it was specifically 'dependent' as a peripheral entity, and there can be no argument that it was 'underdeveloped' by some European power in the process of self-engrandisement. As indicated earlier, it exhibited a well-differentiated core-periphery structure long before European colonialism became a force of structural transformation in South East Asia and its reaction to Western culture in its guise of 'modernization' gives some indication of underlying developmental processes operating in South East Asia as a whole, regardless of the specific forms imposed by the British, Dutch and French elsewhere in the region.

Until the 19th century, Thailand remained virtually a closed cultural system so far as Western influences were concerned. Its monarchical organization reflected the ancient Indian-Buddhist ideals of statehood and the necessity of central direction of a kingdom locked in perpetual struggle for survival against the expansionary forces of its neighbours. The resources of the core of the central plains were organized to that end and the periphery included broad frontier zones with Burma, Laos, Cambodia and Malaya. Western pressures were brought to bear on both core and periphery. The kingdom had to yield peripheral territories to pressures from British Burma and Malaya in the west and the French along the Mekong marshlands in the east until the merits of establishing a buffer zone between themselves came to be recognized by the two major powers. The Anglo-French agreement of 1896 guaranteed the neutrality of the Chao Phraya valley and thus assured the continuity of a Thai state.

By bending rather than breaking under such external pressures, the Bangkok dynasty bought time to transform the internal structures of the state. In the reign of Rama V (1868–1910) in particular, measures were taken to link peripheral areas more firmly with the core. The construction of the railway network for example was undertaken more from political than economic motives, e.g. the Bangkok-Ayuthia-Korat railway completed in 1900 was related to French advances in Laos and the Ayuthia-Chiengmai railway started in 1898 and completed in 1921 to British movements in Burma. However, such railways and their feeder roads had important economic effects. They brought the economies of the Korat Plateau in the north-east, with its rice and cattle farming, the northern region, with its mixed agriculture and forestry, and the southern region, with its peasant-fishing communities into closer relationship with the core area of the Central Plains. In the southern region improved transport facilities encouraged the development of rubber cultivation on Thai, Chinese and Malayan small-holdings and an increase of tin mining.

The gradual opening up of the economy to the influences of international trade had other consequences. The import of Western, and later Japanese, manufactured goods – especially textiles – had a depressive effect on indigenous handicrafts in the core area. Cotton and silk weaving declined in the face of competition from cheap imported European cottons and although the manufacture of tiles, bricks, carts, boats and silver (*niello*) ware continued in the Bangkok region, the role of handicrafts became less important in the core area than it was in the periphery. Most of the small domestic industries in any case were in Chinese hands and it was not until after the 1932 political revolution, with its concomitant policy of national economic planning and import substitution, that Western-style factories, such as cotton, sugar and paper mills, were built in Bangkok. The initial response to meet the growing demand for Western goods was, however, in the expansion of production for export of the one commodity in which the country had a competitive edge, namely rice. By the 1930s Thailand with 1.5 million tonnes per year surpassed French Indo-China and was second only to Burma as an exporter of rice. Some 75% of this came from the delta and was achieved by improvement of flood control and the extension of irrigation. The delta was the scene of much spontaneous colonization by Thai farmers who achieved squatters' rights over areas they reclaimed, and population densities increased to match those elsewhere in the Central Plains (Ingram 1971).

The population of Thailand increased from about 5.5 million in 1855 to 15 million in 1941, partly as a result of natural increase but also due to massive immigration, particularly of Chinese. Initially, the Chinese came as labourers on canal and railway construction, but following a great surge of immigration in the early decades of the 20th century they came to occupy important niches in the economic structure of the country. While the Thai upper classes became increasingly absorbed in politics and administration and the Thai peasants in rice production for export as well as subsistence, the Chinese were quick to exploit the trading opportunities of a developing commercial system. By 1930 90% of rice milling and marketing was controlled and staffed by Chinese; they were greatly involved in market gardening on the outskirts of Bangkok; they were small-holder cultivators of rubber in the south and had built up a sizeable community of tin miners in the Phuket area; they were prominent in the urban retail trade and as financiers; they had established their own chamber of commerce in Bangkok and their own educational system and social services. As they flourished economically in contrast to the increasing indebtedness of the Thai rural population, and as they grew in number to over 2 million, the Chinese minority and its control became a problem of government, but there is no doubt that in the development process they played a vital role. Elsewhere in South East Asia their energy and entrepreneurial capacity were applied within the framework of colonial structures but as the experience of Thailand demonstrates immigration of Chinese into South East Asia would have taken place in any case and their contribution to development has to be viewed alongside the European contribution.

Decolonization and the colonial legacy

By the outbreak of World War II, the ground plan for development in South East Asia was well established, representing in many instances colonial entrenchment of already existing spatial patterns. In essence, it comprised a series of seaboard cores or nuclear areas in which European impact was at a maximum and the processes of development, urbanization and modernization and population growth at their most intense; a series of peripheries around the cores in which enclaves of development, based usually on mines or plantations, were linked to the cores by transport systems expanding from maritime bases; and, beyond these, extensive forest tracts occupied by primitive communities now regarded as politically if not culturally linked to the colonial cores as a result of the European propensity to replace vague frontier zones with internationally agreed linear boundaries. Within such bounded territories, it was anticipated that developmental forces radiating from the cores would ultimately reduce the regional inequalities that had existed for so long and, with due protection for primitive communities exposed to the forces of modernization, raise the general standard of living and equality of life of all the colonial peoples. Reduction of the Malthusian checks of famine, pestilence, and war by European administrations had resulted in a five-fold increase in population (to about 150 million) in South East Asia in the century 1840–1940 and there was every reason to suppose that with continued population growth communities would swarm from the cores and open up new frontiers in the thinly peopled forests of the periphery. It was true that the international economic recession of the 1930s had slowed down the pace of economic growth and expansion and exposed the dangers of over-reliance by colonial economies on a few commodities in the world market, but South East Asia had become one of the most productive areas of the tropical world, contributing much of the world's rubber, tin, copra and palm oil, and there was every expectation that with an upturn of world trade, a greater diversification of production and continued imperial protection of markets and production, progress would continue unabated – to the mutual benefit of metropolitan cores and colonial territories. In more liberal circles it was also visualized that by increasing the contribution of educated native administrators to the government of such territories it would be possible to confer gradually degrees of administrative independence, as well as allow retention of larger percentages of the revenue to satisfy nationalist aspirations for internal development.

Such European anticipations and plans collapsed with the Japanese invasion of the area in 1941–2. At a stroke, the myth of European superiority which had served the colonial powers for four centuries and allowed them to control vast territories with minimum commitment of men and resources was exposed by an Asian power. The apparent ease with which the land and sea forces of Japan

brushed aside power in the Pacific after the bombing of Pearl
Harbour and British power in the Indian Ocean after the destruction
of the British Eastern Fleet and the fall of Singapore, revealed the
geopolitical weaknesses of the European presence in South East
Asia, at the end of long lines of maritime communication. The fact
that the colonial powers were locked in conflict with Germany at the
other end of the world and that the French and Dutch home cores
had already been overrun did little to reduce the perception of
European weakness and inability to combine in opposition to Japan,
nor did the fact of the Japanese defeat in 1945 and the
re-occupation of the area by Indo-British and American forces. The
Japanese interlude of four years' occupation, materially destructive
as it was, confirmed nationalist aspirations present in the colonial
territories and provided the context for a major disconnection of
political links with the West it had taken four centuries to forge. In
two post-war decades, the political destabilization brought about by
the traumatic experiences of World War II, including the decline in
power of former European home cores, resulted in the granting of
independence to former colonial territories. By 1963, the period of
direct Western colonial rule had ended, only Brunei being a partial
exception.

The successor states inherited the bounded territories of former
colonial possessions and, except for Indonesia's take-over of East
Timor from the Portuguese, as shown in Fig. 2.3 there have been no
major revisions of international land boundaries, although as a
result of growing interest in the potential petroleum and other
resources of the South East Asian shelf seas the maritime
boundaries of the states are receiving closer scrutiny and debate.
Along with territories, however, the successor states inherited many
of the development problems of the colonial era. Indeed in several
respects such problems were exacerbated by the exigencies of
new-found statehood. Whereas plural societies with their
complementary economic functions and divisive social structures
could be tolerated, and even encouraged, by alien powers in the
colonial era, they constituted major obstacles to the creation of
national identities. The varied treatment of the substantial Chinese
minorities and less numerous Indian communities in most of the
states is indicative of this problem. Each state had also to face the
problems of integrating culturally distinct communities of the
periphery and extra-periphery with the ideals and policies of central
government, in a situation in which it was increasingly necessary to
demonstrate that the writ of such governments extended to their
international boundaries. Burma, for example, has encountered
enormous difficulties in this respect.

It was also necessary for the new states to replace the professional
cadres of the colonial regimes and seek new sources of aid and
capital investment to restore, run and enhance those services
established in the later colonial era. In most cases, they faced the
same problems of over-dependence on a few commodities to earn

Fig. 2.3 Political boundaries and capital cities

foreign exchange but without the former 'protection' and support of imperial systems. In pursuit of the status symbols of industrial development, many can be accused of ushering in an era of economic 'neo-colonialism' with foreign capital still prominent in new forms of dependency. There is still the need to diversify production and make realistic evaluations of natural resources to replace those wild assertions of untold wealth characteristic of early nationalist statements.

Many of these problems and potentialities are discussed in later chapters in this volume, but in the meantime it is possible to observe that to date most trends and forces have tended to emphasize the power and role of the former colonial cores. Population has tended to be drawn in to such cores and their metropolitan centres, with all the problems attendant on massive and rapid urbanization. The core infrastructures, largely built during the colonial era, have provided the main attraction to new industrial and commercial developments, and central governments, whatever their political policies and ideals, continue to draw most of their financial resources from such cores. With limited resources it has not been possible for them to divert much investment to promote dispersal away from the cores, with the result that regional inequality still remains the dominant characteristic of the geography of South East Asia. Political transformation itself cannot change in the short-term the structural pattern of core and periphery which has been entrenched into the landscape by over two millennia of successive colonization and development.

Further reading

Adas M (1974) *The Burma Delta: Economic Development and Social Change on an Asian Rice Frontier*, Wisconsin University Press, Madison.

Allen G C and **Donnithorne A G** (1957) *Western Enterprise in Indonesia and Malaya*, Allen & Unwin, London.

Andaya B W and **Andaya L Y** (1982) *A History of Malaysia*, Macmillan, London.

Boeke J O M (1942) *The Economic Development of the Netherlands Indies*, Institute of Pacific Relations, New York.

Cady J F (1954) *The Roots of French Imperialism in Eastern Asia*, Cornell University Press, New York.

Cady J F (1958) *The History of Modern Burma*, Cornell University Press, Ithaca.

Coedes G (1966) *The Making of South East Asia*, Routledge and Kegan Paul, London.

Evers H D (ed.) (1980) *The Sociology of South East Asia: Readings on Social Change and Development*, Oxford University Press, Kuala Lumpur.

Fisher C A (1964) *South-east Asia*, Methuen, London.

Furnivall J S (1939) *Netherlands India: A Study of a Plural Society*, Cambridge University Press, London.

Furnivall J S (1948) *Colonial Policy and Practice, a Comparative Study of Burma and Netherlands India*, Cambridge University Press, London.

Geertz C (1963) *Agricultural Involution: the Process of Ecological Change in Indonesia*, University of California Press, Berkeley.

Gourou P (1953) *The Tropical World*, Longman, London.

Groslier B P (1966) *Indo-China*, Archaeologia Mundi, Geneva.

Gullick J M (1983) *The Story of Kuala Lumpur 1857–1939*, Eastern Universities Press, Singapore.

Hall D G E (1964) *A History of South East Asia*, Macmillan, London.

Ingram J C (1971) *Economic Change in Thailand since 1850*, Stanford University Press, Stanford.

Jackson J (1968) *Planters and Speculators: Chinese and European Agricultural Enterprise in Malaya 1786–1921*, University of Malaya Press, Kuala Lumpur.

Mehmet O (1977) 'Colonialism, dualistic growth and the distribution of economic benefit in Malaysia', *South East Asian Journal of Social Sciences* **5**, 1–23.

Osborne M (1985) *Southeast Asia: An Illustrated Introductory History*, Allen and Unwin, Sydney.

Pelzer K J (1945) *Pioneer Settlement in the Asiatic Tropics*, American Geographical Society, New York.

Purcell V (1951) *The Chinese in South East Asia*, Oxford University Press, London.

Ricklefs M (1981) *A History of Modern Indonesia*, Macmillan, London.

Robequain C (1944) *The Economic Development of French Indo-China*, Oxford University Press, London.

Smith R B and **Watson W** (eds.) (1979) *Early South East Asia: Essays in Archeology, History and Historical Geography*, Oxford University Press, New York and Kuala Lumpur.

Swettenham F (1929) *British Malaya*, The Bodley Head, London.

Tate D J M (1970) *The Making of Modern South East Asia: The European Conquest*, Oxford University Press, Kuala Lumpur.

Tate D J M (1979) *The Making of Modern South East Asia: The Western Impact*, Oxford University Press, Kuala Lumpur.

Wales H G Q (1951) *The Making of Greater India*, Quaritch, London.

CHAPTER 3

Concepts of development

David Drakakis-Smith

The previous chapter has clearly indicated the considerable upheaval in the political, social and economic systems of South East Asia wrought by the fluctuations in the balance of political power during the Second World War. Although the colonial powers returned following the withdrawal of Japan, any thought of a rapid restoration of the pre-war political pattern was quickly challenged by a series of vigorous liberation movements. Within the region, independence was mainly achieved by the late 1950s, the bitterness of the struggle being largely correlated with the extent of European settlement rather than the value of commodity exports. Vietnam and Indonesia, for example, experienced very bloody decolonization struggles due to the ferocity with which their extensive French and Dutch immigrant population sought to protect what they regarded as the just rewards of their labours. Few wished to return to a war-devastated Europe.

The surge of independence meant that for the first time the former colonial powers were faced with the fact that many of their traditional supplies of raw materials or cheap food could no longer be assumed to be automatic and that access to these sources would be strongly influenced by the development paths chosen by the newly independent nations. Prior to 1950, 'development' had been discussed only in the context of the European or North American countries. After this date, the concept of development became firmly enmeshed in the growing interest in what was becoming known as the Third World.

At that time, the term Third World was used simply to cover the countries that lay outside the developed capitalist (First) and developed socialist (Second) Worlds. The retention of the term as a convenient way of referring to all developing countries has disguised the fact that over the last three decades massive development gaps and diversification of interests have emerged *within* the Third World. The wealthy oil-based economies, the rapidly industrializing countries (which are well represented in

Table 3.1 Economic and social indicators in South East and East Asia

	GNP per capita		Percentage of labour force in:			Life expectancy at birth (years)
	US Dollars 1985	Average annual growth rate (per cent) 1965–85	Agriculture 1980	Industry 1980	Services 1980	
China	310	4.8	74	14	12	69
Indonesia	580	4.8	57	13	30	55
Philippines	580	2.3	52	16	33	63
Thailand	800	4.0	71	10	19	64
Malaysia	2000	4.4	42	19	39	68
Korea Rep. of	2150	6.6	36	27	37	69
Hong Kong	6230	6.2	2	51	47	76
Singapore	7420	7.6	2	38	61	73
Low-income developing economies	270	2.9	72	13	15	60
Low middle-income developing economies	820	2.6	55	16	29	58
Upper middle-income developing economies	1850	3.3	29	31	40	66

Source: World Bank, *World Development Report 1987*.

South East Asia), the primary commodity producing states and the perenially poverty-stricken countries, all have very different links to the world economy. As a result, exploitative links within the Third World have increasingly occurred – for example, the use of cheap migrant labour from the Indian sub-continent and parts of South East Asia such as the Philippines and Thailand in the oil-rich states of the Middle East. These growing economic differences between developing countries are clearly evident in the World Bank data in Table 3.1, not only in the range of figures themselves but also in the way in which groups of countries are distinguished from one another.

These growing economic contrasts must not be allowed to mask the fact that clear political divisions also exists within the developing nations. Socialists governments emerged very quickly after the Second World War, and nowhere more so than in South East Asia where the defeat of the colonial powers by the Japanese created considerable instability in the late 1940s and early 1950s. The success of Communism in China gave a particularly powerful boost to socialist revolutionary movements in the region, and laid the foundations for lengthy civil wars or insurrections which, although they may have subsided somewhat since the late 1970s, continue to surface throughout the region from time to time.

In many ways, it was the successes of socialist revolutionary movements in countries such as China, Korea and Vietnam, and the threats posed elsewhere, which forced development planners to look more closely at the strategies that ought to be followed in the non-Communist countries of the region. Thus, the newly independent states of South East Asia, like their counterparts in Africa and Latin America, began to receive a stream of advice and aid from individual countries (bilateral assistance) and international organizations, such as the World Bank (multilateral assistance), in order to try to ensure that 'successful' development occurred and so reduce the possibility of a switch to socialism.

Of course, it could be claimed (and often was by socialist strategists) that the advanced capitalist nations were really interested in preserving their monopoly of the cheap commodities produced in their former colonies, and were suggesting development strategies that would in effect perpetuate this situation, but through the medium of individual companies rather than direct colonial rule. Thus, it was claimed, Britain retained considerable influence in Malaysian rubber production in order to maintain a continued supply of cheap latex to companies such as Dunlop.

Not all development planning strategies reflect the interests of the advanced nations, however, for there has been considerable success in South East Asia in promoting diversification from primary commodity production to industrial and service activities (see Ch. 8), thus raising national incomes considerably. Nevertheless, radical critics still suggest that this industrialization has been to the greater benefit of the Western firms involved in it, rather than to the mass of the people of the countries themselves.

This chapter sheds some light on these varying interpretations of development. It takes the form of a review of the major concepts, or schools of thought, that have influenced development planning in the non-socialist economies of South East Asia over the last three decades. The main characteristics of each concept will be discussed and followed by a critique, one which usually derives from the supporters of another school of thought. Although the theories discussed relate primarily to capitalist development, much of the criticism, and some positive concepts, have come from neo-Marxists. Since the realization of the goals of this particular school of thought depends on the overthrow of present governments, it is hardly surprising that such strategies have received scant practical attention. Nevertheless, neo-Marxists have much to offer as critics of capitalism.

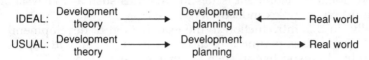

Fig. 3.1 Relationships between development theory and planning

In ideal terms, development planning is supposed to be a response both to the immediacy of the real world and the longer term goals inherent in the strategy being followed, with positive interlinkages with both (Fig. 3.1). Most decision-makers feel they are responding to 'real-world' problems, particularly their material and social aspects. However, they usually do so within a conceptual framework which reflects their own view of the nature of the development process – or at least that which relates to the social group to which they belong. In practice, therefore, the theoretical standpoint can be a very powerful force in determining policies for the real world, with relatively little feedback from the majority of the population being sought or valued except in times of crisis. Analysing the major types of development theory, therefore, enables us to have a closer understanding of the reasons why particular development policies are adopted, and to appreciate how distorted effects can ensue.

It would be a mistake to assume that the examination of post-independence development theory involves the interplay of economies and politics within some sort of historical or cultural constant, even for the small area of South East Asia. The relatively brief post-war period has witnessed extensive changes in the nature of the South East Asia and global economies: from post-war reconstruction, through the international labour migrations of the

1960s, and the multinational company investments of the 1970s, to the world recession of the 1980s. Such events have generated forces that have radically affected attitudes towards development both outside and inside South East Asia, as have changes emanating from within the region itself. In the latter context, one of the most important events has been the schism between the socialist and capitalist nations, the frictions of which have induced massive population movements and related social changes. For example, there was large-scale migration from China during the late 1940s and early 1950s to urban centres throughout the region. The pressures such changes placed on plural societies, such as Malaysia, have very markedly influenced attitudes towards development. Cultural factors have therefore become important in the formulation of development theory. In real terms, this can lead to planning conflicts between the demands for urban-based industrialization, which is often dominated by immigrant groups, and rural development to assist indigenous peoples. Such dilemmas have made regional development issues in countries such as Malaysia and Thailand very complex indeed.

Finally, in this introduction to the examination of development theories *per se*, it is pertinent to discuss what is, or should be, the geographical perspective in such an analysis, even though geography as a discipline has until relatively recently made little positive contribution to the debate as a whole. Originally, few social sciences other than economics were expected to contribute to development theory (particularly by economists!). However, the broader range of analytical perspectives from geography, sociology and political science that have now been incorporated into development analysis have brought both a dynamism and a depth that were previously lacking. The precise role of geography within this broadening of development thinking is still the subject of much debate but little resolution within the discipline (see Rimmer and Forbes 1982, Forbes and Rimmer 1984). Its contribution appears to lie within the analysis of the spatial dimensions of development, both between and within countries. This involves not only description of geographical inequalities but also explanation. Much of the explanation offered in this book may not be geographical in origin but increasingly, through concepts such as core-periphery theory, geographers have sought to make a more positive contribution to the debate.

This emphasis on the multidisciplinary nature of development theory is an appropriate point to introduce the framework within which the examination of the major approaches will proceed (Fig. 3.2). Many reviews of development theories claim that they can be classified into three distinct types: linear or modernization strategy (often incorrectly termed neo-classical theory); dependency theories; and neo-Marxist theories. However, these are only the concepts that have emerged from what might be termed 'macro-approaches' – those which try to interpret global patterns of development based

Plate I Malay wedding, Kuala Lumpur.

Plate II Indian festival of Thaipusam, Penang.

Plate III   Chinatown, Penang.

Plate IV   A typical rice landscape of lowland South East Asia.

Plate V Experimental high-yielding rice, Thailand.

Plate VI Harvesting oil palm, Malaysia.

Plate VII   Oil palm processing mill, Malaysia.

Plate VIII   A Thai village house.

Plate IX Rain-making ceremony, north-east Thailand.

Plate X Village industry, weaving: Thailand.

Plate XI   Village industry, charcoal making: Thailand

Plate XII   Malay-style, coastal fish-trap, Singapore.

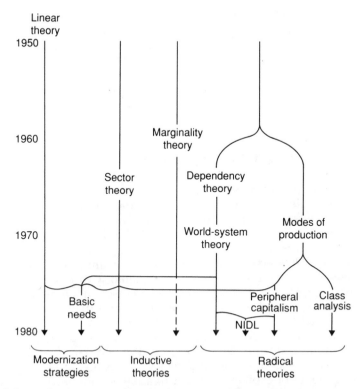

Fig. 3.2 A framework for the study of development theories

on very broad data. In contrast, there have also been 'micro-approaches' to development theory that have been much more strongly based on direct observation and experience of the problems of development itself, as they exist in the Third World. Concepts which fall into this category are sector theory (dualism), marginalization theory, and the 'basic needs' approach.

In essence, therefore, there are both deductive and inductive approaches (Fig. 3.2), and these varied concepts have a range of intellectual and disciplinary origins. However, almost without exception (and irrespective of their political leaning) their origins are 'Western' in nature, they have largely emerged from Western interpretations of development situations. In very few parts of the Third World has there been a substantial attempt to pursue an indigenous development philosophy; examples which come most readily to mind are China, or some Islamic fundamentalist states such as Iran, but in reality not even these countries can pursue a development path totally independent of the advanced industrial nations.

It must not be assumed from the framework illustrated in Fig. 3.2 that developing countries are restricted or restrict themselves to a single developmental strategy. Indeed, this is far from the case and, in response to changing internal and external events of all kinds, individual nations have, quite understandably, proved eclectic in their choice of strategy; the Chinese volte-face of the 1980s, which has reintroduced elements of capitalism, being an obvious case in point. Figure 3.2, therefore, represents a simplification of a far more complex situation, and primarily seeks to place in perspective the major areas of discussion that follow.

Linear theories of development

The main objective of most developing countries, particularly the newly independent nations, in the immediate post-war period was to promote the spread of the life-styles that had been the prerogative of a predominantly expatriate elite during the colonial era. State governments thus became involved in planning for the reproduction of the economic system of the advanced capitalist economies. In short, development was equated with modernization along Western lines and was thought to comprise a simple linear progression from rural underdevelopment to an urban-based industrial society.

Economists in the advanced capitalist nations had divided views about this process in the early 1950s. On the one hand, it was a strategy that appealed to political ideologies anxious to present a successful model of capitalist growth to a developing world apparently vulnerable to Communism. On the other hand, neo-classical economists maintained that the role of Third World nations should continue to lie in providing primary exports, with a cautious reinvestment of earnings into diversified activities.

However, many of the more liberal strategists of the period argued that it was the neo-classical system of free trade and international specialization that had brought about underdevelopment in the Third World, and they began to urge protectionism and industrialization programmes that aimed, through import substitution, at cutting the loss of foreign exchange in unequal trade. Such strategies were largely linked to nationalist euphoria following independence and were not enthusiastically supported by Western governments until the wave of Communism of the late 1950s, typified by the Cuban revolution, escalated insurrections in South East Asia.

Capitalist strategy then began to look more positively at broader growth patterns within the Third World, particularly those which followed a 'similar path' to development in the West – a strategy which had its most explicit and published expression in Rostow's self-proclaimed 'non-Communist manifesto' on the development process. An economic historian, Rostow (1960) postulated that all nations pass through five stages (Fig. 3.3) on the way to development.

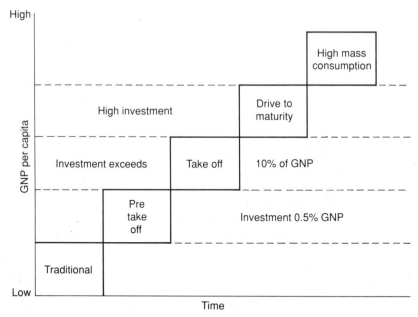

Fig. 3.3 Rostow's stages of growth model

1. Traditional society: oriented around subsistence agriculture and with a few craft industries; strong social stratification but not a society that rewards commercial success with social status.
2. Precondition to take-off: based on a new but small social and political elite with 'modern' ideas on development (this could be expatriate); production and export of valuable local primary commodities emphasized; investment in a supporting infrastructure of roads and railways.
3. Take-off: the stage where dynamic modern attitudes become more widespread and a strong emphasis on a limited set of manufactured products results in economic growth surpassing population growth.
4. Drive to maturity: growth becomes more widespread in the economy; the population becomes more urbanized and educated; social inequality begins to decline.
5. High mass consumption: this stage is only reached if the temptation to 'world domination' is resisted; it comprises large-scale collective consumption of consumer durables, the rapid growth of service industries and a highly urbanized population.

There was widespread support for the overall validity of this model, with discussion revolving primarily around the ways in which 'take-off' (the sudden boosting of economic growth beyond the rate of population growth) would occur: through heavy or light industry, import substitution or export orientation, labour-intensive or high-technology industry. The undisputed indicator of this progression was the level of GNP per capita.

The antecedents of linear development theory were not only economic but also owed much to anthropological and sociological notions on what constituted 'modern' as opposed to 'traditional' attitudes. Particularly close association was made between this dichotomy and the nature of urban and rural communities respectively, with the assumption that an individual 'progressed' as he or she moved further along the line of transition from a rural, traditional way of life to a modern, urban life-style.

Linear development theory began to interest geographers at the same time that the quantitative revolution was affecting the subject. The result was that much of the geographical interest in underdevelopment was concerned with measuring the rate at which traditional societies were changing along Western lines. This diffusion of Westernization was assumed to occur through the medium of the urban system, i.e. that modern attitudes and trends would begin in the capital city and subsequently appear in sequence down the urban hierarchy. In Malaysia, Leinbach (1972) measured 'modernization surfaces' on this basis (Fig. 3.4) and not surprisingly produced a map which clearly reflected colonial urban development and the location of contemporary elites. This did not reflect the overall social condition of the country.

In similar fashion, the cities themselves were examined under the microscope of the new scientific geography and their social ecologies closely scrutinized for the non-Western characteristics that were assumed to be hindering development. There was little attempt to credit Third World cities with inherently different development processes from the West and even less to explain those differences.

By the 1960s the defects in linear development theory were becoming noticeable. Although GNP per capita was rising, much of the growth was illusionary (Table 3.2) and disguised widening inequality both between and within countries where the benefits of industrial development had been soaked up by a small indigenous elite. Such criticism did not worry many linear strategists who were less concerned with alleviating poverty than in building up domestic

Table 3.2 The widening of inequality

	Austria	Philippines
Growth of per capita GNP (1965–85) (average % per annum)	3.5	4.0
Per capita GNP (1985) US $	9120	580
Real growth per capita (1985) US $	319	39

Source: World Bank, *World Development Report 1987.*

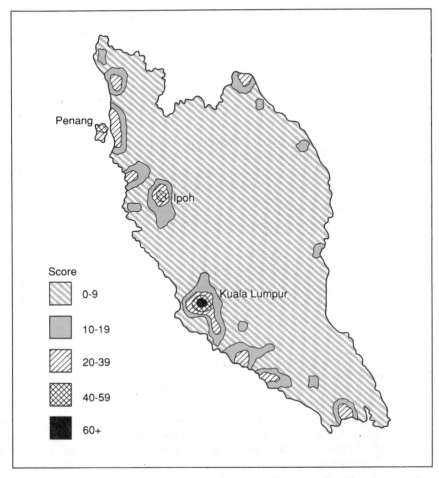

Fig. 3.4 Modernization surfaces in Malaysia. The scored comprise a composite index of features of 'Modernization': the higher the score the more 'modern' the location. (Source: Leinbach, 1972)

savings for reinvestment by an elite, with assumed and consequent trickle-down of benefits.

It soon became apparent, however, that the growing unemployment of the Third World was not a temporary phenomenon. Rostow's model was built on the false foundations of an assumed independence of action by individual nations. It crucially ignored the web of global relationships which conditions the development of individual nations. For the metropolitan powers of Europe and the United States, such relationships have in the past been almost entirely favourable, providing cheap supplies of food and raw materials and also, when necessary, providing reception areas for migrants who had been displaced by the industrial revolution. Over the last three or four decades, of course, these international relationships have been almost entirely unfavourable for most developing nations. Moreover, most of the profitable enterprises of stages II and III of Rostow's model are in the hands of multinational corporations, so that the benefits of economic growth are often siphoned abroad (Table 3.3).

Table 3.3 Percentage of global commodity trade controlled by the 15 largest multinational corporations

Food		Agricultural raw materials		Minerals/metals	
Wheat	85–90	Timber	90	Petroleum	75
Sugar	60	Cotton	85–90	Copper	80–85
Coffee	85–90	Rubber	70–75	Iron	90–95
Corn	85–90	Tobacco	85–90	Tin	75–80
Rice	70	Jute	85–90	Phosphates	50–60
Cocoa	85			Bauxite	80–85
Tea	80				
Bananas	70–75				
Pineapples	90				

Note: In most cases 3 to 6 MNCs account for the bulk of the market. *Source*: Clairmonte and Cavanagh (1983).

By most criteria, the great mass of people in the urban and rural areas of South East Asia was becoming increasingly disadvantaged (see Mehmet 1978). Unemployment in the cities was relatively high even in countries where GNP per capita was rising very rapidly and the fear of many strategists was that these economic problems would cause the urban poor to become fertile ground for revolutionary movements. Policies towards the poor and their activities thus tended to be unsympathetic at best and repressive at worst. Urban squatters, in particular, were harshly treated, with many residents suffering periodic arrests or evictions beyond city boundaries. In Manila, for example, in the early 1960s some 90 000 squatters were removed from the Tondo district near the port in order to permit redevelopment. They were shifted to bamboo huts at a rural site at Sapang Palay, some 25 kilometres from the capital. Within a few years, the great majority had drifted back to the city and were once again in squatter housing. But by the late 1960s, detailed field investigations into the real problems of the Third World were beginning to pave the way for the emergence of concepts very different in scale and orientation for linear theory.

Inductive theories of development

As with several other areas of development theory over the last two decades, the emergence of the two main inductive concepts can be associated with different parts of the Third World. The rise and fall of marginalization theory largely occurred within the context of Latin America, whereas dualism or sector theory was built upon theoretical and investigative work undertaken substantially within Africa and Asia – in South East Asia in particular. It is for this reason, and also because sector theory is currently more influential in development planning, that this discussion is weighted towards that strategy. In addition, it must be noted that much of this theory resulted from urban-based researchers so that the nature of rural underdevelopment is somewhat underemphasized.

Marginalization theory

In Latin America there has long been an established body of anthropological theory in conflict with the basic premises of the linear approach to development. In particular, it was felt by many that individuals do not move smoothly along a continuum of change from a traditional, rural life to that of the modern, industrial city. Instead it was postulated that a large number failed to adapt to the new demands of the industrial economy and remained culturally, socially and economically marginal to it. This was allegedly typified by the squatter settlements that were proliferating around the large cities of the Third World. This approach reached its most formalized structure in the 'culture of poverty' theory of Oscar Lewis (1966), in which a distinctive culture characterized by poverty is seen as a permanent and inevitable way of life for large numbers of migrants to the city, one which is passed on to children without the ability, energy or imagination to lift themselves beyond this condition. Other investigators took up this theme, again largely in an urban context, but it also gradually spread to attitudes about nations as a whole.

The 'advance' made by marginality theory was that it rejected linear notions of automatic progression; its major defect was that it was content primarily to describe but not to account for the extent of such poverty. The resultant impact on development planning was merely to encourage the spread of sympathy and charity for an 'inevitable' and unproductive mass of poor people. However, in recent years the 'myths' of marginality have been effectively dissipated by many investigators, whose researchers have clearly indicated that rather than being apathetic and resigned to their position, most squatters are closely integrated into the productive economy, but in an exploited manner.

Sector theory

Sector theory had its origin in the concept of economic dualism in colonial Indonesia. Until the 1960s, however, such dualism referred primarily to the contrast between rural traditionalism and urban industrialism, and its association with fundamental differences in the Asian and Western ways of life. Subsequent investigations, however, soon began to identify similar contrasts *within* the cities of the Third World. When blended with the contemporary developments in the study of marginalization, this gave impetus to the emergence of the new concept of sector theory.

Essentially, sector theory comprises a general point of view rather than a coherent concept, based as it is upon a range of detailed empirical studies. The basic premise is that urban economic activities in the Third World can be divided into two distinct categories by their internal characteristics. Varied approaches were adopted by the many different researchers; some used occupational categorizations (Hart 1973), others organizational characteristics, each giving the two sectors their own pair of descriptive labels – traditional/modern, lower circuit/upper circuit, bazaar/firm. But the one that has passed into most popular usage is that of Hart's formal and informal sectors.

The main benefit of the empirical studies that began to proliferate in the 1970s was that they established beyond doubt the positive values of what had previously been regarded as an unproductive, disorganized sector of permanent poverty impeding the modernization of the economy. The investigations covered a wide range of activities from small-scale manufacturing (Dwyer 1971), to street-trading (McGee and Yeung 1977) and squatter housing (Drakakis-Smith 1981). In general, these studies all revealed strong traditions and organizational characteristics amongst enterprises and groups of people, together with considerable determination to improve the quality of life for both individuals and their families. In the process, the specific needs of the low-income populations, for cheap goods and services, were met in an efficient manner.

Many of the investigations of informal sector activities occurred in the urban context, particularly with respect to the squatter settlements in which so many were concentrated. One such survey took place in Klong Toey, the largest squatter settlement in Bangkok, the capital of Thailand. Klong Toey contains about 50 000 people and is located on stagnant marshland adjacent to the main commercial port area of the city. At first glance, the squalor and the dilapidation of the settlement is such as to give the impression to Western eyes of it being the home of delinquents, vagabonds and the various un- and under-employed migrants who are illegally occupying valuable land needed for the expansion of the port. In short, the residents of Klong Toey were seen as hindering rather than contributing to the development of the capital city.

However, such an interpretation is very misleading. Research surveys (see Sakornpan 1971) showed that even by the early 1970s the great majority of Klong Toey inhabitants had been resident in Bangkok for over a decade. They had shown considerable initiative in building their homes from recycled materials and (illegally) tapping electrical or water connections. Most of the squalor was the result of inadequate infrastructural provision by the city authorities. Informal but extensive schools had been established in the community by its residents and the great majority of the household heads were gainfully employed in the city. Most worked in or near Klong Toey itself either as street traders or casual labourers. But hours were long (10 to 12 hours per day, seven days per week) and remuneration small (two-thirds of the families had incomes below the poverty line).

In short, the squatters of Klong Toey showed considerable initiative, ingenuity and effort in making a livelihood for themselves in Bangkok despite the lack of concern, even hostility, that the authorities displayed towards their welfare. At the same time, the squatters contributed significantly towards the urban economy as a whole by developing land, creating property, organizing educational services and supplying cheap labour.

It was a multitude of findings such as these which, by the mid-1970s, some development strategists (for example, Sethuraman 1976) began to draw together into a common framework of attributes allegedly typical of all informal sector activities (Table 3.4). The informal sector was characterized as a substantially self-contained but easy-to-enter component of the economy which served a transitional function whereby migrants were gradually incorporated into the formal sector. The obvious value of the informal sector in meeting the basic needs of the poor (particularly employment, income and housing) at little direct cost to the authorities soon resulted in recommendations for policies of support rather than persecution.

Typical of such policy changes was the support for aided self-help

Table 3.4   The attributes of the dual sectors

Formal Sector	Informal Sector
high and middle income	poor and very poor
low unemployment	high unemployment
industry, business, government	artisans, services, petty trade
large scale operations	small scale operations
wage employment	self- and family-employment
high-skill employment	low-skill employment
restricted entry	easy entry
regulated	unregulated
taxed	untaxed
native population	recent migrants
productive employment	residual (unproductive) employment
mainstream	marginal

Source: Sethuraman (1976)

housing that began to appear by the mid 1970s. Briefly, the principle behind such programmes was to put the energies and organizational abilities of squatters themselves to better use by assisting their construction activities. The forms of assistance varied enormously, sometimes they comprised merely the supply of better quality building materials, on other occasions expensive infrastructural services, such as water or sewerage systems, were provided. Often, however, the aid programmes took the form of serviced plots of land on which the squatters could build at their own pace secure in the knowledge that they had legal tenure.

By the late 1970s, research was moving towards a more rigorous testing of the generalized characteristics of the informal sector within the broader framework of development theory (see Rimmer *et al.* 1978, Bromley 1979), and the results indicated that the informal sector was far more complex than had originally been thought. Not only was it quite difficult for migrants or even long-term residents of the city to enter some of the more profitable occupations but, most important, the sector did not and could not function as a separate economic system. The people, enterprises, commodities and capital of the informal sector, or petty commodity production as many radical analysts prefer to call it, were inextricably intertwined with the formal sector in a subordinate relationship. In essence the cheap goods and services of the petty commodity sector were subsidizing the already wealthy, often via activities which were almost entirely controlled by those in formal sector employment. For example, Forbes (1981) and Rimmer (1982) in their studies of pedicab (trishaw) enterprises in Ujung Pandang (Indonesia) and Penang (Malaysia) have shown that the vehicles are usually owned by teachers or other government employees and then rented out to the operators who actually pedal them around the street of the city. Although the operators may work long hours virtually every day of the year, their returns are still far below those of an owner with just two or three vehicles, for whom the income is a supplement to his or her main employment.

On a broader scale, the relationship between the formal and informal sectors is still the cause of heated debate. There seems to be little consistency as to relationships between the proportional importance of the sectors and the level of development. Some strategists claim that growth in manufacturing must inevitably lower the proportion 'languishing' in the informal sector, others suggest that its benefits are confined to a restricted 'labour elite' and that an inflated informal sector helps reduce the demand for higher wages. However, both of these views seem over-simplified and the proportional importance of the informal sector varies in complex ways in individual cities according to their size, economic functions, position in the urban hierarchy and the rate of population growth.

As a result of the conceptual limitations of sector theory, many investigators, particularly those of a more radical disposition, turned to other concepts to explain underdevelopment, However, the theory still retains popularity as an analytical tool, albeit with some modifications. For example, Gilbert (1990) has suggested that the informal sector can be divided into two distinct sub-sectors: the first consists of a set of 'dynamic' enterprises feeding off the formal sector through subcontracts, particularly in export-oriented economies; the second comprises enterprises and activities where survival is the principal objective. Archetypical of the latter would be the scavengers who collect, sell and recycle waste material from individual bins, garbage dumps and street gutters throughout the cities of the Third World. And yet, as Versnel (1986) has shown in a superbly detailed study in Indonesia, the scavengers are no less part of a social network than many other occupational groups and, indeed, can earn more than formal factory employees.

Despite such interpretational difficulties, sector theories have been increasingly incorporated into development strategies during the 1980s. This is in no small measure due to their adoption by major aid-giving agencies, such as the World Bank, as justification for incrementalist programmes of self-help assistance to the poor. Such policies are, however, also the result of liberal tides within the linear approach and will be discussed more fully below. The radical view of such assistance is that as it does not address the fundamental inequalities in society, it should not be supported.

Dependency theory

Dependency theory emerged in Latin America largely within two contexts. The first was the longer history of independence of the Latin American states and their consequently lengthy experience of incorporation into the world economic system. The second was the structuralist stance of the Economic Commission for Latin America in the 1950s which stressed national industrial development. The subsequent extension of this theory was strongly linked to reactions to the diffusionist paradigms of linear strategies. Largely through the work of Frank (1971), it was argued that the unequal exchange of the world economic system promoted the diffusion of underdevelopment rather than development in the Third World. Furthermore, the dependency theorists postulated, the prosperity of the advanced capitalist nations was directly consequential to the transfer of economic surplus from the Third World – an argument given historical validity by the related critique of world systems theory (discussed separately below), of which Wallerstein (1974) has been the leading advocate. In short, development and underdevelopment were seen as two sides of the same coin, a zero-sum equation in which all advances in the West had been at the expense of a passive, exploited Third World.

Spatially this interdependent relationship was expressed in terms of core and periphery, metropole and satellite (Fig. 3.5) where the former exploits the latter by means of interlinked hierarchies of unequal exchange and controls. Within such a framework, urban

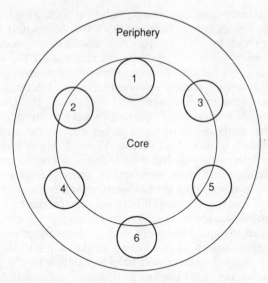

Fig. 3.5 Diagrammatic representation of cores and peripheries in the world-system.

1. Capitalist core countries, e.g. USA and West Germany.
2. Essentially core economies with some peripheral elements, e.g. Australia.
3. Fractional capitalist economies with components of both core and periphery, e.g. Brazil, Hong Kong, Taiwan and South Africa.
4, 5, 6. Peripheral capitalist economies with smaller but crucial core components, e.g. Philippines and Zimbabwe.

systems do not diffuse modernization but further the process of underdevelopment. In this situation of 'dependent urbanization', therefore, urban centres in the Third World were seen as both contributing to, and suffering the consequences of, unequal development. Thus problems such as underemployment, housing shortages and the like were laid at the door of dependent external relationships and national exploitation. One instance of the way this process works may be seen in the import of Western goods, such as processed foods. In Malaysia, for example, many of these goods are reimported from the regional focus of Singapore and are then moved down the urban hierarchy through a series of wholesalers and middlemen to find their way on to the shelves of thousands of small stores. Such foods are bought in both rural and urban areas in preference to locally produced commodities, and effectively transfer profits from the national periphery to the national core and ultimately to the multinational manufacturers in the metropolitan countries. Although it is a land of rich and abundant agricultural potential, food and beverages comprise almost one quarter of the value of Malaysia's total imports.

It is somewhat ironic that despite these newly introduced spatial dimensions to the study of underdevelopment, geographers were poorly represented in the early development of dependency theory, and continue to be uncertain of their role in world systems theory. Since the 1970s, however, both theories have been heavily criticized (Browett 1981a). These critiques make several common points. First, in structural terms, dependency theory is merely a mirror-image of linear development and has no valid concepts of its own; the latter suggests, for example, that urban hierarchies diffuse development, the former that they diffuse underdevelopment. Second, its global perspective is far too general and abstract, divorcing explanation from the realities of societies within the Third World. As Petras (1978, p. 33) has noted, dependency theory suggest that people 'act not for their immediate concrete interests but because the system dictates they act'. The equivalent at the national level is an assumption that all Third World states are helplessly passive victims of global forces over which they have no control.

However, the reality of the Third World is very different. There has been wide variation in the nature of capitalist penetration and in reactions to it. This variation in the pace and nature of underdevelopment cannot be understood without reference to the autonomous histories of individual states during the colonial and post-colonial periods. The complex and changing nature of class relationships, the role of the state and the reaction to capitalist penetration cannot be fitted into the mechanistic, monolithic scheme of dependency and world systems theories.

Within East and South East Asia, in particular, this varied response to Western penetration can be clearly seen. The rapid economic growth of Hong Kong and nations such as Taiwan, South Korea and Singapore appears to refute strongly the image of Third World countries as passive elements within a global system of capitalist exploitation. But even in this context, Asia's newly industrializing countries (NICs) do not exhibit uniform development processes (see Hamilton 1983). All were colonies, Hong Kong and Singapore of Britain, South Korea and Taiwan of Japan, but there the similarity ends. Singapore and South Korea, for example, did not benefit to the same extent as Hong Kong and Taiwan from the flood of refugee capital and labour from China in the 1940s and 1950s. On the other hand, South Korea and Taiwan received considerable aid and investment, both military and economic, from the United States during and after the Korean War of the 1950s. South Korea, Taiwan and Hong Kong have used large amounts of domestic capital to stimulate industrial growth; Singapore in contrast has always encouraged a high level of investment from overseas firms. These and many other variations in the process of political, economic and social change may ostensibly have produced similar results in terms of the present industrialization of the South East

Asian NICs but they seriously question the global validity of world-system theory.

In the wake of such criticism came several types of theoretical response. First, there was a return by the more radical observers towards what might be termed traditionally Marxist notions – a re-emphasis on production rather than distribution. Second, there was a growth of liberal attitudes amongst linear strategists – a 'reformist' (de Souza 1982) tendency which has given rise to what is known as the 'basic needs' approach. Increasingly, however, both trends have been challenged by a persistent and flexible world-systems concept.

The basic needs approach

By the mid 1970s a combination of the continued failure of modernization programmes to bring about a trickle-down of benefits, an increasing appreciation of the criticisms made of world-system theory, and a growing enthusiasm for the incrementalist programmes suggested by sector theory, had caused several major development strategists to reassess the nature of capitalist development concepts (Myrdal 1970, for example). Although adhering to the belief that the continued infusion of capitalist values would eventually remove external and internal inequalities, the more liberal of the linear strategists began to support a restructuring of short-term development goals along more egalitarian principles to meet targets related to the alleviation of the basic needs of the poor. They further believe that this can and does occur within a variety of political systems, citing China, Cuba, Taiwan and South Korea as cases in point (see Streeten et al. 1981).

The basic needs in question primarily encompass income, health, nutrition and housing. Within each field, a series of statistical indices is used to indicate the extent of underdevelopment and so establish appropriate remedial targets. As these data obviously relate to the poorest households, the programmes are frequently geared to the setting of minimum acceptable standards. These are relatively straightforward as far as income, health, nutrition, education and even housing are concerned, but are much more difficult to establish in relation to employment, for which incrementalist schemes aimed at increasing opportunities in 'the informal sector' are usually proposed.

Criticism of the basic needs approach has come from both left and right wings of the political spectrum. On the one hand, for example, are the conservative capitalist strategists who argue that investment in basic services deflects valuable capital away from vital growth-inducing sectors such as manufacturing. Basic needs protagonists have countered with the economically robust but morally weak argument that, in fact, their policies improve 'the quality of human resources. A healthy well-fed labour force is capable of greater physical and mental effort than one that is ill, hungry and malnourished' (Streeten *et al.* 1981, p. 105). The basic needs literature is thus full of depressing arguments about how to alleviate the worst basic needs without sacrificing too much growth – in effect making basic minimum standards of living the maximum likely to be achieved under capitalism.

In this context, many basic needs strategies place great emphasis on the need to identify the extent of poverty and the number of poor (see Richards and Thompson 1984). Whilst such statistics in theory have some use in making decisions on investment priorities, their main findings are often ignored because in most Third World countries they reveal far greater poverty in rural rather than urban areas. In contrast, most basic needs programmes are targeted at the more accessible urban poor and even then seldom deal with their poverty in its entirety but rather with elements of it (housing, health, nutrition, transport, etc.) in isolation from each other. Radical critics, such as Browett (1981b) and de Souza (1982), have been particularly vigorous in their condemnation of this aspect of basic needs programmes. They point out, for example, that investment in basic health care is wasted without an improvement in living conditions which in turn is dependent on regular incomes. It is alleged that the basic needs approach encourages excessively narrow targeting and is problem- rather than people-oriented. In short, it is remedial and cosmetic, and leaves unchanged the basic inequalities in society.

As de Souza (1982, p. 124) has commented, geographers who support the basic need thesis seem to 'view space, spatial patterns and spatial structures as autonomous from social organisation'. This leads to regional development policies which appear to be using basic needs investment to counter disparity and poverty but which in reality accentuate the inequalities within such regions by favouring those currently in control of the distribution systems viz. the already better-off. For example, such consequences have clearly occurred in programmes in Thailand designed to encourage development in the north east. In that dry, unproductive and politically vulnerable border region of Thailand the introduction of irrigation systems and associated tenure reforms have consistently favoured the larger farmers, who are put in control of the reorganization and take the best irrigated land for themselves. Despite the public promises, therefore, many small farmers find themselves little better, or even worse, off than before (see Lightfoot and Fuller 1983). The radical critique is thus based not so much on the distributive (rather than production) emphasis of the basic needs concept, but on the fact that its interpretation of both those systems is devoid of class perspectives.

Despite such criticisms, the movement towards the basic needs approach has gained great momentum in recent years as increasing amounts of aid from the principal agencies, particularly the World Bank and United Nations, have become tied to projects of this nature.

Radical theories of development

It is not easy to synthesize the recent radical, neo-Marxist contribution to the debate on development theory because it has been so disparate and subject to eclectic criticism. This is particularly true in attempting to identify the contribution of geography to radical development theory and vice-versa. Indeed, the relationship between the two is still the focus of considerable unresolved debate (see de Souza 1982, Rimmer and Forbes 1982, Forbes and Rimmer 1984, Taylor 1986). Essentially the debate amongst radical development geographers concerns the role of space and spatial analysis, not merely in descriptions of underdevelopment, but in explanations of its origins.

It must be admitted at the outset that the geographical or spatial factor has not yet emerged as a crucial element within the two principal approaches that can be identified within the complex of neo-Marxist development theory, viz. mode of production analysis and class analysis. However, both these approaches have refocussed attention away from the abstract global level of analysis towards that of particular social organizations within the Third World itself. In this context, geographers have been able to make an increasingly important contribution to the theoretical debate by using their empirical tradition in analyses that seek to verify theoretical ideas (Forbes and Rimmer 1984).

In its most basic and original form, the Marxist theory of development assumes that all societies begin with a form of simple Communism with no private property and societal egality. As economies become more complex so two distinct classes evolve – an underclass involved in production and a dominant ruling class that exploits the former and appropriates most of the surplus value produced. This results in class conflict which, according to Marx, is the principal force for social change and results in a succession of economic systems. As the inherent conflict between the bourgeoisie and proletariat continues, Marx would argue, social revolutions will overthrow industrial capitalism and restore a Communism in which, once again, there will be egalitarianism and no private property.

Marx wrote little on what we now call the Third World. His writing relates primarily to the role of the British in India, as well as to the then colonial periphery of Ireland. Neo-Marxists have interpreted his analysis into two alternative development paths: the first, in which industrialization in the periphery is blocked by metropolitan countries in order to retain the periphery's emphasis on primary, mainly agricultural, production; and second, a deliberate promotion of industrial development in the periphery, leading eventually to a proletarian revolution.

In historical terms, it is the former path which dominated development in the periphery during the colonial period and which has persisted very strongly in the contemporary Third World through the forces of neo-colonialism. However, there has clearly been a major restructuring of the world economy in the post-colonial period and rapid industrialization is now a feature of many Third World countries, particularly in South East Asia. It is in the interpretation of these recent changes that radical analysts have varied so much.

Although all would claim to focus their analysis on events and relationships at the global level, it is clear that some theories emphasize more than others that full appreciation of the present situation can only come from an analysis of the world economic system. The more 'traditional' Marxists on the other hand tend to discuss more willingly the nature of underdevelopment within geographically smaller social formations.

Using such a division as a simple guide, this overview of radical theories of development first discusses the neo-Marxist theories of peripheral capitalism and class analysis, then examines contemporary world-systems theory.

Neo-Marxist approaches

Although some early neo-Marxists assumed that extended Western interest in the Third World promoted development through industrialization, as suggested for India by Marx, post-colonial analysts, with one or two exceptions discussed below, have been more prone to accept the alternative view that the consequence of neo-colonism was 'blocked' or distorted development in which the peasantry and proletariat were exploited by foreign capitalist interests.

This argument has become more incisive with the emergence of the concept of *peripheral capitalism*. This interpretation is itself based on the rise to prominence of mode of production analysis in which the nature of the Third World is seen as a reflection of the articulation of two particular modes of production – the capitalist and pre-capitalist. Modes of production have several basic components (Fig. 3.6), of which most neo-Marxists would argue, the social relations of production are the most important. They further

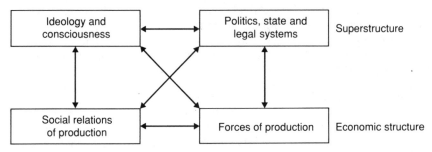

Fig. 3.6 Mode of production components. (Source: Peet 1980)

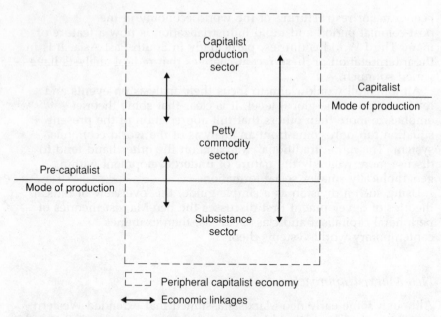

Fig. 3.7 Relationships within peripheral capitalism

suggest that within the Third World the articulation or co-existence of the capitalist and pre-capitalist modes of production has produced a dual economic system. Unlike sector theory, however, the emphasis is strongly on the relationship between the two systems not on their distinctiveness (Fig. 3.7). Indeed, a central tenet of peripheral capitalist analysis is the emphasis placed on the way in which the petty commodity sector is manipulated and controlled to support the reproduction of capitalism.

Rimmer (1982) has provided a useful account of the ways in which this relationship is maintained in his studies of trishaw and pedicab operations in Penang, Malaysia. He found that there were 2445 trishaws in operation in the town. Of these less than 20 per cent were directly used by their owners, and five owners held 42 per cent of the total trishaw fleet. Many of the multiple owners had other business interests ranging from retailing, to associated transport operations (such as taxis), to manufacturing. The annual rental (of M$360) charged for each trishaw by the owners to the pedallers creates a flow of economic surplus or profit to the capitalized sector, since many owners had their major occupations within that sector. In addition, there are many other ways in which the operations of the trishaw industry creates wealth for capitalism. The demand for spare parts, for example, benefits the importers of metal and hardwoods as well as the manufacturers themselves; the imposition of licence fees on owners and operators, as well as periodic traffic fines, all transfer cash from the trishaw riders to the state. In all of these and similar ways there is, therefore, what has come to be known as a transfer of value from the poor to other classes in the economy. It is by its emphasis upon these exploitative links within the social relations of production that the peripheral capitalist approach differs from sector theory which would be content simply to stress the existence and positive, even self-contained, nature of the 'economy of the poor'.

When the concept of peripheral capitalism is extended from the urban or national to the global scale, it also offers a more plausible explanation of the diversification of economic growth patterns which have occurred amongst Third World countries. The main theoretical concept which has been used in this context is that of the new international division of labour (NIDL). Underpinning this concept is the massive movement of capital from developed to developing countries which has occurred since 1970, primarily due to the contrast in labour costs. Manufacturers have been able to take advantage of this cheap Third World labour because of technological developments, such as containerization and new communication links, which have enabled a fragmentation of the production process. Thus General Motors boasts that it produces a world car with most of the components being manufactured in dozens of different Third World locations and assembled wherever major markets exist in the advanced nations.

The relationships are basically the same as for the trishaw industry described above, with there being a transfer of value from the Third World to the advanced nations by means of multinational corporations using cheap labour (usually female) to produce commodities that realize large profits in the markets of the west. Ninety per cent of labour force in the electronics factories in South East Asia, for example, comprise girls between 15 and 25 years of age. On average, their wages are about 40 per cent of comparable occupations in Britain but they work almost twice the hours. In addition to all these criticisms, it has also been pointed out that many multinational firms in South East Asia use imported materials and energy, so there are few backward linkages into the local economy (for further discussion of the operations of multinational companies see Dixon *et al*. 1986).

Despite its perceptive analysis, the peripheral capitalism approach has itself been subject to criticism. For many investigators such as Petras (1978) and Henderson (1986), mode of production analysis is still too economically deterministic and ignores the local forces, class relationships and state intervention which have shaped contact within Western capitalism. Indeed, even the international division of labour is not new. Walton (1985) claims that there have been three such divisions altogether. The first occurred during the colonial phase of commodity export from colony to metropolitan factory; the second occurred when some newly independent countries began to encourage import-substitute industries (often heavy industry or branches of international firms) to serve domestic markets; the third is the current growth of export-oriented manufacturing plants. NIDL theory ignores these various stages and also the regional and national linkages that occur within the periphery, i.e. between Third World nations. In short, it is claimed that semi-independent and fully independent economic growth has occurred in many developing countries, sometimes in parallel, sometimes in conflict, with Western-sponsored industrialization.

Fitzgerald (1981, pp. 15–16) thus identifies three major development strategies based on these internal and external class alliances. First,

> 'the neo-colonial strategy where in the national regime participates with the core bourgeoisie in exploiting the indigenous labour force. Second, the national regime may undertake "national development" strategy based on exploitation of the indigenous labour force and the limitation or elimination of the share going to imperial firms. Third, the regime may ally with the indigenous labour force, nationalise foreign and even indigenous enterprise, redistribute income, and generally undertake a "national populist" strategy as against (sic) core capital'.

Clearly, Fitzgerald and others value developmental class analysis because of its re-emphasis on the nation state as the principal unit of analysis, thus enabling empirical investigation to take up the specifics of development dynamics; for example, in developmental histories or the changing role of the state. However, an important criticism of both NIDL and class analysis is that they tend to neglect not only the spatial details of the underdevelopment process but also some of the social issues as well, particularly the gender or ethnic dimensions of the processes under scrutiny (Henderson 1986). Moreover, by moving away from broad global theory and more towards an analysis of the real world, both the peripheral capitalism approach and class analysis have been laid open to the criticism that their approach is primarily a critique of the past or contemporary development with little predictive accuracy of future trends. In particular, critics point to the fact that the Communist revolution has occurred in so few parts of the world and seems increasingly unlikely to occur as the various underclasses of the Third World are assiduously assuaged with crumbs from the development table. As a result there has been a reappraisal of neo-Marxist arguments and an appeal for more realistic attitudes towards the continued problems of underdevelopment (Corbridge 1986).

World-systems theory

For many critics during the 1970s, world-systems theory was synonymous with dependency theory and was subject to much the same criticisms, i.e. that it was too broadly based, too vague, too all-embracing and simply too crude to explain the complex socio-political variations within the Third World. And yet, despite such comments, world-systems theory is experiencing a new lease of life as many economic and political geographers have been forced by the realities of the changing world-economic system to look beyond the boundaries of the core to explain the changes that have taken place within countries such as Britain and the United States (see Hamilton 1986, Thrift 1986).

The current protagonists of world-systems theory (see Taylor 1986) argue that it is a more suitable vehicle for understanding the present situation in the Third World than is neo-Marxism. The differences between the two are complex and sometimes difficult to discern but may be summarized as follows. First, world-systems theory recognizes the fact that capitalist penetration of the Third World cannot be equated with the spread of wage-labour 'as probably less than half of the world's labour is proletarian' (Taylor 1986, p. 281). It is suggested that instead, there is a single mode of production, capitalism, superimposed over varied forms of labour control. Thus, the role of subsistence and petty-commodity households or enterprises is to shoulder the burdens of the reproduction of labour in both rural and urban areas.

The second contrast with neo-Marxism relates to the notion that feudalism progressed into mercantilism, then through a bourgeois revolution into industrial capitalism and will eventually, through a proletarian revolution, progress into socialism. Wallerstein (1983) argues that the bourgeois revolution never occurred and that the aristocracy became the bourgeoisie. The continued influence of the old school tie, Oxbridge and club membership in Britain perhaps supports this argument.

The third difference closely follows this argument in that whilst both world-systems theory and the neo-Marxist approach concur that there is currently a crisis of capitalism (this does not equate with the global depression of the 1980s which is merely a cyclical downturn within capitalism), the advent of state socialism, particularly since 1945, is not seen as heralding the imminent arrival or world socialism. Indeed, Wallerstein (1983) would argue that state socialism is a nineteenth century strategy and that socialist states have easily been incorporated into the world capitalist economy and are subsequently modified as a consequence, willingly or unwillingly.

Another argument, often put forward by geographers in favour of world-systems theory (see Corbridge 1986), is that it tries to bridge the artificial contrast between core and periphery by recognizing the existence of a semi-periphery. The notion of the semi-periphery has proved difficult to conceptualize (see discussions in Drakakis-Smith 1990 Smith and Nemeth 1990). In many ways it is not surprising that there has been no agreement on the list of features, or states, that allegedly comprise the semi-periphery, for the notion is more symbolic than real, representing as it does the full range of features between the extremes of core and periphery. However, discussion of the existence of a qualitatively, if not quantitatively, recognizable semi-periphery does serve a useful purpose in drawing attention towards processes which gave rise to social formations that were neither core nor periphery.

Overall, however, it is not so much the detail of world-systems theory that has attracted most criticism but rather its abstract, highly generalized nature. In particular, critics have pointed out that, although very detailed on present patterns of unequal exchange, the theory is very uninformative about the ways in which the world economy ought to develop in the future. Wallerstein (1983) himself simply asserts that the role of world-systems theory is to ensure the disintegration of capitalism into a classless society overseen by some unspecified form of trans-state government. For many critics vague statements such as these not only seem to demand almost a blind faith in the future from supporters of world-systems theory, but also undermine the credibility of its other dimensions.

Conclusion

To most non-Marxists, particularly those in government, the principal criticism of neo-Marxist or world-systems theories is that they offer little in the way of immediate alternatives to the present situation other than through revolution – a proposition that seems increasingly unrealistic to some and unwelcome to others. This is why the liberal wing of linear development strategy, the acceptable face of capitalism through the basic needs approach, has been so much more influential in shaping actual policy decisions in the Third World.

Perhaps, this is only to be expected in a capitalist world system. However, even the basic needs approach seems radical in the context of many development priorities and has been incorporated into planning strategies only because it is backed by the massive funds of the international agencies. Left to their own devices, most South East Asian governments would undoubtedly prefer to follow conservative linear strategies in concert with the international capitalism that is increasingly coming to dominate the development process through the global division of labour, the export of core capital to the periphery, and the communications revolution that enables multinational firms to control fragmented global production from a metropolitan base. Clearly, at this point the discussion of development theory must necessarily give way in the chapters that follow to that of the development process *per se* and the impact it has had in South East Asia.

Further reading

Bromley R (1979) *The Urban Informal Sector*, Pergamon, Oxford.

Browett J G (1981a) 'Into the cul-de-sac of the dependency paradigm with A. G. Frank', *Australian and New Zealand Journal of Sociology* **17**, 14–25.

Browett J G (1981b) 'On the role of geography in development geography', *Tijdschrift voor Economische en Sociale Geografie* **72**, 155–61.

Clairmonte F and Cavanagh J (1983) Multinational corporations and the struggle for the global market', *Journal of Contemporary Asia* **13**, 446–80.

Corbridge S (1986) *Capitalist World Development*, Macmillan, London.

Dixon C, Drakakis-Smith D W and Watts D (eds.) (1986) *Multinational Corporations and the Third World*, Croom Helm, London.

Drakakis-Smith D W (1981)*Urbanization, Housing and the Development Process*, Croom Helm, London.

Drakakis-Smith D W (1990) 'The built environment and social movements in the semi-periphery of Northern Australia' in D. W. Drakakis-Smith (ed.), *Economic Growth and Urbanisation in Developing Areas*, Routledge, London, pp. 196–237.

Dwyer D J (1971) 'Problems of the small industrial unit', in D. J. Dwyer (ed.), *Asian Urbanisation: A Hong Kong Casebook,* Hong Kong University Press, Hong Kong, pp. 123–36

Fitzgerald F T (1981) 'Sociologies of development' *Journal of Contemporary Asia* **11**, 5–18.

Forbes D (1981) 'Production, reproduction and development: petty commodity producers in Ujung Pandang, Indonesia', *Environment and Planning A* **13**, 841–56.

Forbes D and Rimmer P (eds.) (1984) *Uneven Development and the Geographical Transfer of Value*, Monograph H16, RSPACS, Australian National University, Canberra.

Frank A G (1971) *Capitalism and Underdevelopment in Latin America*, Pelican, Harmondsworth.

Freidman J (1972) 'The spatial organisation of power in the development of urban systems', *Development and Change* **4**, 12–20.

Gilbert A G (1990) 'Urbanisation at the periphery: reflections on the changing of dynamics of housing and employment in Latin American cities', in D. W. Drakakis-Smith (ed.), *Economic Development and Urbanisation in the Third World*, Routledge, London, pp. 73–124.

Gottmann D (ed.) (1980) *Centre and Periphery*, Sage, London.

Hamilton C (1983) 'Capitalist industrialisation in East Asia's Four Little Tigers', *Journal of Contemporary Asia* **13**, 35–73.

Hamilton F E I (ed.) (1986) *Industrialization in Developing and Peripheral Regions*, Croom Helm, London.

Hart K (1973) 'Informal income opportunities and urban employment in Ghana', *Journal of Modern African Studies* **2**, 61–89.

Henderson J (1986) 'The new international division of labour and American semiconductor production in Southeast Asia', in C. Dixon et al. (1986), pp. 91–118.

Leinbach T (1972) 'The spread of modernization in Malaya: 1895–1969', *Tijdschrift Voor Econ. En Soc. Geografie*. **63**, 267–77.

Lewis O (1966) 'The Culture of Poverty', *Scientific American* **215**, 19–25.

Lightfoot P and **Fuller T** (1983) 'Circular rural-urban movement and development planning in Northeast Thailand', *Geoforum* **14**, 277–87.

McGee T G and **Yue-man Yeung** (1977) *Hawkers in Southeast Asian Cities*, International Development Research Centre, Ottawa.

Mehmet O (1978) *Economic Planning and Social Justice in Developing Countries*, Croom Helm, London.

Myrdal G (1970) *The Challenge of World Poverty*, Pantheon, London.

Peet R (ed.) (1980) *An Introduction to Marxist Theories of Underdevelopment*, Monograph HG14, Research School of Pacific Studies, Australian National University, Canberra.

Petras J (1978) *Critical Perspectives on Imperialism and Social Class in the Third World*, Monthly Review Press, New York.

Richards P and **Thompson A** (1984) *Basic Needs and the Urban Poor*, Croom Helm, London.

Rimmer P (1982) 'Theories and techniques in Third World settings: trishaw pedallers and towkays in Georgetown, Malaysia', *Australian Geographer* **15**, 147–59.

Rimmer P J and **Forbes D** (1982) 'Underdevelopment theory: a geographical perspective, *Australian Geographer* **15**, 197–211

Rimmer P J, Drakakis-Smith D W and **McGee T G** (eds.) (1978) *Food, Transport and Shelter in Southeast Asia and the Pacific: Challenging the Unconventional Wisdom*, Monograph HG12, Research School of Pacific Studies, Australian National University, Canberra.

Rostow W W (1960) *The Stages of Economic Development: A Non-Communist Manifesto*, Cambridge University Press, London.

Sakornpan C (1971) *Klong Toey: A Social Work Survey of a Squatter Slum*, Thammasat University, Bangkok.

Sethuraman S (1976) 'The urban informal sector: concept, measurement and policy', *International Labour Review* **114**, 69–81.

Smith D and **Nemeth R** (1990) 'Dependent Urbanization in the contemporary semi periphery: deepening the analogy', in D. W. Drakakis-Smith (1989) pp. 8–36.

de Souza A R (1982) 'Dialectic development geography', *Tijdschrift voor Economische en Social Geografie* **73**, 122–8

Streeten P *et al.* (1981) *First Things First: Meeting Basic Human Needs in Developing Countries*, Oxford University Press, London.

Taylor P (1986) 'The World-systems project' in R. J. Johnston and P. Taylor (eds.) *A World in Crisis?* Blackwell, Oxford, pp. 269–88.

Thrift N (1986) 'The geography of international economic disorder', in R. J. Johnston and P. Taylor (eds.) *A World in Crisis?* Blackwell, Oxford, pp. 12–67.

Versnel H (1986) 'Scavenging in Indonesian cities', in P. Nas (ed.) *The Indonesian City*, Foris, Dordrecht, pp. 206–19.

Wallerstein I (1974) *The Modern World System*, Academic Press, New York.

Wallerstein I (1983) 'An Agenda for world-systems analysts', in W. R. Thompson (ed.) *Contending Approaches to World System Analysis*, Sage, Beverly Hills, 299–308.

Walton J (1985) 'The third 'new' international division of labour' in J. Walton (ed.) *Capital and Labour in the Urbanized World*, Sage, Beverly Hills, 3–14.

World Bank (1987) *World Development Report 1987* Oxford University Press, New York.

CHAPTER 4

Environmental resources

Chris Barrow

Topography and structure

Discussion of environmental and natural resources is made easier if South East Asia is divided into *Mainland* (including peninsular) and *Island* subregions. Some authorities would include Papua New Guinea in Island South East Asia, but it is arguably too-closely associated with the Australian environment to truly be part. Island South East Asia is a particularly appropriate term, for there are over 13 000 islands forming Indonesia alone, and over 7000 in the Philippines. Buchanan (1967, p. 14) aptly described the situation when he noted that the south east margin of tropical Asia is fragmented into '. . . a complex of peninsulas and islands, of land-girded seas and sea-connecting straits' (Fig. 4.1). The present outline of South East Asia reflects post-glacial rises in sea-level; most of the Malacca Straits, Sunda and Java Seas, which today are seldom more than 45 m deep, were dry land for long periods during the last two million years. Present patterns of plant and animal species distribution reflect 'land bridges' which existed before sea-level began to rise roughly 18 000 years ago.

The degree of marine influence over the environment, settlement, communications and the development of resources, both in Mainland and Island South East Asia, is probably unmatched in any other world region. The rivers of South East Asia are also more easily navigable than those of tropical Africa or America, although sand bars often obstruct their estuaries (Fisher 1967, p. 59; Buchanan 1967, p. 13).

Drawing on geological geophysical and topographical evidence, South East Asia may be divided into three basic physical units (Dobby 1973, p. 17–27; Phoon Phon Asanachinta, 1974).

1. The stable and ancient Indo-Malayan Massif plus the Sunda Platform.
2. The stable Sahul Shelf.
3. The Young (Tertiary) Fold Mountain and Island Arcs.

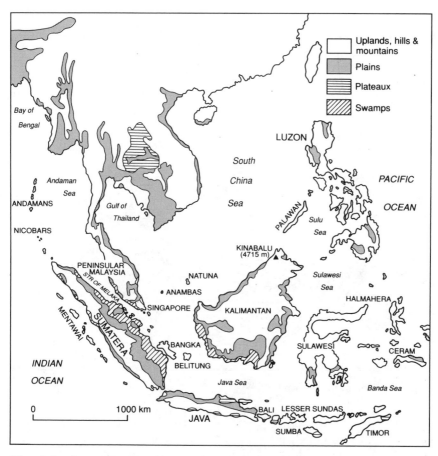

Fig. 4.1 Generalized landforms. (Source: Swan 1979)

The Indo-Malayan Massif/Sunda Platform (Fig. 4.2) consists of ancient crystalline rocks folded along north-south lines in the early Mesozoic. Subsequent erosion reduced much of the block to an area of subdued relief; today it is either covered by shallow seas or alluvial plains, or moderately elevated hills or plateaux with mature topography. Locally, for example in northern Burma, uplift during the Tertiary mountain-building period has resulted in more rugged relief. The Indo-Malay Massif/Sunda Platform has no active volcanism and little or no earthquakes, although shocks originating from Sumatra may sometimes be felt in peninsular Malaysia.

The Sahul Shelf has a similar morphology to the Sunda Platform and is also free of recently active volcanoes. Partly covered by shallow seas and terminating on the north east along the main east-west range of mid-Papua New Guinea, it is best considered as a northern extension of the Australian continental mass.

79

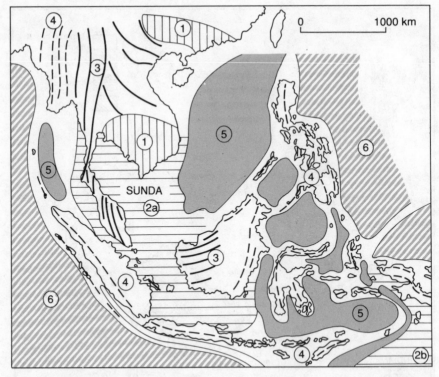

Fig. 4.2 Generalized Structure.

1. Ancient crystalline Indo-Malayan Massif.
2. (a) Sunda Platform.
   (b) Sahul Shelf.
3. Zones of Tertiary folding.
4. Volcanic areas and regions of earthquake activity.
5. Marine basins and troughs.
6. Oceanic abysses
   (Source: Swan 1979)

The younger fold mountain and island arcs are looped around the Indo-Malayan Massif/Sunda Platform and are composed mainly of sandstones and limestones. They are believed to have been uplifted during the Alpine (Tertiary) mountain-building period. The greater part of the island arc system is below sea-level; submarine peaks rise steeply for thousands of metres from the floors of Indian and Pacific Oceans, sometimes emerging to over 5000 m above sea-level. These emergent peaks form 'strings' of islands which Dobby (1973, p. 20) subdivided into the following.

1. *The Burma-Java loop* stretches from the India-Burma border to Sumatra and South Java, although not always as a distinct topographic feature, and then eastwards as a string of islands.

2. The *arcs on the Pacific margins* run through the Philippines and north-western Irian. This appears as a line of islands encircling small, deep, square-shaped, ocean basins. On the Pacific side of these arcs there are troughs or trenches, much deeper than the average depth of that ocean and similar to the oceanic trenches off Sumatra and Java.
3. The *complex of short arcs and knots of islands of the Suluwesi/Moluccas group* which enclose deep oceanic basins, such as the Molucca Sea.

The Indo-Malayan Massif/Sunda Platform is mainly composed of metamorphic rocks: quartzites and shales, which have been weathered to produce subdued relief. Sediments from early Mesozoic fold mountains were redeposited in shallow marine and estuarine environments and uplifted in the late Mesozoic, especially the Cretaceous, and in the early Tertiary to form sandstone plateaux like the Korat Plateau (Thailand), East Java, the now fragmented plateaux in Kelantan, Pahang and Johore States (West Malaysia), and a large part of Borneo. In some areas massive limestone was uplifted and is now dissected into striking karst-type topography, examples of which may be seen in West Malaysia near Ipoh and in parts of Thailand and Sabah.

The younger (Tertiary) fold mountains and island arcs which enfold the Indo-Malay Massif/Sunda Platform are much contorted, fractured and affected by volcanic intrusions or flows. The associated contact metamorphism has in many areas resulted in the deposition of metallic minerals including tin. The island arcs have continued to develop during the Quaternary, consequently there is considerable volcanic activity and earthquakes are common. Indonesia alone has about 70 active volcanoes. The catastrophic 1883 eruptions of Krakatoa and, in 1963, Agong were particularly forceful; and the Philippines have had at least a dozen eruptions in recent years (Hill 1979: pp. 1–5). Although not especially high the ranges of South East Asia are quite rugged and together with thick forest cover, frequently deeply incised rivers, swamps and mangrove-choked coasts, have made communications difficult except up and down river courses.

Much of lowland peninsular Malaysia, Eastern Sumatra and Kalimantan is covered with deposits of recent (Quaternary) river or coastal sediments. In contrast, the Philippines have few such alluvial deposits, primarily because there are only two extensive Philippines lowlands – the central plain of Luzon and the Bicol Lowlands. Rapid erosion of the South East Asian landmasses means that coastal sedimentation can be considerable; spits, bars and offshore beaches are common, especially on the inner coasts of the Sunda Platform but less so on those coasts facing the Indian and Pacific Oceans. It is rather an irony that, while rivers of South East Asia tend to be more navigable than those of tropical Africa or South America, entry to their courses from the sea is often difficult. Spits, bars and offshore beaches play a significant role in the siting of many coastal settlements near river mouths.

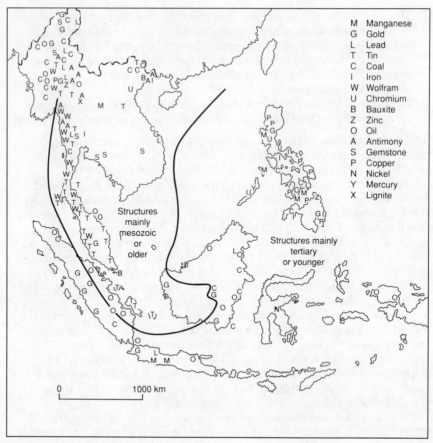

M	Manganese
G	Gold
L	Lead
T	Tin
C	Coal
I	Iron
W	Wolfram
U	Chromium
B	Bauxite
Z	Zinc
O	Oil
A	Antimony
S	Gemstone
P	Copper
N	Nickel
Y	Mercury
X	Lignite

Structures mainly mesozoic or older

Structures mainly tertiary or younger

0   1000 km

Fig. 4.3 Principal mineral deposits. (Source: Courtenay 1979)

South East Asia is one of the more mineral-rich regions. Most of the minerals are associated with Mesozoic igneous intrusions and their surrounding metamorphic complexes or occur in sediments eroded and redeposited from them. Malaysia has long been the world's leading tin producer, Indonesia is second. Gold, silver, and lead are mined in Burma, Indonesia and in the Philippines – where chromium, copper, cobalt and molybdenum are also exploited. Tungsten is produced in Burma; there are diamonds in South East Borneo, and coal in Java and Sumatra (Fig. 4.3). There are also lower-value minerals, which are nevertheless useful: limestone, bauxite, kaolin and laterites. Petroleum resources are increasingly being discovered and exploited. Indonesia has considerable reserves. Malaysia has within the last few years become self-sufficient and there appears to be much more awaiting discovery, particularly in the South China Sea.

Climate

Few world regions have so uniform a temperature regime as South East Asia. With the exception of highlands above a few thousand metres altitude, and northern and central Burma and northern Thailand where continental Asian influences are felt, South East Asian temperatures seldom fall below 26°C. This uniformity of warmth is due to the distribution of land about the Equator and to the moderating influence of the seas, which are roughly four times more extensive than the islands and peninsulas they surround (Fisher 1967, p. 11). It is rainfall, rather than temperature, which determines the rhythm of the natural environment, agriculture and social life. Though distinct differences in annual rainfall volumes make it possible to recognize distinct climatic subdivisions of South East Asia, it is the incidence of rainfall rather than its volume that is important (Dobby 1973, p. 32).

To understand the distribution and periodicity of rainfall in South East Asia, one must examine the way in which seasonal air movements impinge upon the land. Two similar air masses move across South East Asia: (a) The Northern Tropical Air Mass which manifests itself as the North East Trade Winds, normally blowing from the Tropic of Cancer toward the Equator; and (b) The Southern Tropical Air Mass, originating over the Tropic of Capricorn and moving towards the Equator as the South East Trade Winds. Both these air masses have similar physical characteristics: each, having blown across tropical seas, is warm and damp. Where the Northern and Southern Tropical Air Masses meet, roughly along the Equator, what is commonly called the Intertropical Front (Dobby 1967: pp. 32–7), or Intertropical Convergence Zone (ITCZ) is formed. Upward air movements occur in the zone and this tends to produce convective rainfall.

This relatively simple pattern is complicated by two other influences. First, the two air masses are seasonally displaced across the Equator by the movement of the Earth relative to the Sun. In the southern hemisphere summer the front moves south, and in the northern hemisphere summer it moves north. When the front moves into the northern hemisphere the South East Trade Winds become south-westerly winds and North East Trade Winds become north-westerlies when the front moves into the southern hemisphere. Second, low-pressure areas developed over the Asian continent and Australia cause seasonal deviations of the air streams over South East Asia. Summer in Australia (November–January) distorts the front south to Northern Australia. Summer in India, Central Burma and Thailand has an even more marked effect. The North India Low Pressure System formed over these lands draws part of the Southern Tropical Air Mass over the Equator during the late northern hemisphere spring and this flow of air is maintained in a parabolic path curving across the northern Indian Ocean until November. Another low-pressure system is established over eastern

Asia a few months after the North India Low Pressure System. This draws flows from the Southern Tropical Air Mass northward through the South China Sea during June-July-August (Dobby 1973, p. 35).

As a consequence of these movements of moisture-laden air, West Burma, South East Thailand and west-coast Sumatra experience a rainy season between April and September – the South West Monsoon. Peninsular Malaysia, in Sumatra's lee, gets relatively little South West Monsoon precipitation. During the northern hemisphere winter (mid-October to March), Vietnam, peninsular Thailand and peninsular Malaysia get considerable North East Monsoon precipitation. In the west of the aforementioned regions, and in Island South East Asia, it is the early part of the North East Monsoon (November to February) which is the wettest period. From southern Sumatra to West Irian the principal rainy season is November to March, and for this region June to September is a markedly dry season. The rest of Sumatra and most of Borneo have a rainfall fairly uniformly distributed through the year. In Java there is a sharp distinction between dry and wet seasons, especially in East and Central Java. High ground denies large areas of North East Monsoon rainfall, particularly in eastern peninsular Malaysia.

In Mainland South East Asia in addition to the two main seasons in the year (the South West Monsoon season, established between late May and June and ending in September and the North East Monsoon season which begins in late October or November and ends in March) there are two 'transitional seasons', the first from April to May and the second during the month of October. During these two transitional seasons conventional storms are frequent and heavy. In peninsular Malaysia transitional season storms cause two peaks in the year's daily rainfall records, in April/May and in October. Between April and November, often violent line squalls known locally as 'sumatras' develop in the Malacca Straits at night and move upon west-coast peninsular Malaysia, mainly between Port Swettenham and Singapore. Their arrival is marked by heavy thunder storms mainly between midnight and noon the following day. There is a similar strongly diurnal variation of rainfall in eastern peninsular Malaysia during the North East Monsoon (Koteswaram 1974, p. 73).

Except where the influence of continental Asian air masses is felt, or in the rain-shadow of high ground, few parts of South East Asia have a season sufficiently rainless to be considered 'dry' (i.e. at least one month during which less than about 60 mm of precipitation falls), but there is considerable seasonal variation in degree of wetness.

Total precipitation tends to progressively lessen eastwards across Indonesia, particularly east of central Java (Dobby 1973, p. 37). The regions with relatively low rainfall are the 'dry belt' consisting of upper Burma, central Cambodia and central Thailand, which are shielded from moist winds by highlands, and east and central Java. The 'dry belt' has at least one dry month and east and central Java has up to five practically rainless months. Most of the Philippines, except for the east and central regions, gets the bulk of annual precipitation between May and October, and in general the percentage of yearly rain falling in this period increases as one goes north. During the North East Monsoon season, Luzon and adjoining regions get very little rainfall due to anticyclonic conditions at that time of year. The east and central Philippines get their maximum rainfall during the North East Monsoon season, due to monsoon air flows striking the highlands which flank the east and central coasts. From March to August this region has a drier season. The southern Philippines get some rainfall during the North East Monsoon.

Rainfall ranges form about 400 mm per year in central Burma and a few other rain-shadow regions to 5900 mm per year (recorded in Taiping, peninsular Malaysia) (Koteswaram 1974, p. 74). At its simplest, large areas of South East Asia have more than 2000 mm per year precipitation, on western-facing uplands the rainfall is likely to be between 2000 and 3000 mm per year, and to the east of high ground is likely to be less than 2000 mm per year. Certainly the Malay Peninsula north to about seven degrees from the Equator, Borneo, Sumatra, western Java and the eastern Philippines receive an excess of precipitation over evapotranspiration (Koteswaram 1974, Furtado 1980, p. 74). Rainstorms can be very intense, for example a fall of 430 mm in 24 hours was recorded at Penang (peninsular Malaysia). Such heavy storms often cause severe local flooding, erosion and landslides especially where the forest cover has been removed from sloping land.

Relative humidity is generally high. In peninsular Thailand, peninsular Malaysia and Island South East Asia, except for east and central Java, it typically ranges from 65 to 75 per cent during the day to 90 per cent or more at night. High humidity reduces solar radiation reception over much of South East Asia, especially in regions near the Equator. Typically cloud cover builds up during the morning and reduces the amount of sunlight reaching the ground, especially during the afternoon. In Jakarta, for example, less than 70 per cent of the possible sunshine hours are sunny. Coastal lowland and island conditions are generally moderated by the influence of the sea, in particular by on-shore and off-shore breezes.

Typhoons (tropical cyclones experienced in parts of South East Asia) and tropical depressions develop in the area between the Philippines and longitude 160°E between May and November, and especially between July and September. These storms tend to move roughly west-north-west to strike the Philippines (especially Luzon), Indonesia and coastal Vietnam. A few typhoons and depressions form in the South China Sea, usually west of the Philippines, but only a fraction of these take a path which affects South East Asia; seldom do these storms affect South East Asia south of Cambodia.

Agriculture in South East Asia depends heavily on the monsoons and the monsoon rainfalls are important in maintaining streamflow and many ground-water supplies. South East Asia's long-term climatic trends have not yet been established, so it would be wise to exercise caution when undertaking any development of water resources or agriculture in case there are future reductions, increases or changes in the timing of precipitation. In spite of considerable progress, there is still a need for better medium- and short-term weather forecasting in South East Asia.

Soils

The soils of any locality is a function of a complex of factors: parent rock, climate, relief, aspect, vegetation, drainage, human and animal activity and the time span over which deposits have developed. Dobby (1967, p. 75) noted that although most soils in South East Asia have been evolved under conditions of plentiful moisture they vary considerably. The term 'laterite' has frequently been used to describe South East Asian soils but it is so broad that it misleads. There is a risk of generalizing but, with reasonable safety one may describe the bulk of South East Asians soils as 'mature' or even 'senile', free-draining, relatively infertile (calcium or phosphorus deficiency is particularly common), and unstable in that they are prone to erosion and what fertility they have can be difficult to maintain if they are cultivate. Since periods of high rainfall and high evapotranspiration do not always coincide, even in the humid 'equatorial' parts South East Asia, the most important characteristic of a soil, as far as plant life is concerned, is not necessarily its fertility but its ability to retain moisture. This is especially important in regions away from the Equator where rainfall is more irregular (Williams and Joseph 1973, p. 20, Furtado 1979, p. 75).

The tendency of many South East Asian soils to erode easily is often exacerbated by the presence of impermeable deposits beneath the uppermost layer (the A-horizon) which hinder drainage and cause the ground above to become waterlogged, in which state it can easily flow down even slight gradients. Slumping and landslides are common wherever sloping ground is disturbed by cultivators, loggers, road or railway routes or building activity. Great care is needed when developing slopes of more than ten degrees gradient, if such land is to be cropped then perennial crops with ground cover plants beneath them are desirable to protect the soil. Slopes of greater than 20 degrees are best left undisturbed. It should be noted that erosion not only results in the loss of soil which could support crops or wildlife, it also causes problems when the soil enters streams and when that soil is deposited by streams.

In the warm, moist environment of South East Asia, chemical rather than mechanical weathering predominates, and is relatively fast so soils tend to 'mature' very quickly. One consequence of this is that similar soils can sometimes be produced from quite different parent rocks; for example, red oxisols can form above granite or limestones. There is, however, a tendency for acidic rocks like granite or sandstones to yield less fertile soils and for intermediate or basic rocks, such as balsats or andesites to weather and form soils relatively rich in plant nutrients. Not only is soil formation comparatively fast, South East Asia escaped severe Quaternary climatic conditions which in many other parts of the world stripped away deposits that had accumulated during the Tertiary. Those regions of South East Asia away from the Equator which have a markedly seasonal rainfall, for example the drier areas of Burma, Thailand and Central Java, for part of the year have little surface evaporation and for part of the year experience considerable surface evaporation. The soils which form in these regions frequently resemble those formed in desert environments.

In those parts of South East Asia where there are humid tropical conditions, rainwater – well charged with carbonates and the products of vegetative decomposition as it passes through the atmosphere, forest canopy and litter – dissolves clay, iron and aluminium compounds to form silica-rich, bleached soils with iron or aluminium precipitated at depth (i.e. podsolized soils). If rainwater is low in organic acids, silicates may be broken down into salts which leach away; clays and iron compounds accumulate as a result of further chemical processes to form red and yellow soils. The greater the proportion of iron or aluminium hydroxides to silica, the more 'laterized' the soil is deemed to be. Hard, impervious deposits (indurated layers), formed by deposition or redeposition of organic matter, clay minerals and hydrous oxides of aluminium and/or iron, are especially common in regions with marked dry seasons and in more humid regions where forest cover is removed. These layers can hinder drainage and make cultivation difficult.

In Island South East Asia, where there is active volcanicity, soils may be formed directly by ash 'fall-out', or indirectly where ash-charged streams deposit their load. Light ash falls can modify existing soils and ash may be carried to farmland suspended in irrigation water. Volcanic or volcanicity-modified soils tend to be fertile. Geertz (1963, p. 12) noted that a mere nine per cent of the Indonesian archipelago supports two-thirds of the total Indonesian population, and that area corresponds to the occurrence of volcanic soil. In central Java, for example, population distribution closely reflects the occurrence of volcanic soils.

South East Asia's alluvial soils are generally fertile, and like the volcanic soils attract settlers. Cultivation tends to be practised in lowlands and river valleys where riverine or marine alluvium has been deposited. Some lowlands have a cover of peat. When drained this can be fertile; but it is prone to shrinkage and often requires the application of lime and/or artificial fertilizers before cropping is possible. Peat formation is also common at altitudes above about

1370 m (Whitmore and Burnham 1969, p. 102). Alluvial soils and peatlands, especially in coastal lowlands may sometimes have excess sulphate content, sufficient to hinder rice cultivation. Nevertheless deltaic soils, tidal marshlands, mangrove swamps and coastal lagoons are farmed or used for pisciculture in many regions, and there is a vast untapped potential for aquaculture, forestry and coastal swampland rice cultivation (ESCAP 1978: pp. 68–148). A biological problem, rather than one of soil chemistry or structure, but one which greatly hinders the cultivation of some crops (including potatoes and tomatoes), is the presence of large numbers of nematode worms in many South East Asian soils.

There is a considerable range of possible soil classifications, since much depends on the purpose of the classification. A reasonably useful general classification, based on FAO information and drawn up by Hill (1979, pp. 14–15) is given in Table 4.1.

Table 4.1 South East Asian soils

Soil order	Great soil Group	Description	Location
Entisols (recent soils)	Alluvial soils	Young water-borne sediments without prominent horizons except at top. Generally poorly drained.	Flood plains, valley bottoms, coastal flats.
	Regosols (rhegos: blanket)	Sediments other than alluvium. No prominent horizons. Includes beach and dune deposits, slope colluvium, volcanic ash in areas of active volcanism.	Coastal zones, foot-slopes, volcanic areas. Flat, undulating terrain.
Vertisols (verto: turn)	Grumosols	Dark, heavy clay soils of tropics. 40–80 per cent of clay is montmorillonitic, which expands when wet, and shrinks and cracks when dry. Well-developed horizons. Parent materials largely of	East Java, Madura, Lesser Sundas, central plain of Thailand, basalts and old alluvium in Cambodia, southern Vietnam, Luzon.

		basic composition or calcareous clays.	
Inceptisols (incepto: begin)	Andosols (an: dark)	Thick black to very dark grey-brown, organically enriched surface horizon. Medium-light texture. Parent material < 1 metre from surface. Mainly from volcanic ash weathered under humid conditions.	Volcanic areas of Sumatra, Java, Bali, South Luzon, islands off West Burma.
	Acid brown forest soils	Young, shallow soils < 80 cm thick, on steep slopes and youthful volcanic ash, in areas subject to continual erosion. Of patchy distribution. *B* horizon slightly illuviated.	Mountains and steep hills.
Spodosols (spodos: woodunder) or ash)	Podzols (pod: (zola: ash)	Very acid, bleached light-textured soils; eluviated *A*, and illuviated *B* horizon rich in organic matter and iron oxides. Developed in areas of poor or once poor internal drainage, especially on level or undulating terrain, in wet climates.	Lowlands. West and South Kalimantan, Sarawak, Sabah, Brunei, coasts of South Cambodia, South Thailand, peninsular Malaysia, East Sumatra.
Alfisols (alf: ped*alfer*: aluminium and iron compounds)	Grey podzolic soils	Acid soils formed on old transported materials, e.g. old alluvial terraces, on flat to undulating terrain in monsoonal climates. Eluviated *A*, illuviated *B* horizon.	Chief type in Mekong and Irrawaddy terraces.

	Lateritic concretions or laterite evident. Clay predominantly kaolin.	
Red-yellow podzolic soils	Acid soils on flat to mountainous terrain, where parent materials non-basic. Eluviated *A* horizon, illuviated *B*. Kaolinitic clays. Laterite or lateritic concretions present.	Dominant type in wetter South East Asia, where non-basic parent materials occur, e.g. Vietnam, non-volcanic West Java and Sumatra.
Low humic gley and grey hydromorphic soils	Acid to neutral soils, developed on low-lying, poorly drained, alluvial or colluvial material. Eluviated *A*, illuviated *B* horizons. Older river terraces, and small patches on lower slopes.	Especially along Mekong River. Widespread.
Red-brown earths	Neutral to mildly acid soils developed on limestone and some basalts, in seasonally dry environments (annual rainfall < 1500 mm). Moderate to good drainage. Clays, kaolinitic especially; some little quartz, also present.	Central Burma, Cambodia, Laos, South East Vietnam, central plain of Thailand, central and East Java.
Non-calcic brown soils	Acid soils developed on an acid to intermediate rocks in seasonally dry environments (annual rainfall c. 1500 mm). Kaolin dominant clay mineral.	Central Burma, South East Vietnam, East Java, Lesser Sundas.

Oxisols (ox: oxide)	Dark red and reddish-brown latosols	Deeply weathered, leached, reddened, acid to mildly acid soils; very clayey and with insignificant amounts of sand. Concretions in *B* horizon. Developed on basic rocks. Excellent physical properties and efficient surface drainage.	Lower volcanic ranges of West Sumatra, Java, Bali, Maluku, Philippines. High plateaux of Burma and West Thailand.
Oxisols (ox: oxide)	Red and yellow latosols	Acid. Less-clayey, more sandy, developed on less-basic parent materials. Illuviation in *B* horizon. Wide range of landforms.	Especially in Kalimantan, North East Sumatra, Sarawak, Brunei, Sabah, Mindanao.
Histosols (histos: tissue)	Organic soils	Acid to very acid soils > 30 cm thick, · with organic matter > 30 per cent. Peaty soils in ill-drained situations. Many overlie a highly-weathered white to grey mineral horizon.	For example east coast of Sumatra, east and west coasts of peninsular Malaysia, Sabah, Sarawak.
Special soils (not in seventh approximation)	Anthropomorphic (anthropos: man)	Develop a bluish-grey surface gley with iron mottling due to artificial flooding for rice-growing: a temporary characteristic.	Rice-growing alluvials or other artificially terraced terrain.

Source: Swan, (1979, p. 14).

Water resources

Other than the Irawaddy, Salween, Chao Phraya and Mekong, few South East Asian rivers are more than a few hundred kilometres long. Most are short and turbulent. Streams tend to fluctuate in level rapidly as a consequence of heavy convectional rainfall, for example the Salween can rise nearly 20 m in only a few hours. Forest cover intercepts rainfall and moderates the rate at which runoff reaches streams and the amount of detritus carried. In areas where forest is undisturbed, streams are usually clear, acidic, low in dissolved oxygen and rich in silicates. Deforestation can result in erratic flows (indeed streams may even dry up between rains) and increased silt loads. It is not only deforestation which results in heavy silt loads in streams, volcanic ash falls and the erosion of unconsolidated volcanic deposits also add much detritus. In Sumatra and Java rivers are especially silty for the latter reasons; the Solo River (Java) carries an average of 2000 g/m^3, roughly sixty-times the load of the Rhine at Paris though it is about one-sixtieth as short (Dobby 1973 pp. 50, 83). Mining waste also contributes silt to rivers, especially in peninsular Thailand and Malaysia; this silt is mainly sterile silica sand, and if spread by floods it can ruin productive land (as has been the case along the Malacca River, peninsular Malaysia).

Because of the abundant silt, deltas are commonly well developed, and in many regions are growing at ever-increasing rates, e.g. there are large deltas on the Irawaddy, Mekong (Thailand/Kampuchea/South Vietnam), Solo, and Tjimanoek (Java). Sediment washed down by streams is becoming a problem in many coastal and estuarine waters of South East Asia, particularly the Straits of Malacca, and around Java, Sumatra and Kalimantan. Estuaries are filling up, hindering shipping; and inland deposition along stream courses is causing flooding more often and to higher levels than was once the case (for example, Kuala Lumpur has had to invest in costly flood mitigation engineering). Irrigation systems become choked; bridges may be carried away when floodwaters can no longer freely discharge; and communications routes and urban settlements get inundated. Even when it reaches the open sea, silt may cause problems; coral reefs off the shores of many parts of South East Asia are being damaged by it.

In South East Asia, Quaternary climatic conditions did little to create lakes. Rapid siltation and the ability of vegetation quickly to colonize shallow water have also conspired to ensure that there are comparatively few large bodies of still water to be found in this part of the world. There are many seasonal swamps, e.g. Tasek Bera (peninsular Malaysia) or Tonle Sap (Kampuchea) but the only sizeable lakes are volcanic crater lakes in Indonesia (e.g. Lake Toba, Sumatra) and the Philippines. Despite the scarcity of large, natural lakes there are an estimated 500 000 hectares of smaller ponds, mining pools, man-made fish ponds and paddy-fields. These, together with brackish lagoons, mangrove swamps and estuaries offer tremendous potential for aquaculture (Ruddle 1982), but the lack of large natural lakes and consequent dearth of information on South East Asian limnology has made the prediction of environmental impacts associated with large hydroelectric reservoirs more difficult.

Most South East Asian nations have some potential for hydroelectric generation and, in particular, the Mekong could become one of the world's most impressive examples of integrated river basin development for power, flood control and navigation improvement, if political stability and agreement between nations sharing its waters could be reached and maintained. Because most South East Asian rivers vary quite a lot in flow, large-scale generation of hydroelectricity usually requires big storage reservoirs which are costly, can generate unwanted environmental and socioeconomic side-effects (Goldman 1979, Szekely 1982), and are vulnerable to having their useful life shortened by siltation resulting from difficult-to-control logging and cultivation activities. In more remote areas, mini-hydroelectric plants (generating less than five megawatts), which can harness small, fluctuating, silt-charged streams may offer a cost-effective, less environmentally damaging, means of rural electricity supply. The adoption of mini-hydroelectric generation may even help to reduce deforestation by reducing some people's need to collect fuel wood.

On the whole, groundwaters are under-utilized in South East Asia, reflecting inadequate exploration for the resource, past availability of surface water supplies and the cost and difficulty of sinking and pumping wells or boreholes. Groundwater exploitation will increase in the future, but already legislation controlling exploitation is inadequate in most countries. There are already problems of saltwater contamination of aquifers (water-bearing rocks) and aquifer collapse leading to ground subsidence and damage to buildings in the Djakarta metropolitan region which result from ill-regulated extraction. In many suburban areas of Thailand, Malaysia, Indonesia and the Philippines, shallow groundwaters are increasingly contaminated by sewage and industrial pollutants, and in rural areas pesticide and fertilizer contamination of groundwater supplies is increasingly common.

Forests and deforestation

A humid tropical, or seasonally moist tropical environment, wide variety of soils and relief and relatively limited climatic disturbances during the Quaternary have ensured South East Asia has a particularly rich flora and fauna. Indeed, the Indo-Malesian Floristic Region (Indonesia, Malaysia, Brunei, the Philippines, Singapore, eastern Timor, eastern Papua New Guinea, plus the rather less species-rich Solomon Islands) has more flowering plants (at least

35 000 species) than any other floristic region (Dobby 1967, p. 64). Tropical West Africa, for example, has a mere 13 000 species.

The natural vegetation cover of most of South East Asia is, or was, forest. Two principal types can be distinguished, although it must be stressed that these are generic and within each there are considerable variations caused by factors such as soil, local microclimate, aspect, altitude, drainage and disturbance by man or animals. The first is tropical 'evergreen' rainforest (tropical moist forest), the climax vegetation of lowland South East Asia wherever there is pronounced dry season and a rainfall in excess of about 2000 mm per year. Undisturbed South East Asian lowland rain forest can have a very high species diversity. Two hundred species of tree per hectare have been recorded, but generally trees of the family Dipterocarpaceae, which includes commercially important species like *Shorea* spp. and *Dipterocarpus* spp., are dominant (Whitmore 1975).

Above roughly 700 m, *Dipterocarpus* spp., progressively become less common and are replaced by tropical evergreen oaks and other highland species. Above roughly 2000 m on the highest mountain ranges (e.g. Mt Kinabalu, Sabah), Ericaceous species (including *Rhododendron* spp.) become quite common and above the 'tree line' which, for example, on Mt Willhelm (Papua New Guinea) lies at about 4000 m, only low-growing shrubs, grassland and alpine herb associations flourish. On isolated peaks these vegetation zones may lie at somewhat lower altitude than on mountain ranges.

Wherever soils dry out seasonally, or where soils are very poor, the climax vegetation is likely to be seasonally humid tropical forest (also called, depending on the criteria used to describe it, deciduous tropical forest, seasonally dry forest or 'monsoon' forest – the latter is a poor name because it emphasizes rainfall as a formative factor, whereas soil and historical factors are probably really as important). Structurally, this forest is less complex than the tropical 'evergreen' forest of lowland equatorial South East Asia. The canopy (the more-or-less continuous surface formed by the treetops) is lower and more open and the species diversity is less; although typically with 40 to 60 tree species per hectare, it is still very much richer than forests at higher latitudes. Leguminous trees are particularly common but dipterocarps are also present, notably *Shorea* spp. and the commercially important teak (*Tectona grandis*). Teak is most common and best-developed where there is a rainfall of 1000 to 2300 mm per year and a clearly defined 'dry' season. Regions with such conditions include: Burma, North Thailand, Laos and parts of Indonesia (notably Java) (Richards 1964, pp. 327–33). It is, however, not easy to generalize on the occurrence of forest type as it is a function of several variables, as Table 4.2 seeks to indicate.

There are estimated to be 5.4 million hectares of mangroves in South East Asia (Davidson *et al.* 1985, p. 13). The more-sheltered coasts of Thailand, the west coast of peninsular Malaysia, Sumatra, Java and Burma all have extensive areas (ESCAP 1978, p. 83). Varying in width from one to several kilometres mangrove swamps

Table 4.2 Relationship between soil, climate and vegetation cover in South East Asia (simplified)

Climate	Rich soil	Poor soil
Dry	S	S
Seasonal-wet	R	S
Ever-wet	R	R

Key: R = tropical 'moist' rainforests;
S = seasonally humid ('monsoon') forest.

contain between 20 and 40 tree species, some related, some unrelated, but all of broadly similar physical appearance. Some of these trees are of economic value (e.g. *Rhizophora* spp. and *Avicenna* spp.) providing fuel wood, charcoal, tan-bark and building materials and chips for rayon manufacture. Where the water is less salty and silt is abundant, the nipa palm (*Nipa fruticans*) flourishes, and provides thatch and palm sugar.

Some attempts have been made to protect and regenerate South East Asia's diminishing mangrove stocks, but in general too little is being done to combat this problem. Along the west coast of peninsular Malaysia there is a long history of intensive management; nevertheless, between 1963 and 1978 an estimated 58 per cent of peninsular Malaysia's mangroves were cut and not replaced (Barrow 1980, p. 11). A major cause of this deforestation has been cutting for charcoal production; Furtado (1979, p. 99) reported that 566 000 m^3 of mangrove charcoal was produced in peninsular Malaysia alone in 1970. The loss of mangrove and nipa swamp forest exposes coasts to increased erosion and storm damage and reduces the breeding and feeding grounds available for commercially valuable fish, crustacea and other wildlife. Recent research suggests that mangrove swamps have considerable potential for effective, low-cost sewage and industrial effluent disposal: the effluent can be simply channelled through a maze of mangrove roots where it is rapidly decomposed by bacteria before it reaches the sea.

Perennial crops like rubber, oilpalm and spices have long been important to South East Asian economies. However, there is considerable scope for crop diversification and for increased exploitation of forest products from areas where forest removal is inadvisable. The market for furniture and woven goods made from rattan, produced from species of forest liana and undergrowth shrubs, has increased in recent years (Caldecott 1988, Ives and Pitt 1988).[†] Other forest products which can be collected without total

[†]In 1988 the world market for rattan products was worth about U.S. $ 3000 million, and was virtually all met from wild material. Some twenty-five rattan varieties are collected in South and South East Asia and processing industries employ many people. Because rattan is often collected from undisturbed forest – and even national parks and reserves, causing considerable damage to other plants and to the soil – and because of the demand, there is now an urgent need to encourage the cultivation of rattan.

destruction of forest cover include wood oils, bamboo, honey, ornamental plants and waxes.

Freshwater swamp, peatland and sandy soil environments in South East Asia have potential that could be better developed. Freshwater swamps could be used for aquaculture and/or wider cultivation of sago palm (the main sago producing palm is *Metroxylon sagu*, but sago can also be obtained from *Metroxylon rumpii, Cayota urens, Arenga saccharifera* and *Phoenix acaulis*). Cellulose, alcohol and other products can be made from sago palm (Staunton and Flach 1981). Inland from coastal mangrove swamps, there is often a belt where soils tend to be poor and sandy, woodland, dominated by casurina trees (*Casurina equisetifolia*) and coconut palms (*Cocus nucifera*), covers these soils and can yielded commercially valuable timber, nuts, fibre and oil. Deforested land often develops poor, sandy soils on which it is only worthwhile to establish hardy crops like cashew nuts (*Anacardium occidentale*) or coconuts. It should not be overlooked that one of South East Asia's most successful crops – oilpalm (*Elaeis guineensis*) is a native of tropical West Africa, and that there are a number of other palm species, potentially as valuable as *Elaeis*, growing wild in the forests of South East Asia, Africa and Amazonia. Commercial production would not compete for the same market as *Elaeis* because their oil or seedpulp has different qualities.

Throughout South East Asia, man has altered the natural cover, probably for 8000 years or more, 'slash-and-burn' shifting cultivation (called *ladang* in Malaysia/Indonesia: *taungya* in Burma) has caused considerable forest disturbance (Spencer 1966). But in recent decades the extent and degree of damage has increased. A report has estimated that highland dwellers and migrant 'squatter settlers' in the lowlands are presently responsible for roughly 60 per cent of South East Asia's forest losses each year (Davidson *et al.* 1985, p. 14). An increasing percentage of South East Asia's remaining forests are in highlands, largely because commercial logging interests find it more difficult to operate there, and this is where shifting cultivation is being practised. There is a need to develop forest management/conservation strategies which involve highland peoples. At present highland peoples are often ignored by South East Asian governments so forest protection efforts miss what should be their main target.

Collection of wood for domestic fuel is probably another major cause of forest loss in South East Asia. In Thailand, the Philippines and Indonesia, at least 50 per cent of the people use wood as their primary fuel source, and in Burma, lowland Sumatra and Indochina most cooking is done on wood fires. Small-scale industries, for example brick, tile and pot manufacture, sugar refining, smelting and smithying, use wood or charcoal. As fuel wood supplies dwindle, South East Asian countries will sooner or later have to develop suitable, renewable, affordable alternative energy sources such as fast-growing tree plantations, the use of solar energy,

Table 4.3 South East Asia: forest cover

Mainland South East Asia	123 million hectares of forest which covers about 63 per cent of total land area; 34 million hectares of total land area are used for agriculture.
Island South East Asia	187 million hectares of forest which covers about 65 per cent of total land area; 46 million hectares of total land area are used for agriculture.

Note: Furtado (1979) estimated that 38 per cent of South East Asian moist tropical forest had been cut by 1979, leaving 187 million hectares and 81 million hectares of other types of forest.
Source: Tran Van Nao (1974).

Table 4.4 Land area, forest area and area of United Nations-recognized national parks or equivalent areas in four major South East Asian nations

Country	Total land area (km²)	Forest area (km² estimated)	Reserve areas (km²)
Indonesia	1 904 000	800 000	13 259
Malaysia	330 000	170 000	5 143
Philippines	300 000	160 000	2 179
Thailand	514 000	130 000	10 759

Source: McNeely, in Furtado (1980, p. 337).

hydroelectricity and biogas. Tables 4.3 and 4.4 give an indication of South East Asia's forest cover.

Logging is another major cause of deforestation. The export of timber began to 'boom' in the 1960s when commercial timber interests recognized that large profits could be made and when South East Asian governments realized how lucrative and rapid an earner of foreign exchange government-controlled logging or the taxation of commercial logging could be. Large areas of South East Asia, especially at altitudes below about 500 m, have been logged and timber extraction is now tending to shift to land as high as 1000 m or even higher. In 1985 it was estimated that about 1.4 million hectares of 'undisturbed' forest were being lost through logging each year (Davidson *et al.*, 1985, p. 15).

In favourable localities, regeneration of forest after logging, or clearance for other reasons, can be rapid and the results after, say, 20 years can *look* almost like the original cover. Unfortunately, the degree and extent of forest disturbance usual today means that such complete regeneration is increasingly rare. It is also likely that regenerated forest which *looks* like the original cover, will in fact not re-acquire a natural species diversity for centuries. For forest to become properly re-established, it is necessary for the cleared land to be left undisturbed for it to escape severe soil damage, and for there to be suitable areas of the original forest cover left to permit re-seeding. Often these conditions are not satisfied. Much of South East Asia is now covered with poor-quality regrowth (belukar – Malaysia/Indonesia), of lower stature than the original cover, with far fewer species per hectare and dominated by a few light-loving 'pioneer' shrubs or trees. In the earliest stages of forest regeneration, or if tree cover cannot get re-established, bracken and grasses, especially lalang grass (*Imperata cylindrica*), are common. Large areas of South East Asia which were once forested are now clothed with lalang, which has little value as forage for livestock and is difficult to clear for cultivation.

Three main logging strategies are used in South East Asia: *selective cutting, shelterwood-cutting* and *clear-cutting* (or uniform-cutting). Selective-cutting is the removal of the best specimens of commercially valuable timber from a forest, hopefully with minimum damage to other species. At present only ten or twelve species of tree are easily marketed and in practice this means roughly 30 to 40 per cent of forest is actually selected. More is damaged by access tracks and the felling of trees. Selective cutting should offer the best chances of forest regeneration of the three strategies. Shelterwood-cutting involves making several cuttings over a period of time, each of varying intensity. These open up gaps in the forest canopy where timber is extracted. When the cleared areas have regenerated satisfactorily, the surrounding 'shelterwood' trees which provided the seeds and shade for regeneration are cut (if they are of value). Clear-cutting involves the more-or-less complete removal of trees, usually to make woodchips for paper pulp or compressed chipboard. Immature specimens of commercially valuable trees may be left after clear cutting, but the degree of clearance means that soil erosion and/or weathering is likely, so satisfactory natural forest regeneration is rare. Even careful selective cutting is likely to cause 'genetic erosion': the removal of the best specimens of a given species of tree leaving behind an inferior breeding stock to regenerate it.

Organized, government-sponsored, agricultural development, transmigration and settlement schemes are a major cause of deforestation, especially in Malaysia and Indonesia. In peninsular Malaysia vast hectarages have been converted from forest to oilpalm and rubber plantations, so much so that the view greeting anyone journeying down the Malay peninsula today is a somewhat monotonous one. Hydroelectricity development often involves the creation of a reservoir which floods hundreds of square kilometres. This has destroyed much forest, especially in Thailand. The forest destroyed by hydroelectric development tends to be the most diverse in species because it is on the rich alluvial valley soils and these are the areas inundated. In Indochina, especially Vietnam and

Kampuchea, the use of defolient chemicals during the Vietnam conflict resulted in the loss of large areas of forest. Once defoliated, these forests have tended to be settled and cleared for cultivation so regeneration of a large portion of the affected land is unlikely.

Rates of deforestation

Estimates of how much of the world's tropical forest has been lost, how much remains and what the rates of deforestation are can be unreliable and have been subject of much bitter argument (Myers 1980, 1986; Simon 1986). Certainly, deforestation in South East Asia has been rapid and extensive, and it is likely that the loss is taking place faster than in any other tropical forest region of the world (Lal et al. 1986, p. 189). About 23 per cent of the world's tropical forest, roughly 180 million hectares in late 1980s according to the IUCN (Davidson et al. 1985, pp. 5–13), is in South East Asia. About 43 per cent of South East Asia as a whole is still forested. Indonesia is still about 60 per cent forested, mainly in Sulawesi, Maluku and Irian Jaya. † Roughly 60 per cent of Malaysia is still covered with forest (but some of this cover is disturbed forest). Between 1981 and 1985 about 1.5 million hetares of forest were cleared from South East Asia each year – the most rapid removal being in Thailand where it was estimated around three per cent of forest cover was removed annually.

Not only is the rate of forest destruction alarming, the rate of replanting and the record of conservation are generally poor, for example less than one per cent of what is cut from Indonesia each year is re-afforested (Davidson et al. 1985, p. 29). The critical question that needs to be asked now is how much forest will be left undisturbed when the timber industry shifts from logging natural forest to harvesting intensively managed tree plantations or replanted forests, and when will governments cease forest clearance for agriculture and urban development? Will what is left be enough to ensure that representative forest reserves can be maintained indefinitely?

United Nations recognized national parks or equivalent areas comprised a mere one per cent of the four major South East Asian nation's total land area in 1980 (see Table 4.4). Many of the reserves that have been established are too small, and poorly policed or are simply unrepresentative of lowland moist tropical forest and therefore provide inadequate protection for representative sample of South East Asia's vegetation and wildlife.

---

†Between 1982 and 1984 roughly 3500 square kilometres of Eastern Kalimantan was destroyed by fire. Same authorities suggest drought related to the El Niño-Southern Oscillation events was in part responsible; however, there are also indications that logging and settlement predisposed the region to damage (Adriawan and Moniaga 1986; Malingreau et al. 1985).

The environmental consequences of deforestation

Considering the vast areas cleared of forest, knowledge of the structure and function of tropical forest/humid tropical ecosystems is weak. Much more information is needed on forest productivity, resilience of forest reserves, on the subtle, close relationship between environment, pest animals, disease-carrying organisms and creatures which pollinate forest plants and crops. For example, pollination of the popular durian fruit trees (*Durio* spp.) over a wide area of Malaysia depends on bats which roost and feed in restricted localities which are under threat of destruction. Commercially valuable timber trees like *Shorea* spp., seem to be pollinated by thrips which require plots of relatively undisturbed forest close to the *Shorea*. For plantations to flourish or timber to continue to regenerate may require careful, planned conservation of forest areas. It is common for fungi to form a symbiotic (mutually beneficial) relationship with the roots of forest trees. Without these fungi, trees have difficulty obtaining nutrients. If forest is cut and the soil 'sterilized' by exposure to weathering, even if an area is replanted, the correct fungi may not be easily re-established and so the reforestation fails.

The hydrological effects of deforestation are reasonably well established but the climatic impacts are rather less well established. Widespread cutting alters an area's albedo (reflection of sunlight) and reduces the amount of water vapour being transpired and evaporated. Local air temperatures, downwind airflows and humidity may well be altered. If fire is used and/or a lot of dust is liberated following forest clearance, this might influence the formation of precipitation on a regional scale.

A major impact of deforestation is that flora and fauna are lost. Apart from the moral aspect of such loss, some of this wildlife might well have commercial potential as food plants, sources of drugs or industrial raw materials in the future. There are large areas of once-forested scrubland in South East Asia. Priority should be given to getting these to yield crops of timber or useful forest products so that less undisturbed forest is cleared.

Cropping strategies

The clearance of forest so that annual or perennial crops can be planted is often far from satisfactory. It is something of a cliché that nutrient cycling in tropical forests is 'close', i.e. that plant debris are rapidly broken down, are reabsorbed by the vegetation and the soil holds little reserve of plant food. Plantations of perennial crops like oilpalm, rubber, cocoa and coffee have replaced forest over considerable areas of South East Asia. Whether these crops can repeatedly be replaced as they become senescent without a reduction in yields is not really established. Though the aforementioned crops (except coffee) make relatively light demands

on soil fertility, few plantations have been established for much
more than 80 or 90 years.

Much of the forest cleared by smallholders or shifting cultivators is
planted with annual crops. Un-irrigated, rainfed hill rice is widely
grown where there is a suitable rainfall, but the yield is fairly low
(seldom over 700 kg per hectare per year) and soil exhaustion and/or
weed growth allows further crops only for two or three years. The
land must then be left fallow for many years before cropping again.
Irrigated, paddy or wet-rice (sawah – Malaysia/Indonesia) will grow
satisfactorily on a very wide range of soils, provided that the field
can be bounded and kept flooded to a satisfactory depth. In parts of
Java, Thailand and Malaysia paddy-fields have yielded rice crops
every year for centuries without the application of any fertilizer
(typically giving between 1000 and 2000 kg per ha per year). Two
broad strategies of paddy-rice cultivation have been traditionally
practised: *rice monoculture* under which most of the cultivator's land
is planted to rice in both the wet and dry seasons and two crops a
year are obtained; and *mixed rice farming* under which at least one
crop other than rice is planted during the drier season or in rotation
with the rice. Rice monoculture is generally only possible where the
monthly rainfall is at least 200 mm for six to eight months, or where
there is adequate irrigation. Without such water supplies,
paddy-fields could not be kept adequately flooded.

Traditional wet-rice cultivation succeeds even where soils are of
quite poor fertility because the throughflow of irrigation water
deposits sediment-containing plant nutrients on the paddy-field, and
because bacteria and algae among the flooded rice stems and in the
mud fix nitrogen. Geertz (1963, p. 30) described wet-rice cultivation
as '. . . an ingenious device for agricultural exploitation of a habitat in
which heavy reliance on soil processes is impossible . . .'. Not only
can wet-rice cultivation be sustained on poor soil with no addition
of fertilizer, yields can be 'stretched' with the input of what is often
the only thing that is available to many South East Asia smallholders
– labour. Better weeding, better tillage, transplanting of seedlings
from 'nursery beds' and improved water control can significantly
boost rice crops.

During the 1960s agricultural innovations, many of them
originating from the International Rice Research Institute at Los
Banos in the Philippines, changed traditional rice cultivation over
large parts of South East Asia and other parts of the tropics. New
rice, maize and barley varieties were developed which, given
adequate water, artificial fertilizer (and usually the application of
pesticides), produced (and continue to produce) more grain in
relation to stem and leaf and were less likely to lodge (fall over if
exposed to wind or heavy rain and become difficult to harvest).
Some of the new 'miracle rices' mature much faster than traditional
varieties, and some are also able to grow near the Equator, where
little variation of day length interfered with the maturation of
traditional varieties. Rice could be grown in parts of South East Asia

where previously conditions were unfavourable, and in many areas, two or even more crops a year could be obtained (giving previously impossible yields of at least 3000 and sometimes as much as 7000 kg per hectare per year). The catch phrase 'green revolution' was used to describe the impact of these innovations. Very quickly countries like Malaysia and the Philippines became self-sufficient in rice for the first time in decades. Unfortunately, not all farmers benefited from the 'green revolution'. Some suffered social and economic disadvantages, and there have been environmental repercussions. Some of the more severe resulted from increased use of pesticides and herbicides. Whether the new fertilizer, hybrid seed and pesticide-dependent rice cultivation will be as sustainable as traditional cultivation methods have been remains to be seen. Hopefully it will be, as many millions of South East Asians now depend on it for their food supplies.

Environmental degradation and progress towards environmental management

The economies of South East Asia (with the exception of Singapore) depend, and will continue to depend for some time to come, more on natural resource exploitation and agricultural development than manufacturing. There is a need in most of the countries of South East Asia for effective environmental planning and legislation to ensure environmental damage, especially deforestation, soil erosion and pollution, is minimized and, wherever possible, renewable resources exploitation is sustained. In practice economic growth has had, and still has, priority over environmental concerns. Even the richer countries like Malaysia have hesitated to implement environmental controls or planning measures which might be seen to discourage investment or hinder development activities, but there have in the last few years been promising trends towards better environmental management, especially in Malaysia and Singapore.

Since the 1960s agriculture throughout South East Asia has come increasingly to rely on inputs of agrochemicals: chemical fertilizers, pesticides and herbicides. These agrochemicals can greatly increase crop yields, especially if applied to high-yielding varieties; unfortunately they can also contaminate the environment, and man. With high rainfall and high temperatures it is difficult to prevent agrochemicals from contaminating streams, ponds groundwaters and even marine environments. The effects may be rapid and spectacularly obvious – fish kills and chronic poisoning of wildlife, livestock and people – or slower and insidious, but not necessarily less dangerous, such as long-term contamination of groundwater and soil.

Unfortunately, organisms in the food web tend to concentrate pesticides, herbicides and heavy metal pollutants. For example, paddy-field plankton may acquire a dangerous level of contamination by concentrating poisons from an environment where agrochemicals have been diluted to a 'safe' background level. Fish, prawns or crabs feeding on the plankton further concentrate the pollutant; if higher organisms including man consume these animals they may become ill or die. There have already been cases of death or illness from this process of 'biological magnification' of pollutants and it is poor people who depend on fish, shellfish or crustacea from paddy-fields, streams or coastal waters, who are most vulnerable, and their children even-more so. People suffering from agrochemical pollution may be wrongly diagnosed. There have been cases where health authorities have for years vainly tried to identify 'mystery' illnesses amongst villagers and then it has come to light that a landlord has been using pesticide which has then contaminated fish eaten by the local people. Increasingly, the protein sources available to the poor are becoming contaminated and reduced through pesticide pollution. In Malaysia, for example, between 1972 and 1982 fish yields from paddy-fields have declined considerably; the same is true in the Philippines, Thailand and probably many other parts of South East Asia, due it seems to the application of agrochemicals associated with double-cropping rice. Aquatic organisms are especially vulnerable to agrochemical pollution, and amongst the wildlife which is killed there may well be species of fish or other organisms with potential for domestication for aquaculture/pisciculture. In the high temperatures of South East Asia agrochemicals may be unstable. A 'safe' compound may break down before or after application to the crop or soil into compounds which are dangerous. Many agrochemicals or their decomposition products are very persistent and are not rendered safe by decay. It is unfair to blame modern farming trends entirely. Health authorities have used pesticides to try to control disease-carrying mosquitoes and, in urban areas especially, the public are increasingly using pesticides to try and combat household insect pests. In general, however, there has been minimal official encouragement of the use of non-persistant pesticides in South East Asian agriculture.

The countries of South East Asia, apart from Singapore, have relatively high rates of population growth. One consequence of this is a steady stream of rural-urban migrants seeking employment and better conditions in the cities. Providing enough low-cost housing for these migrants is difficult and a result has been the growth of squatter settlements or shanty towns around or within many cities. Crowded, poor quality housing, usually lacking in services, highly insanitary and often located on land that is waterlogged, prone to flooding, steep or polluted (and for these reasons is available to the poor because no one else wants it) means that the inhabitants of these settlements face considerable environmental health hazard. Inadequate urban road networks, poor public transport services and an increase in the number of private cars owned by richer citizens results in traffic jams and local air and noise pollution (McAndrews and Chia Lin Sien 1979; p. 7). Again it is the likely to be the poor, and especially their children, who suffer from such pollution. The combustion of leaded petrol by South East Asian road traffic continues without restriction it is likely to have serious long-term effects on the health of children in many cities.

The Malaysian environment: a case study

With an environment and population growth rate reasonably 'typical' of South East Asia, but having made much more progress than most South East Asian countries in exploiting natural resources and in trying to manage that exploitation to minimize environmental damage, Malaysia is interesting, for what happens in Malaysia now may well happen more widely elsewhere in South East Asia in the future. (Aiken and Moss 1976, Barrow 1980).

At the start of the twentieth century, peninsular Malaysia was largely clothed in tropical rainforest and the population was probably about one million. Under British colonial administration was begun a *laissez-faire* exploitation of the country's tin deposits. Rubber was introduced from Brazilian Amazonia and, later, oil palm from West Africa. Environmental degradation was well under way before independence in 1957. Between 1957 and 1968 Malaysia expanded exploitation of these resources (Hill 1982) and embarked on developing export-substitution industries. Large areas of forest were also being cleared for land development and resettlement schemes. The population had grown at about 2.7 per cent per annum (and in the 1980s exceeded 13 million). By the mid-1970s the impact of development on the Malaysian environment were becoming obvious. Some of the earliest reaction to this came from non-governmental bodies, and these played a key role in promoting media, public and government interest in environmental problems (Barrow 1981). The most active of these non-governmental bodies were the Malaysian Nature Society, the Selangor Environmental Protection Society, Sabat Alam Malaysia and the Consumer's Association of Penang.

After deforestation, the major cause of damage to the Malaysian environment has so far been pollution of the stream and river systems. The main sources of pollution have been agro-industrial effluent, human sewage and silt, and dust or toxic wastes from mining operations. Agrochemical pollution, especially by pesticides, has increased since the 1960s and is already a major problem.

One reason for widespread oil palm effluent pollution is that the processing has had to be done close to the plantations because the fruit deteriorates fast after gathering. Many small- or medium-sized factories rather than a few large ones have been built and this makes pollution control more costly and difficult to monitor. But in the last five years or so palm oil producers have been forced by government legislation and public pressure greatly to improve their effluent disposal. But the damage had already been done. By 1978 fisheries had been ruined by palmoil effluent and sewage pollution in 42 rivers and in 16 little had been left alive. In recent years there has been a marked decline in catches from Malaysia's paddy-fields and smaller streams, most probably because of pesticide pollution (Bull 1982, p. 65). Catches of hardy, but commercially undesirable, fish like *Anabantids* and catfish have increased, while carp and osteoglossid species, which are generally preferred by Malaysians, have declined.

Industrial pollution has become more frequent since the mid-1970s, and has affected several rivers and some coastal waters. Before about 1977 heavy metal pollution was rare; now it is increasingly reported, especially on the west of the peninsula. By 1978 things were bad enough for a Malaysia newspaper to call the country an 'Effluent Society' (Consumers' Association of Penang 1978, p. 22).

Malaysia's mining laws failed to prevent the deposition of over 80 000 hectares of tin tailings (infertile silica sand), mainly in the west of the peninsula, nor the removal of environmentally sensitive areas of limestone outcrops for cement manufacture (Barrow 1980, 1981). In East Malaysia (Sabah) worries have been voiced about the Japanese-owned Mamut Copper Mine (on the Sugut River about 120 km from Kota Kinabalu) which has caused siltation and toxic copper waste problems (Ziauddin Sardar 1980: pp. 700–1). Development of offshore petroleum resources in the South China Sea has led to some forest clearance and pollution in the east of peninsular Malaysia where pipelines, refining and petrochemical plants have been constructed. The risk of marine pollution from these installations is added to the danger of spillage from oil tankers navigating the Straits of Malacca, where they run the risk of collision or running aground (Consumers' Association of Penang 1978, p. 37).

In parts of Malaysia, Thailand, Singapore and Indonesia (especially Bali), tourism has become important. Hotel development, especially if not associated with adequate, well-maintained sewage disposal facilities, can pollute coastal waters causing fish kills. There has been considerable pollution of the coastal waters of Penang by hotel-derived effluent, for example. Un-regulated sport fishing and SCUBA diving can damage marine wildlife, and the sprawl of hotel buildings, associated roads, increased road traffic and airport facilities means pollution, forest clearance and loss of farmland. As well as damaging the physical environment, tourism, if not carefully managed, may have deleterious effects on the socio-environment of the host country.

In 1974 an Environmental Quality Act became law in Malaysia, and in 1975 a Division of the Environment was established within the Ministry of Science, Technology and Environment. These were 'landmarks' in South East Asian environmental legislation. Before, Malaysian environmental protection and management had been managed by a number of governmental agencies with little coordination. Malaysia is the first South East Asian country (Singapore has a much smaller land area, and is best considered a city-state) to have made real efforts to get effective environmental management.

In 1981 the Fourth Malaysia Plan paved the way for the introduction of obligatory environmental impact assessment for all new developments deemed likely to significantly affect the Malaysian environment. Under the Fifth Malaysia Plan (1986-90), environmental impact assessment procedures will be enforced. Government standards and environmental monitoring services have been improved considerably in the last few years (Ziauddin Sadar 1980, author's personal observation August 1986). Things are by no means perfect in Malaysia, but the country does have a media, public and government which does more than simply pay lip-service to minimizing natural resource and industrial development problems.

Other South East Asian countries are following, or will probably follow, Malaysia's lead. For example, Indonesia established a Co-ordinating Ministry of Development and Environment in 1978, and the Philippines (influenced by American environmental legislation developments – notably the USA's National Environmental Policy Act of 1970) established a National Pollution Control Commission in 1976. Various United Nations (UN) agencies have been active in promoting environmental management in South East Asia. The UN Environment Program (UNEP) has established a Regional Office in Bangkok, Thailand. The UN Economic and Social Commision for Asia and the Pacific (ESCAP) established a Regional Centre for Technology Transfer, which seeks to promote environmentally sound alternatives of development (also in Bangkok). A UN Asian and Pacific Development Centre has been established in Kuala Lumpur and has been active in promoting environmental assessment of South East Asian development projects (UNAPDC 1983); and a UN Asian and Pacific Development Institute, established in Bangkok, has promoted the incorporation of environmental dimensions into South East Asian development planning.

Although there has been much despoilation of the South East Asian environment, there now seems to be growing awareness of the problems, sufficient to give one hope that the future is not one of total ruination. However, there are key problems which South East Asian countries will have to master in the not too distant future. These are:

1. control of logging, collection of forest products, forest clearance and destruction of wildlife;
2. creation of adequate sized, well-protected, forest reserves which hold representative selections of flora and fauna and which are viable in the long-term;
3. reforestation and the establishment of tree plantations, especially on degraded land;

4. provision of alternative energy supplies or sustainable supplies of fuelwood;
5. development of strategies which help farmers in highlands, especially those practising shifting cultivation, to avoid or control soil erosion;
6. control of agro-industrial (above all pesticide), sewage and industrial pollution;
7. increase of public awareness about and concern for the environment;
8. whenever possible, pursuit of *sustainable*, non-polluting development projects, programmes and policies.

Further reading

Adriawan E and **Moniaga S** (1986) 'The burning of a tropical forest: fire in East Kalimantan. What caused this catastrophy?', *The Ecologist* **16**, 269–70.

Aiken S and **Moss M** (1976) 'Man's impact on the natural environment of peninsular Malaysia: some problems and human consequences', *Environmental Conservation* **3**, 273–83.

Barrow C J (1980) 'Development in peninsular Malaysia: environmental problems and conservation measures', *Third World Planning Review* **2**, 9–25.

Barrow C J (1981) 'A review of the environmental problems and environmentalist activity in peninsular Malaysia', *Environmental Education and Information* **1**, 49–61.

Buchanan K (1967) *The Southeast Asian World: an introductory essay*, Bell, London.

Bull O (1982) *A Growing Problem: pesticides for the Third World*, Oxfam Public Affairs Unit, Oxford.

Caldecott J (1988) 'Climbing towards extinction', *New Scientist* **118** (1616), 62–6.

Consumers' Association of Penang (1978) *The Malaysian Environment in Crisis: selections from press cuttings*, The Consumers' Association of Penang, Penang, Malaysia.

Courtenay P P (1979) 'Minerals, mining and power' in Hill, R. D. (ed.). *South East Asia: a systematic geography*, Oxford University Press, Kuala Lumpur, 133–49.

Davidson J, Tho Yow Pong and **Bijveld M** (eds.) (1985) *The Future of the Tropical Rainforest Areas in South East Asia*. Commission on Ecology Paper No. 10, International Union for Conservation of Nature and Natural Resources, Gland, Switzerland.

Dobby E H G (1973) *South East Asia*, University of London Press, London.

ESCAP (1978) *Deltaic Areas (Water Resources Series No. 50)*, (Proceedings of the Third Regional Symposium on Deltaic Areas, Bangkok, Thailand, 22–8 November, 1977), Economic and Social Commission for Asia and the Pacific, United Nations, New York.

Fisher C A (1967) *South East Asia: a social, economic and political geography*, Methuen and Co., Singapore.

Furtado J I (1979) 'The status and future of the tropical moist forest in Southeast Asia', in McAndrews.

Furtado J I (ed.) (1980) *Tropical Ecology and Development* (Proceedings of the Vth Intervational Symposium of Tropical Ecology, 16–21 April, 1979, Kuala Lumpur, Malaysia), 2 vols., The International Society of Tropical Ecology, Kuala Lumpur.

Geertz C (1963) *Agricultural Involution: the process of ecological change in Indonesia*, University of California Press, Berkeley.

Goldman C R (1979) 'Ecological aspects of water impoundment in the tropics' *Unasylva* **31**, 2–11.

Gradwohl J and **Greenberg R** (1988) *Saving the Tropical Forests*, Earthscan Publications, London.

Hill R D (1979) *South-East Asia: a systematic geography*, Oxford University Press, Kuala Lumpur.

Hill R D (1982) *Agriculture in the Malaysian Region*, Research Institute of Geography, Hungarian Academy of Sciences: Geography and World Agriculture II, Akademiai Kiado, Budapest.

Ives J and **Pitt D C** (eds.). (1988) *Deforestation: social dynamics in watersheds and mountain ecosystems*, Routledge, London.

Kartawinata K, Adsoemarto S, Riswan S and **Vayda A P** (1981) 'The impact of man on a tropical forest in Indonesia', *Ambio* **10**, 115–119.

Koteswaram P (1974) 'Climate and meteorology of humid tropical Asia, in UNESCO, *Natural Resources of Humid Tropical Asia*, Natural Resources Research XII, UN Educational, Social and Cultural Organization, Paris, 28–85.

Lal R, Sanchez P A and **Cummings R W jr** (eds.) (1986) *Land Clearing and Development in the Tropics*, Balkema, Rotterdam and Boston.

McAndrews C and **Chia Lin Sien** (eds.) (1979) *Developing Economies and the Environment: the Southeast Asian experience*, McGraw-Hill, Singapore.

Malingreau J P, Stephens G and **Fellows C** (1985) 'Remote sensing of forest fires, Kalimantan and North Borneo in 1982–83', *Ambio* **14**, 314–21.

Myers N (1980) 'The present status and future prospects of the tropical moist forests', *Environmental Conservation* **7**, 101–14.

Myers N (1986) 'Tropical deforestation and species extinctions: the latest news', *Futures* **17**, 451–63.

Richards P W (1964) *The Tropical Rain Forest: an ecological study*, Cambridge University Press, Cambridge.

Ruddle K (1982) 'Brackish water aquaculture in South East Asia', *Mazingira* **6**, 58–67.

Simon J L (1986) 'Disappearing species, deforestation and data', *New Scientist* **110**, 60–3.

Spencer J E (1966) *Shifting Cultivation in South East Asia*, University of California Press, Berkeley.

Staunton W R and Flach M (1981) *Sago – the Equatorial Swamp as a Natural Resource*, Martinus Nijhoff, The Hague.

Swan B (1979) 'Geology, landforms and soils', in Hill (1979), pp. 1–15.

Szekely F (1982) 'Environmental impacts of large hydroelectric projects in tropical countries', *Water Supply and Management* **6**, 223–42.

Tran Van Nao (1974) 'Forest resources tropical Asia' in UNESCO, *Natural Resources of Humid Tropical Asia*, UN Educational, Scientific and Cultural Organization, Paris, **XII**.

UNAPDC (1983) *Environmental Assessment of Development Projects*, United Nations Asian and Pacific Development Centre, Kuala Lumpur.

Whitmore T C (1975) *Tropical Rainforest in the Far East*, Clarendon Press, Oxford.

Whitmore T C and Burnham C P (1969) 'The altitudinal sequence of forests and soils on granite near Kuala Lumpur', *Malayan Nature Journal* **22**, 99–118.

Williams C N and Joseph K T (1973) *Climate, Soil and Crop Production in the Humid Tropics*, Oxford University Press, Kuala Lumpur.

Ziauddin Sardar (1980) 'The fight to save Malaysia', *New Scientist* **87** (1217), 700–3.

CHAPTER 5

Human resources

Chris Dixon

This chapter provides an overview of the human populations of South East Asia. The region, although in general typical of much of the less-developed world in terms of its population structure, is in human resource terms probably the most complex and varied in the world. It should be borne in mind that it is not strictly meaningful to discuss 'population' in isolation from natural resources (see Ch. 4) and the wider aspects of directions and levels of development. Thus the material presented here should be seen as providing a background for the discussion of developmental and resource issues.

South East Asian population data

In common with the rest of the less-developed world South East Asian population data suffers from serious shortcomings. For much of the region there is a lack of coverage at sufficiently frequent and regular intervals to provide complete census and vital statistics returns. In Laos the first full census took place in 1985; Vietnam conducted its first in 1979; in 1981 Kampuchea carried out a partial census which provided the first meaningful population estimates since 1963; and in 1973 Burma conducted its first census since 1931. For a number of countries the census coverage is not only incomplete but this to an unknown extent. In both Burma and the Philippines, for example, uncertainty over government control leads to serious questioning of the returns. Overall, even where full census material is available, problems of data collection, processing and publishing limit its utility. Changes in procedures and definitions, mainly in the interests of improving reliability and coverage, make for problems in studying even short-term temporal changes. In consequence much of the data presented in this chapter is drawn from the estimates of a variety of national and international agencies.

General demographic characteristics: South East Asia in the less-developed world

In 1986 the population of South East Asia was 408.3 million; this was 10.8 per cent of the population of the less-developed world, 52 per cent of that of India and 38 per cent of that of China. Despite a general slowing of the region's growth rates since the mid-1970s, the population is continuing to grow at the very rapid rate of 2.3 per cent a year compared to 2.0 per cent for the less-developed world as a whole. Malaysia, Burma, Laos and Vietnam all have growth rates above the average for their World Bank income class (Table 5.1). Since 1980, only Singapore has experienced a lower growth rate than India.

Table 5.1 Average annual growth rates of population by World Bank income group[1] (percentage)

	1960–70	1970–80	1980–86
Burma	2.2	2.2	2.0
Kampuchea	2.6	−0.6[2]	3.7[3]
Laos	1.8	1.0	2.0
Vietnam	3.1	2.8	2.6
All low income	2.2	1.9	1.9
Indonesia	2.1	2.3	2.2
Philippines	3.0	2.7	2.5
Thailand	3.0	2.5	2.0
All lower-middle income	2.6	2.6	2.5
Malaysia	2.9	2.5	2.7
Singapore	2.4	1.5	1.1
All upper-middle income	2.5	2.2	1.9
China	2.3	1.5	1.2
India	2.2	2.0	1.8
All less-developed	2.4	2.1	2.0
South East Asia	2.8	2.7	2.3

Notes:
1. For 1986 the income groups were: low income, per capita GNP less than US$400; lower-middle income, per capita GNP of over US$400; and upper-middle income, per capita GNP of over US$16 000.
2. Based on a population estimate of 1970 at 7.1 m and the returns for the 1982 census which gave a total population of 6.7 m.
3. 1982–4. In 1981 the growth rate was estimated to be 4.6–5.2 per cent, one of the highest in the world (Kiljuen 1984, p. 34).
Sources: World Bank (1988), United Nations (1986).

Table 5.2 General demographic characteristics 1986

	Population (million)	Crude death rate per 1000	Crude birth rate per 1000	Infant mortality rate per 1000	Percentage of population less than 15	Life expectancy at birth
Brunei	0.2	3.5[1]	29.3[1]	16.0	37.0	66
Burma	37.7	10.0	33.0	70.0	36.7	59
Indonesia	168.4	11.0	28.0	84.0	33.2	57
Kampuchea	6.4	19.7[2]	38.0[2]	160.0	33.0	37
Laos	3.7	15.0	39.0	122.0	43.0	50
Malaysia	16.1	6.0	29.0	30.0	39.0[1]	68
Philippines	57.3	7.0	35.0	51.0	40.3	63
Singapore	2.6	5.0	16.0	9.0	24.4	73
Thailand	52.6	7.0	25.0	48.0	38.0	64
Vietnam	63.3	7.0	34.0	59.0	42.5[1]	65
South East Asia	408.3	9.0	28.0	64.0	37.0	59
All less-developed countries	3761.0	10.0	30.0	109.0	43.5	59

Notes: 1. 1985
: 2. 1983

Sources: United Nations (1986), World Bank (1988).

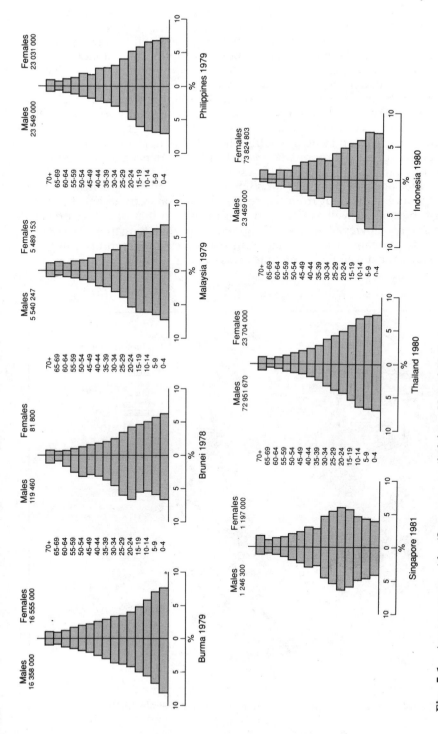

Fig. 5.1 Age–sex pyramids. (Source: compiled from United Nations, *Demographic Year Book 1982*)

As may be seen from Table 5.2, although the region's CBR (crude birth rate) and CDR (crude death rate) are slightly below those prevailing in less-developed countries in general, South East Asia has a substantially lower CDR (crude death rate), IMR (infant mortality rate), and a smaller proportion of the population less than 15 years old. The age–sex distributions (Fig. 5.1) for the region, and for the individual countries other than Singapore and Brunei, are also very close to the broad-based less-developed norm, with 50 to 53 per cent of the population less than 20 years of age. The Brunei population structure suffers from the distortions to be expected in a small population which has experienced a high degree of international migration. At present, the data is insufficient to produce age–sex pyramids for Vietnam, Laos and Kampuchea. Singapore stands out from the rest of the region with an age–sex distribution, as well as CBR and CDR, closer to that of a developed economy. However, it is worth noting that while Singapore in 1986 had 34 per cent of its population in the under-20 age group, the average for developed economies was 17.

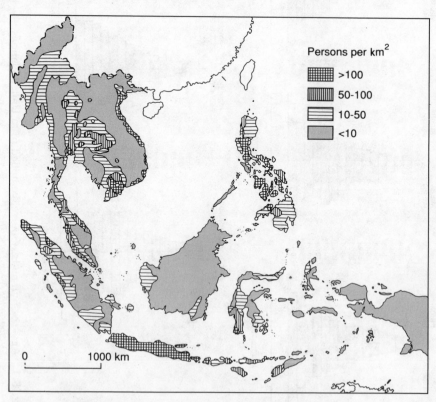

Fig. 5.2 Generalized population density. (After Neville 1979)

Overall, the structure of the South East Asian population is one that implies a continued high growth rate. This has serious implications for development projects and planning in the region over the next twenty years.

The distribution of population

The population of South East Asia is extremely unevenly distributed (Fig. 5.2). Java, for example, in 1985 accounted for 30 per cent of the region's population on less than 3 per cent of the region's area. In contrast, over 85 per cent of South East Asia, with less than 50 persons per square kilometre, may be considered thinly peopled. These regional level comparisons do not tell the entire story, there being considerable variations in density at all scales. As may be seen from Fig. 5.2, the contrast between the inner islands of Indonesia (Bali, Java and Madura) and the outer islands is very considerable. Within Java itself the variations are by no means negligible, however.

Crude density figures can give only a very general picture of the distribution of population. Table 5.3 provides, in addition, cultivated area and agricultural densities for comparison. These latter measures, although less accurate than the crude densities, provide considerably more insight into man–land relationships within the region. The coefficient of variability for the measures of density decreases from 66.7 per cent for the crude statistic to 43.6 per cent

Table 5.3 Population density 1985

	Crude density		Cultivated area density		Agricultural population[1] density	
	Persons/km^2	Rank	Persons/km^2	Rank	Persons/km^2	Rank
Burma	54.7	5	369.1	5	184.4	5
Indonesia	87.4	4	797.1	2	388.9	2
Kampuchea	40.2	7	238.4	7	172.2	6
Laos	17.8	8	457.4	4	337.1	4
Malaysia	47.2	6	355.9	6	123.9	8
Philippines	181.7	2	689.8	3	337.8	3
Thailand	100.0	3	262.0	8	167.7	7
Vietnam	184.9	1	797.2	1	563.2	1
Coefficient of variability $\left(\frac{50}{\bar{x}} \times 100 \right)$	66.7	–	43.8	–	49.2	–
South East Asia	87.2	–	530.1	–	287.2	–

Note: 1. Cultivated area and the population directly dependant on agriculture.
Source: FAO (1986).

for the cultivated density, indicating that densities become more homogeneous when expressed in terms of cultivated land. In addition, the rankings of some countries alter markedly depending on choice of measure (Table 5.3). These are largely explained by differences in the proportion of the surface area cultivated and the nature and intensity of agricultural activity.

The destruction resulting from the Second World War, the protracted Vietnam conflict, conditions of almost civil war in Burma, and such shorter-term disturbances as the 'Malayan Emergency' had profound, if mainly short-term, effects on the distribution of population. In the Malayan peninsula between 1950 and 1952 forced population movements, a part of the British security measures, resulted in approximately one million rural people being resettled in some 600 compact settlements (Neville 1979, p. 58). During the Vietnam War and its aftermath, large-scale planned and spontaneous movements of people took place in all three of the former Indo-Chinese states (Nhuân 1984, p. 83; Kiljuen 1984, pp. 12–13).

Over the last thirty years the general level of population densities has not only increased sharply but has reduced in variability. For South East Asia as a whole the average crude density rose by 48 per cent between 1970 and 1985. More significant are the sharp increases in cultivated density, a direct result of a slowing down and in some cases a cessation of the expansion of the area cultivated (Table 5.4). Resettlement and land-reclamation schemes have contributed to the reduction in density variability between, and more especially within, countries, especially in Thailand, Malaysia and Indonesia. The opening up of land unsuited to rice cultivation, for example the clearance of large tracts of land in North East Thailand since 1958

Table 5.4 Percentage change in cultivated and agricultural density

	Cultivated density			Agricultural density		
	1970–75	1975–80	1980–85	1970–75	1975–80	1980–85
Burma	17.4	10.3	9.7	11.3	3.9	3.9
Indonesia	1.6	12.6	3.0	−4.5	3.6	−5.8
Kampuchea[1]	2.4	−8.6	13.5	0.3	−7.0	10.2
Laos	12.7	3.2	9.2	7.0	1.4	6.4
Malaysia	6.6	8.9	11.5	−6.0	−5.5	−2.0
Philippines	8.1	6.9	9.8	5.7	3.8	5.9
Thailand	5.8	2.5	3.0	−11.7	−4.3	−2.2
Vietnam	8.9	−1.9	6.0	2.9	0.3	1.8
South East Asia	25.3	0.4	6.0	4.1	5.1	0.8

Note:
1. The figures for 1970 to 1980 are based on estimates and are suspect. This is particularly the case for cultivated land: the FAO makes the unrealistic assumption that there was no change during the period.
Source: FAO (1986).

for the production of maize, cassava and kenaf, and large-scale irrigation and flood-control schemes have either resulted in an extension of the area cultivated or more significantly increased the carrying capacity of previously rain-fed cultivation. If the ambitious Indonesia 'transmigration' programmes or the grandiose plans to develop the water resources of the Mekong Basin were to be successfully implemented, major long-term changes in the distribution of population in the region would result.

The origins of the present pattern of population distribution

The uneven distribution of population in South East Asia has been a feature of the region since very early times. This pattern has been accentuated since the early nineteenth century by the rapid growth of population and the imposition of Western economic and political patterns.

There is a strong environmental element behind both the past and present distribution. Early settlement in South East Asia favoured the alluvial lowlands and their fringes. These 'ecological niches' were occupied by pioneer agriculturalists and, later, wet-rice cultivators (Ng 1979). The major river valleys of the mainland became the foci of dense populations based on wet-rice production and formed the 'hearths' of a number of the region's major cultures and the cores of successive early states (Fisher 1971, p. 102). In general, population growth, accompanied by associated technical and organizational developments, both made possible and necessary a seaward movement of population foci into the more extensive, potentially more productive but environmentally more difficult lower valleys. The seaward movement of the core of the Thai state from Sukothai to Ayuthia and then to Bangkok is one of the clearest examples of this long-term trend. In contrast, the extensive interfluve areas remained essentially negative ones for cultures based on wet-rice cultivation. These areas were sparsely settled by various 'relic' peoples displaced from the lowland areas by the expansion of successive major cultures. The lower deltas of the Mekong and Irrawaddy were similarly avoided by settlement prior to the mid-nineteenth century, the environmental difficulties of these areas remaining beyond the organization and technical abilities of the mainland peoples (Adas 1974, p. 17).

In contrast to the mainland, in island South East Asia early populations were heavily concentrated on a series of generally small alluvial coastal plains while the interiors remained extremely thinly peopled. The small, fragmented wet-rice cores of the islands are largely responsible for the fluid nature of the pre-colonial political pattern. Peninsular Malaysia represents a zone of transition between the mainland and island settlement patterns. Early wet-rice cultivation was concentrated in the north east in the alluvial lower deltas of Kelantan and Trengganu, in contrast settlement along the west coast was closer to the island pattern.

Developments associated with the imposition of colonial rule did not bring any fundamental change to the general pattern of population distribution. New crops, often associated with plantation production, were important in opening up previously thinly peopled areas, for example in the Annamite ranges of Indo-China. In the Mekong and Irrawaddy deltas, the French and the British colonial administrations were responsible for major flood control and irrigation schemes which resulted in a rapid buildup of population densities.

Elsewhere the negative zones of the pre-colonial period remained largely untouched, the main impact of the colonial economies being confined to the already comparatively densely peopled areas. It was in these areas that technical improvements, health measures and new food and commercial crops had their main impact. The resultant increases in density from the early nineteenth century were, in certain areas, striking. In Java and Madura, population density increased from 96 persons per square kilometre in 1817 to 940 by 1940 (Gourou 1961, p. 110). Some locally important increases in density resulted from mining developments after 1860, most notably in peninsular Malaysia, South Thailand, north western Burma and parts of Borneo. Overall, however, during the late nineteenth and early twentieth centuries, the '. . . juxtaposition of densely occupied areas and sparsely peopled expanses . . .' (Gourou 1961, p. 106) remained largely undisturbed, the contrasts, if anything, becoming more marked.

Early cultural influences

South East Asia contains a remarkable variety of peoples and cultures. In this respect, Indonesia is representative of the region with over 300 recognized ethnic groups, speaking variations on 31 groups of languages and living under conditions ranging from primitive hunter-gathering to modern urban manufacturing, the other smaller countries of the region exhibiting similar, if lesser, degrees of variability. A very general impression of the human mosaic of South East Asia is presented in Figs. 5.3 and 5.4 and Table 5.5. Fisher (1971, p. 63) summarized the background to this diversity.

As in relief so in human geography South East Asia is characterized by great diversity and complexity, and these traits are intimately related to its position and configuration which has made the region both a bridge and a barrier to the movement of peoples and ideas.

The region's character as a focus of converging land and sea routes has brought long-term movements of people overland from the north, north east and north west, together with coastal and sea movements from both the east and west.

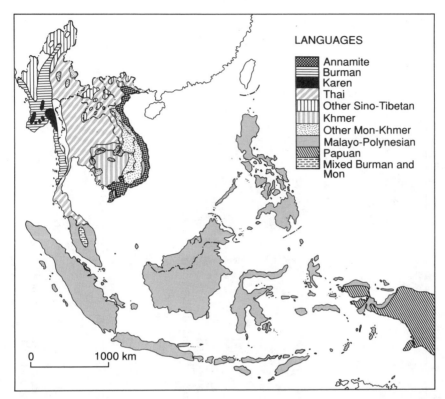

Fig. 5.3 Languages of indigenous peoples. (After Fisher 1971 and Neville 1979)

It is generally accepted that in the pre-colonial period the peopling of South East Asia took place through a series of long-term southerly migrations. In the main these had their origin in the continental interior, percolating into mainland South East Asia and then 'island hopping' into present-day Indonesia. These movements were probably supplemented by movements along the off-shore island arcs. Some of the early movements continued beyond the region's limits into the Pacific. In general, the successive waves of migrants have occupied the favoured 'ecological niches' in the coastal and riverine alluvial lowlands, displacing or assimilating earlier settlers in the process. The displaced groups moved either into the upland areas or further south into the outer islands. However, considerable intermixture has taken place with new migrants being absorbed by the older, and in the long term perhaps only adding linguistic, technical and cultural elements to those of the established population.

The island interiors contain scattered groups, comprising relics of very early migration. Elsewhere the interiors of the islands are

Table 5.5 Estimates of ethnic composition[1]

	> 80%	> 60%	> 40%	> 20%	> 10%	> 5%	> 1%
Brunei			Malay	Chinese		Kadazan Kadayan	
Burma		Burmese				Karen[2] Shan[2]	Chinese Kachins Other hill people
Kampuchea	Khmer					Vietnamese	Chams Chinese
Indonesia		Javanese			Sudanese	Madurese	Chinese Balak Buginese Dyaks
Laos			Lao[2] Montagnard[3]				
Malaysia			Malay	Chinese		Indian	
Philippines				Cebuano[4] Tagalog	Iloko Panay-Hiligaynon	Bikol Bisaya	
Singapore		Chinese			Malay	Indian	
Thailand	Thai[2]				Chinese	Lao[2] Montagnard[3] Chinese	
Vietnam	Vietnamese						Hill people[5]

Notes:
1. These proportions are based on a variety of estimates.
2. Thai-Lao-Shan belong to the same ethnic group.
3. Montagnard is a French collective term for hill peoples, in particular the Muong, Meo, Yao, and Thai.
4. Cebuano is the first language of 24 per cent, Tagalog of about 21 per cent.
5. The Thai hill people population is probably less than half a million, over 50 per cent of which are Karen and 20–25 per cent Meo. Present in small numbers are the Yao, Lisu, Lahu, Akha, Lawa and Khamu.
6. Including Chin, Lolo, Naga and Palaung Wa.
7. There are in total 28 minor groups with estimated populations of 1–2 million.
Source: Based on Neville (1979).

120

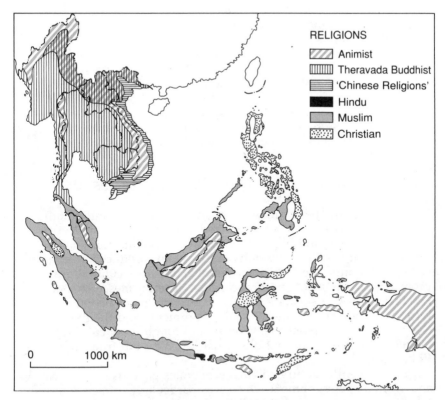

Fig. 5.4 Religions of indigenous peoples. (After Fisher 1971, and Neville 1979). 'Chinese religions' is a far from satisfactory category, comprising an amalgam of Confucian, Buddhist, Taoist and even animist ideas

peopled by Nesiot peoples who migrated into the region between 3000 and 4000 years ago. The lowlands of the islands are occupied by Pareoean people who have been settled there for over 2000 years. Inter-island migrations, Arab and more importantly Indian and Chinese influences have added to the complexity of the islands but the Pareoean migrations remain the dominant element of the ethnic composition.

The situation in the mainland is far more complex, a result of a continuation of major movements until the twentieth century and very minor ones until the present. Of the earlier Pareoean peoples, only the Khmer survive in significant numbers; the Mon similarly exist in identifiable groups in Burma and to a lesser extent in Thailand. These earlier groups have been overlain by the later Tibeto-Burmese and Thai-Lao-Shan peoples. Of the third major mainland grouping, the Vietnamese, little is known of their origins. They possibly represent a fusion of Malay and Thai strains, with early origins in the Red River Basin and later moving slowly south between 2000 BC and 1400 AD to occupy the approximate area of modern Vietnam.

As well as these major ethnic groups, there were a variety of lesser groups, again almost entirely of Mongoloid stock, from which the present hill peoples, notably the Karens, Chins, Nagas and Kachins of Burma, and the Mans, Miaos and Lolos of Laos and Vietnam are descended. Some of these, for example the Karens of the Thai-Burmese border area, have been present for over 1000 years, while others have only migrated during the last twenty years.

The development of the plural societies

On the eve of the penetration of the Western powers into South East Asia, the region already exhibited a mosaic of cultural types and a fragmented and highly fluid political structure (Fisher 1971). The nineteenth-century divisions of the region produced a series of states with boundaries which frequently cut across cultural groups producing a pattern of 'cultural clusters'. The Burmese, Thai and Lao boundaries all separate homogeneous groups of peoples, with resultant problems for the colonial governments and their successors (see Ch. 2). In addition, the colonial era brought new cultural influences to the region not only from the West but from India and China as well.

The nineteenth-century colonial economies gave rise to a social and economic environment which was particularly attractive to the inhabitants of the densely peopled provinces of southern China and India. In general, the colonial powers saw the indigenous population of South East Asia as unsuited to employment in the 'modern sector' of the economy. This view was reinforced by the low population densities of most of the region and the resultant limited possibilities of transferring labour wholly out of the traditional sector; the unwillingness of certain indigenous groups to abandon traditional work; the existence of enclaves of Chinese, Indian and Arab traders and entrepreneurs who controlled the trade and much of the production; and the favourable views of the Chinese, in particular as workers and entrepreneurs, held by various influential colonial administrators. Not all the colonial powers were as overtly enthusiastic over immigrant workers as the British in the Straits Settlements but, except in the Philippines, where restrictions were imposed by the new US administration in 1900, there were few restrictions on immigration until the 1930s.

The great increase in Chinese immigration after 1800, although largely induced by Western activities, was also stimulated by the internal conditions in China. Southern China in particular was experiencing rapid population growth, a consequence of two centuries of stability under the Manchu dynasty, with limited expansion of food production. In the first half of the nineteenth century the deterioration of living conditions was aggravated by natural disasters and repeated outbreaks of political unrest. From the early 1800s, migration flowed not only into the areas of Western penetration but also into areas such as Thailand largely bypassed by the West until after 1855. In the 1820s, Chinese were entering the Kingdom at a rate of 15 000 a year and coming to exercise considerable control over the economy.

The rate of migration increased as the century progressed and Chinese flocked to the mines, plantations and ports of South East Asia with the introduction of steam shipping speeding and cheapening the process. In general, the migrants travelled under appalling conditions and their treatment on arrival was usually equally unfavourable, the excesses of this traffic led many to label it the 'pig trade'.

Until 1929 the Chinese population of South East Asia remained essentially transitory (Purcell 1951). After this date increasingly strict control of movements resulted in a greater degree of permanency in the communities. The majority of South East Asian Chinese remained poor, but a significant number came to control the internal trade and commerce of many of the region's economies. Southern China, the area of origin of the vast majority, had a long history of trade both internal and overseas. This background, and the movement from an area of restricted opportunities, seems a sufficient explanation of the 'success' of the Chinese, without recourse to dubious racial theories. It is difficult to quantify the contribution made by the Chinese to the South East Asian economy. By the late 1930s, outside of the Netherlands East Indies or Burma, Chinese investment represented about 40 per cent of the regional total. In Burma the role of the Chinese in the economy was taken by Indian migrants, while in the Netherlands East Indies the scale of Dutch investment and the more restrictive economic policies of the colonial administrations resulted in an over-shadowing of the Chinese. In all the countries of the region other than Burma the Chinese came to dominate petty trading, money-lending, small-scale production and the collecting and processing of much agricultural produce. It has been estimated that by the late 1930s, 80–90 per cent of the rice mills in Thailand, over 80 per cent of those in Indonesia and 75 per cent of those in the Philippines, were owned by Chinese (Fisher 1971, p. 182).

In contrast to the widespread Chinese migrations into the region, the Indian presence after 1834 was mainly confined to the British territories and comprised a smaller number of people with generally less economic power. As with the Chinese, most of the Indians arrived as labourers. The majority of the migrants into Burma and the Malay states were Tamils from Madras who generally failed to rise from the level of labourers because of their conditions of employment. The average Indian labourer came as a recruit to work in specific plantations. From the 1880s this recruitment was the subject of increasing official supervision. Thus short-term contracts, controls over movement, and the imposition of strict discipline and regimentation prevented these groups from establishing themselves permanently in the region. The Indian migrants, although less varied in language and culture than the Chinese, were by no means homogeneous; the dominance of Tamils should not be allowed to obscure the variety in caste, language and religions of this group. The Chettyars, among the best known and documented of the smaller groups, originally a money-lending caste, came to dominate many parts of the Burmese economy. By 1930 there were one million Indians in Burma; they held the largest share of investment, owned over half the cultivated land and dominated trade and commerce.

Not only did the immigrant communities make their own contribution to South East Asia's rich cultural pattern, but so also did the colonial powers under whose auspices they migrated. The Dutch, French, British and to a lesser extent the Spanish, imposed elements of language, education, architecture, administrative structures and attitudes on their colonial possessions. Throughout the colonial period, the contact between the rival colonial structures was kept to a minimum, thus emphasizing their distinctiveness. In many countries the experience of the colonial period and the often protracted and turbulent process of decolonization resulted in a strong anti-Western feeling being generated. Despite this, the colonial contribution to the region's rich cultural pattern remains considerable.

The multiracial societies in the post-colonial period

The process of decolonization further complicated the cultural pattern of South East Asia. In Malaysia, and to a lesser extent Burma, independence was given to territorial units which had been created to facilitate the withdrawal of British rule. The inability of successive Burmese governments either to control or to accommodate the minority peoples of those outer regions of the country which had been separately administered under the British, lies at the root of Burma's ongoing economic and political problems. As may be seen from Table 5.5, the Burmese only represent 60 to 65 per cent of the total population. The creation of the Federation of Malaysia in 1963 represented an attempt to produce a politically independent unit in which the immigrant communities, particularly the Chinese, would not outnumber the indigenous Malays (Fisher 1968, pp. 75–146).

During the immediate post-colonial period the role of the 'intermediate layer' between the colonial powers and the indigenous population, played by the Chinese in most parts of the region, the Indians in Burma and to a lesser extent the Vietnamese in Laos and Cambodia, was one fraught with contradictions. In many countries, while the wealthy members of the immigrant groups had much in common with the leaders of the nationalist movements, the antipathy felt towards the colonial powers became directed at the Vietnamese, Indian and Chinese communities as a whole.

In striking contrast to the comparatively homogeneous and generally well-defined Indian and Chinese immigrant populations of the region, the upland minority groups of mainland South East Asia are both extremely heterogeneous and poorly enumerated. Continued long-term movements of peoples from Yunnan, migrations due to the Vietnam War and its aftermath, together with movements between Thailand and Burma due to persecution and unstable conditions, have produced a complex and poorly recorded pattern of population. Policies towards many of these peoples, especially in the so-called 'Golden Triangle' of northern Thailand, North East Burma and south western China, have been complicated by the opium trade. Thai policy, in particular, towards the 'hill tribes' has been variable with, at different periods, indifference, genocide and assimilation being the dominant ingredient (McCoy 1972).

The topography of the area has played a major part in maintaining ethnic and linguistic divisions both between upland peoples and between them and the lowlanders. Although some assimilation and resettlement of certain groups, notably the Lawa of northern Thailand, have taken place (Kunstadter 1967, p. 48) the majority of these peoples have become increasingly marginalized posing major problems of social welfare, economic development and security for the mainland states.

The plural societies of South East Asia provide a rich and varied culture but they also give rise to serious problems for the region's governments. Long-term internal conflicts, as in Burma, positively discriminatory economic policies, as in Malaysia, and officially multi-lingual societies, as in Singapore, are costly and dilute scarce developmental resources. For many South East Asian countries the problem remains one of welding the plural society into a national whole.

The changing rates of population growth

As in less-developed countries in general, the largest single determinant of the changes in the rate of natural population growth in South East Asia over recent decades has not been the level of fertility but a decline in mortality rates. However, the pattern and timing of the decline in mortality varies considerably within the region.

The pre-nineteenth-century pattern of population growth in South East Asia appears to be one of overall very slow growth, with more localized centres of increase and decrease. Longer-term cycles which affected larger areas were associated with major mainland empires. It is likely that at times some of these early states supported population densities well above those present in the first half of the nineteenth century. In general, political instability in the pre-colonial period, internecine warfare, endemic and epidemic diseases and unreliable environmental conditions kept population growth at a low level (Fisher 1971, p. 176). While Western penetration of the region may well in its early and more brutal stages have reduced population locally, the main demographic consequence of colonization was the mitigation of some of the continuous and periodic checks on population growth. The imposition of Western rule brought comparative peace and order to the region, introduced new crops, opened up new land to cultivation and by improving transport reduced the effects of harvest failures. There appears to be a general relationship between the date at which European territorial control was established and the acceleration of the rate of population increase. The decline in mortality rates within countries was by no means even, being influenced by the extent and degree of European control as well as the environmental conditions. To a considerable extent the persistence of higher mortality rates and lower population densities in the outer islands of Indonesia are due to the more difficult and generally less healthy environmental conditions. Similarly the striking upland-lowland population densities that characterized early nineteenth-century Burma may be related to the incidence of malaria, as may the striking east-west contrasts in peninsular Malaysia.

If the general thesis that population growth rates accelerated following the imposition of European control is correct we would expect the Philippines and Java to have exhibited this tendency first. By 1800 they were almost the only territories where European control had expanded beyond the immediate coastal trading centres. In Java from the early 1800s the Dutch, and between 1811 and 1816 the British, maintained peace (often by no means gently), improved agricultural techniques, particularly with respect to water control, drained swamps, and implemented sanitation and hygienic measures. The Dutch were pioneers in tropical medicine, introducing smallpox inoculation and quinine, during the 1830s. Forest clearance and swamp drainage further reduced the breeding grounds for mosquitoes and parasites. A fall in the Javanese death rate between 1800 and 1830, with a constant birth rate, resulted in a probable expansion of population from 4.5 million in 1815 to 10 million in 1850 (Carr-Saunders 1936, p. 92).

In the Philippines, the general rate of population growth also began to accelerate during the early nineteenth century. A study by Smith and Ng (1982) suggests that growth rates increased sharply from an annual average of 0.54 per cent between 1571 (the beginning of Spanish control) and 1817 to 1.54 per cent between 1817 and 1876. While the growth rate during the first two centuries of Spanish rule appears low compared to later standards, it is very likely that it was much higher than elsewhere in South East Asia. Outside of the Philippines and Java, the acceleration of population growth probably took place after 1850. Thailand's incorporation into the world market from 1855 appears to have been accompanied by a fall in mortality rates as the cash economy penetrated the rural areas (Mounge 1982, p. 130). In general, although there was considerable regional variation, mortality rates appear to have fallen from 40 per 1000 in the second half of the nineteenth century to 30 by the 1930s. The sparse evidence that is available for other parts of South East Asia, suggests that these levels of mortality were at or near the regional norm.

The unstable conditions that affected much of the region during the immediate post-war period kept death rates at or above 25–30 per 1000. Where stability was established, death rates fell and life expectancy at birth increased. In Thailand, for example, CDR (crude death rate) fell from the 30 per 1000 rate of the 1930s to 20 per 1000 by the mid-1950s. Throughout the region, the sharp falls in death rates since the 1940s can be attributed to the imported foreign medical technology and government-based expansion of public health provision. Malaria control by DDT spraying was started on a large scale in peninsular Malaysia in 1947 and Thailand in 1949, for example; by 1954 most of the South East Asian countries had eradication schemes. Other mass campaigns, notably for smallpox, cholera and plague, had by 1960 virtually eradicated these diseases from South East Asia with the exception of Burma where smallpox remained a significant cause of death until the mid-1960s.

Table 5.6 Infant mortality rates and life expectancy at birth, 1960–86

	Infant (age 0–1 years) mortality rates per 1000 live births			Life expectancy at birth		
	1960	1986	Percentage change 1960–85	1960	1986	Percentage change 1960–85
Burma	158	70	−55.7	46	59	+28.3
Indonesia	150	84	−44.0	41	57	+39.0
Kampuchea	146	160	+ 9.6	46	37	−19.6
Laos	155	122	− 2.2	43	50	+16.3
Malaysia	72	30	−58.3	65	68	+ 4.6
Philippines	106	51	−51.9	53	63	+18.9
Singapore	36	9	−69.4	72	73	+ 1.4
Thailand	103	48	−53.4	52	64	+23.1
Vietnam	157	59	−62.4	43	65	+51.2

Sources: United Nations (1986), World Bank (1983, 1988).

Other factors influencing mortality levels, such as improved sanitation, clean water supplies, health education, access to medical services and nutrition are much more difficult to quantify. The rapid fall in mortality since 1960 (Table 5.6) however, points to a considerable contribution from these sources. The expansion of low-cost public health and medical services, often utilizing only partially trained personnel, has increased access for many rural communities in many cases where public expenditure on health has shown scant increase. Improvements in nutrition are difficult to measure; but a general increase in calorie intake has taken place between 1979–80 and 1987–88 for all countries except Kampuchea. The growth in calorie intake in predominantly rice-eating communities needs to be offset against the declining nutritional value of rice where high-yielding varieties and mechanical polishing have been adopted. However, the improvements in South East Asian calorie consumption are particularly impressive when viewed in relation to the static situation in India and the overall decrease in Africa south of the Sahara.

Table 5.7 Crude death rates

	1960	1965	Crude death rates (per thousand) 1970	1975	1980	1986	Percentage change 1960–86
Brunei	n/a	n/a	5.5	4.1[1]	3.6	3.5	–
Burma	21.0	18.0	17.4	n/a	13.0	10.0	−54.2
Indonesia	22.0	20.0	19.4	15.1[2]	13.0	11.0	−50.0
Kampuchea	19.0	20.0	15.6	40.0[2]	n/a	19.7[3]	+ 3.6[5]
Laos	23.0	n/a	17.2	n/a	21.0	15.0	−30.0
Malaysia	16.0	12.0	10.8	6.2[1]	7.0	6.0	−37.5
Philippines	15.0	12.0	12.0	9.1[2]	8.4[4]	7.0	−53.3
Singapore	8.0	6.0	5.2	5.1[1]	5.0	5.0	−37.5
Thailand	15.0	10.0	10.4	8.3[2]	8.0	7.0	−53.3
Vietnam	21.0	17.0	16.1	11.4[2]	8.0	7.0	−66.7

Notes:
1. 1976
2. 1978
3. 1983
4. 1982
5. This refers to 1960–83 but in the context of recent Kampuchean demographic history is not very meaningful.
Sources: World Bank (1983, 1988), United Nations (1976, 1986).

The variability in CDR and life expectancy in South East Asian countries has declined since 1960, but still remains high (Table 5.7). Perhaps more significantly, considerable spatial and personal variation remains within countries. In particular, IMR (infant mortality rate), generally accepted as one of the most sensitive measures of the health environment, shows rather more variation than the crude death rate (Table 5.6). Striking rural-urban differences still exist: in Peninsular Malaysia for example, the 1985 rural IMR was 19.9 per 1000 live births compared to 13.6 in urban areas. Similar variations are to be found elsewhere in the region. The existence of lower rates in urban areas is accepted as a normal feature of less-developed countries, but in some cases the poor living conditions found in urban areas coupled with successful rural health programmes have reversed the situation. In Thailand the rural IMR has been below the urban since the early 1970s; in 1984 the urban rate was 27.1 per 1000 compared to 7.1 in rural areas. Fragmentary evidence also points to marked class differences, with the overall lower urban rates masking the prevalence of rates higher than those of rural areas in the most deprived central slums and peripheral squatter settlements. The existence of similar variations in rural areas is even less well documented.

Changing levels of fertility

Fertility levels in South East Asia have, with the possible exception of Laos and Kampuchea, fallen over the last twenty-five years. As may be seen from Table 5.8, while in Indonesia, Singapore, Thailand and Brunei, the CBR has fallen below 30 per 1000; only in Singapore does the rate approach that of the developed economies (approximately 14 per 1000). The pattern of decline has varied, the steady fall in the Philippines between 1960 and 1986 contrasting with Indonesia where the slow fall of the 1965–75 period was followed by a rapid decline. These crude national figures conceal wide spatial, ethnic and class differentials, examination of which

Table 5.8 Crude birth rate: live births per 1000 population 1960–86

	1960	1965	1970	1975	1980	1986	Percentage change 1960–86
Brunei	n/a	41.8	37.0	31.7	30.5	29.8	−28.7[1]
Burma	43.0	42.3	40.3	39.4	37.0	33.0	−23.3
Indonesia	46.5	46.0	43.0	42.5	35.0	28.0	−39.8
Kampuchea	45.0	44.9	43.9	39.9	30.9	45.5	+ 1.1[2]
Laos	44.0	44.0	44.9	44.6	43.0	39.0	−11.4
Malaysia	44.9	41.6	36.5	32.3	31.9	29.0	−35.4
Philippines	47.0	44.2	40.6	38.3	34.0	35.0·	−25.5
Singapore	38.0	32.1	27.0	17.7	17.1	16.0	−57.9
Thailand	44.0	43.7	41.9	37.0	30.0	25.0	−43.2
Vietnam	47.0	42.5	41.4	40.8	35.0	34.0	−27.7

Notes:
1. 1965–85
2. In the context of recent Kampuchean demographic history this is not a very meaningful figure.
Sources: United Nations (1976, 1979, 1986), World Bank (1983, 1987, 1988).

provides considerable insight into the processes of, and prospects for, the region's economic development.

The past pattern of fertility

It is generally assumed that CBR of 43 to 47 per 1000 prevailing over much of South East Asia in the 1940s was little below the long-term historical level. However, information on the birth rate of even the 1930s is scant. In 1934 the CBR of Cochin-China was reported as 39 per 1000 and that for Hanoi 25 per 1000 (Naval Intelligence Division 1943, pp. 236-7). It is highly probable that these represented some of the lowest rates in French Indo-China rather than the general situation. The 1930 Dutch census returns for Java suggest a rate of 40 per 1000 for Java and a slightly higher rate for the Outer Islands. These figures may well be unrepresentative of the region as a whole, although there is evidence in certain areas for a temporary decline in the CBR during the 1930s in response to the international recession. It may be that the rate of 45 per 1000 that is believed to have prevailed in Thailand from the late nineteenth century until the early 1950s is more typical of the region.

Prior to the 1950s, however, fertility was not uniformly high in South East Asia. Spatial and temporary variations in birth rates, particularly where large-scale migration took place, were common. A number of studies of Chinese migration reveal that in the early stages the communities were transient and predominantly male with a consequent low birth rate, fertility subsequently rising as the sex balance was restored by increased female immigration. More significant for our understanding of population processes in the region is the existence of traditional methods of fertility limitation. The desire to limit births cannot be merely equated with modernization. Fertility is limited by a variety of means, for example late marriage, the existence of large celibate groups and taboos on re-marriage, as well as by 'folk' methods of birth control. There is little doubt that reductions in fertility took place in some South East Asian communities in response to land shortage, unstable conditions or economic recession. However, important as the existence of these practices are, they have made no long-term impact on the region's demographic structure.

The process of fertility decline

Studies of the adoption of family planning in South East Asia and less-developed countries in general have tended to stress the importance of urban dwelling and high education (Prachubmoh et al. 1973). A number of studies suggest strong associations between the education of mothers, their participation in regular employment outside the home and the number of children in the family, these relationships holding in both urban and rural areas, but in rural areas there are less opportunities for female employment outside of the home and education levels are lower (Neville 1979, p. 63).

In Malaysia, where the rates of fertility are amongst the lowest for less-developed countries in the tropics, there remains appreciable variation between ethnic groups and classes. Prior to 1970, the fertility rates were much higher for the Indians than for the other groups. Since 1965 the decline in fertility rates has been greatest for the Indians and least for the Malays.

In Singapore there are sharp differences in fertility levels between the dominant Chinese population and other groups. Striking overall variations in the marriage rate and size of family according to women's levels of education have been revealed. In the 1980 census, for example, the average number of children ranged from 3.5 for mothers with no qualifications to 1.65 for those with tertiary education. Similarly, the percentage of women remaining unmarried increased from 4.5 per cent of those over thirty with no qualifications to 17.8 per cent for those with tertiary education. This has resulted in a much-discredited government campaign to encourage the more educated women to marry and have children, and the less educated to restrict their fertility.

Studies such as that of Jones (1977) in Indonesia confirm the existence of higher rates of adoption of family planning amongst urban communities as against rural. However, not only can rural adoption rates be above urban ones, as in densely populated eastern Java, but there are wide variations between rural areas. It is important to stress that rural fertility rates in South East Asia are by no means uniformly high.

In a study of the northern Thai province of Chingmai by Pardthaisong (1974) it was found that the majority of acceptors of birth control were the wives of landless agricultural workers with only primary level education. Similarly, a sample survey of the whole of rural Thailand conducted by Prasithrathsin (1973) revealed that the adoption rate for family planning was higher for landless than for landed families. Additionally, the study found that there was no significant differences in contraceptive practice amongst landed families when examined against their size of land holding. A further, more detailed study of rural contraception practice in northern Thailand by Mounge (1982) suggests, however, that the situation is actually more complex. For the most impoverished group with no chance of gaining access to land, children perhaps provide the only means for the family to improve its economic standing. In contrast, the landless tenant with limited land might, by hard work and limiting family size, eventually improve the family's economic standing and acquire land. These studies, then, question the views that family planning in rural South East Asia depend on reproducing urban levels of education and patterns of employment.

The labour force

The conventional definition of labour force, those aged between 15 and 64, is of limited utility for rural South East Asia. Those aged below 15 play a vital role in agricultural activities; in wet-rice growing communities, children are playing a full part in transplanting by the age of 9 or 10 and boys are ploughing at 12. Many tasks are undertaken by the young and the old that release more productive members of the community; this is the case, for example, with tending livestock (especially water buffalo) and producing handicrafts. The contribution of those outside the 15–64 age band is difficult to quantify but it brings into question the significance attached to the dependency ratio (the percentage of the population aged less than 15 and more than 64). Even in Singapore, with a dependency ratio in 1985 of 33 per cent which was equal to the developed world mean, the 'dependants' remain an appreciably important part of the production system. Throughout the South East Asian region this contribution, especially of the under-15 group, to the urban-industrial economy remains considerable, if impossible to quantify. Child labour in South East Asian manufacturing is widely attested to and plays an important role in maintaining viable family incomes and keeping labour costs low; this is particularly the case in the 'informal sector' of the urban economy.

The dependency ratio for South East Asia as a whole has increased since the 1960s, a reflection of high population growth rates and increasing life expectancy. This change was by no means uniform, however (Table 5.9).

It is important to qualify the regional generalizations with respect to the socialist economies. For Laos and Kampuchea, the problem in the 1980s remains one of overall labour shortage, while in Vietnam there are shortages in certain rural areas. The progress of rural reconstruction in all three countries is highly labour-intensive, for example the rehabilitation of the rice lands of the Mekong Delta, and to this end, and for ideological reasons a general policy of 'de-urbanization' and resettlement has been enforced (Nhuân, 1984, pp. 83–4). However, in Vietnam, and to a lesser extent in Laos, policies of population limitation have been implemented since the late 1970s.

Agriculture remains the main source of employment for every economy in the region except Singapore and Brunei. A general reduction in the level of rural residence has been accompanied by a decline in the dominance of agriculture as a source of employment. However, the particularly sharp decline in the share of agriculture in the total employment of the Philippines and Indonesia and the accompanying rapid expansion of the service sectors (Table 5.9) has to be seen in terms of continued rapid population growth, fundamental changes in the structure of the agricultural economy and consequent large-scale migration to urban areas.

While Santos (1979) has stressed the importance of the 'informal' service sector in developing countries, the rapid growth of service sector employment in many South East Asian countries is a highly debatable indicator of the degree of modernization. The continuing high rates of population growth that most of South East Asia is

Table 5.9 Labour force 1965–85

| | Percentage of population in the 15–64 age group | | Percentage of the population engaged in | | | | | | Average annual percentage growth rate of the labour force | |
| | | | Agriculture[1] | | Industry | | Services | | | |
	1965	1985	1965	1985	1965	1985	1965	1985	1965–80	1980–85
Burma	57	54	64	53	14	19	23	18	2.2	1.9
Indonesia	53	56	71	57	9	13	21	30	2.1	2.4
Kampuchea	52	n/a	80	n/a	4	n/a	16	n/a	1.2	n/a
Laos	56	53	81	76	5	7	15	17	1.6	1.8
Malaysia	50	59	59	42	13	19	29	39	3.4	2.9
Philippines	52	56	58	52	16	16	26	33	2.5	2.5
Singapore	53	67	6	2	27	38	68	61	4.2	1.9
Thailand	51	59	82	71	5	10	13	19	2.8	2.5
Vietnam	n/a	55	79	68	6	12	5	21	1.8	n/a

n/a not available.
Note: 1. Includes forestry and fishing.
Source: World Bank (1987).

experiencing, coupled with the high proportion of the population who are less than 15 years old, implies a rapid expansion of the region's labour force. Signs of the economy's inability to absorb the 'new entrants' and the increasingly large numbers displaced from agricultural work are now widespread; evidence for this is particularly strong in the Philippines, Indonesia and Thailand. The unemployment statistics for the region are partial and far from reliable. In 1986 the official rates for the Philippines, Thailand and Singapore were 6.1, 6.3 and 9.5 per cent, respectively. Particularly for Thailand and the Philippines these rates are almost certainly seriously underestimated. In Thailand some 700 000 people are entering the official job market every year. Considerable attention has focussed on the level of unemployment amongst graduates. In 1986 some 70 per cent of unemployed Thais aged 15 to 24 were graduates. The level of unemployment and underemployment in South East Asia has been increased during the 1980s by economic recession which has in many instances curtailed the growth of manufacturing sectors and the demand for primary exports (Dixon 1984).

For many governments in the region a large, unskilled, quiescent urban labour force underpinned by cheap-food policies and informal sector services is basic to industrial development policy. This type of labour supply situation, especially when allied to the development of free trade zones, has been particularly attractive to multi-national corporations. However, since the late 1970s a number of countries have become more concerned with producing a more skilled and highly trained workforce.

Education and training

Education and training in the context of South East Asia are complex and, for many countries, poorly documented. While the view of 'investment in man' and the 'quality of population' has become a basic tenet of the international agencies and national development policies, there are a number of other conflicting motives. National education schemes have been, and still are, used to impose uniform nationalist views on the population. Regional languages and cultures have been suppressed by the imposition of compulsory primary education in the national language. In Thailand and Indonesia, for example, private Chinese schools were banned in the 1950s in the interests of increasing 'national cohesion'. Additionally, education schemes have been used to enhance loyalty to the government, particularly where insurgency was present. Education has been, and indeed still is, an important factor in the whole process of 'internal colonialism' within the region. Many of these policies had their antecedents during the colonial period, with active suppression of indigenous cultures and limited and highly selective education programmes.

Education, particularly at primary level, has been heavily emphasized in the development plans and public expenditure programmes of most South East Asian countries since the 1960s. Progress since 1960 has been considerable with, by 1985, only Laos and Kampuchea reporting less than 100 per cent of the eligible 5 to 11 age group in primary education. In contrast, despite substantial increases in secondary education, only in Malaysia, Singapore and the Philippines were more than 50 per cent of the eligible 12 to 17 age group receiving education in 1985.

Population planning in South East Asia

Since the late 1950s, population policy has come to occupy a significant position in the various regional and national planning institutions established by South East Asian governments. In general, the prevalent view of governments outside of the socialist group, and of the international agencies, is that population growth both stems from development and dilutes development efforts, so that even greater levels of growth are necessary. These views provide the rationale for the widespread adoption of policies aimed at reducing the rates of population growth. Between 1964 and 1974 the lead given in Asia by India, Pakistan, South Korea and China, was followed by Indonesia, Laos, Malaysia, the Philippines, Singapore and Thailand. By 1979 Kampuchea, Vietnam and Burma had followed suit. However, more recently there have been dramatic reversals of policy by both Malaysia and Singapore (Dwyer 1987).

In the non-socialist countries policy statements linked programmes aimed at fertility reduction to development efforts. The need for the population policy to be an integral part of development policy has been emphasized repeatedly by the international development agencies. During the United Nations Second Development Decade (1970–80) the interrelationship between economic, social and demographic variables was frequently stressed.

In practice, this integration has not occurred, as Whitney (1976) has noted:

Understandably, but nevertheless naively, government leaders and planners as well as demographers frequently assumed that all that was needed to reduce high birth rates was to provide the technology.

Thus, in practice, population policy was treated rather like vaccination programmes. Neither Whitney's (1976) review of South East Asian population policy in the early 1970s nor an examination of national programmes in the 1980s reveals any major change in practice as against policy statements. This is despite the writing-in of population growth-rate targets in national plans and the establishment of, for example, a Family Planning Board in Malaysia and a Population and Manpower Planning Division in Thailand.

Despite the shortcomings of many of the region's schemes, the reduction of fertility in a number of the region's countries since 1960 has been impressive. The rate of growth of population in South East Asia is declining, but whatever assumptions are made about continuing falls in the birth rate, the region's population will increase considerably in size. Population projections are mainly exercises in speculation; Table 5.10 is representative of the more optimistic views. Even under these assumptions, the region's population is likely to increase by over 40 per cent before the end of the present century and that of Indonesia by 65 per cent. The latter figure is disturbing, even frightening, given the conditions that prevail in parts of Indonesia today.

Table 5.10   South East Asian population projections 1986–2000

	Projected average annual growth rate (%) 1986–2000	Projected population (millions) 2000
Burma	2.3	52.0
Indonesia	1.8	207.0
Kampuchea	6.4	n/a
Laos	2.8	5.0
Malaysia	1.9	21.0
Philippines	2.3	76.0
Singapore	0.8	3.0
Thailand	1.6	65.0
Vietnam	2.4	88.0
South East Asia	1.9	536.0
China	1.4	1279.0
India	1.8	1002.0
All less-developed countries	2.0	4926.0

Source: United Nations (1982), World Bank (1988).

While it is clear that much remains to be done with respect to the promotion of family planning in South East Asia, it is important to see that the problem is a much wider one than that of distribution and education. Coale (1973) proposed three prerequisites for a sustained major fall in fertility. First, fertility must be considered a matter for rational choice; couples must be both aware of the possibility of controlling family size and find it an acceptable form of behaviour. Second, reduced fertility must be seen as advantageous within the perceived social and economic circumstances; and third, effective and acceptable techniques of birth control must be available. In Coale's view, the first and third are present before the onset of the decline and it is the change in the second factor which is crucial. Knoedel (1977) in a review of Asian fertility changes sees the 'innovation' and 'adjustment' processes as frequently operating concurrently. While in practice the prerequisites for sustained fertility decline may well be less clear-cut than Coale has suggested, the second factor remains crucial. It is the failure of South East Asian population planning programmes to appreciate this that is disturbing. Almost all criticisms are couched in terms of shortcomings in the distribution or the acceptability of the 'technology'. It is vital that the failure of large numbers of the region's inhabitants to adopt family planning is not seen in terms of ignorance and lack of opportunity but in the main the result of economic conditions which make, at least in the short-run, a large family economically advantageous.

Further reading

Adas M (1974) *The Burma Delta: Economic and Social Change on an Asian Rice Frontier, 1852–1941*, University of Wisconsin Press, Madison.

Carr-Saunders A M (1936) *World population*, Clarendon Press, Oxford.

Coale A J (1973) 'The demographic transition', *Proceedings of the International Population Conference*, Vol. 1, Leige, pp. 53–72.

Dixon C J (1984) 'The Far East after the boom years', *Geographical Magazine* **56**, 61–6.

Dwyer D J (1987) 'New population policies in Malaysia and Singapore', *Geography* **72**, 248–50.

FAO (1982) *Production Year Book*, Rome.

FAO (1986) *Production Year Book*, **40**, Rome.

Fisher C A (1968) 'Malaysia: a study in the political geography of decolonization' in Fisher, C.A. (ed.), *Essays in Political Geography*, Methuen, London, pp. 75–146.

Fisher C A (1971) *South-east Asia: A Social, Economic and Political Geography*, Methuen, London.

Geertz C (1963) *Agricultural Involution: the Process of Ecological Change in Indonesia*, University of California Press, Berkeley.

Gourou P (1961) *The Tropical World*, Longman, London.

Hull T H (1981) 'Indonesian population growth 1971–80', *Bulletin of Indonesian Economic Studies* **17**, 114–20.

Ingram J C (1971) *Economic Change in Thailand 1850–1970*, Stanford University Press, Stanford, California.

Jones G W (1977) 'Fertility levels and trends in Indonesia', *Population Studies* **31**, 29–41.

Kiljuen K (1984) *Kampuchea: Decade of the Genocide*, Zed Books, London.

Knoedel J (1977) 'Family limitation and the fertility transition', *Population Studies* **31**, 219–49.

Kunstadter P (ed.) (1967) *Southeast Asia: Tribes, Minorities and Nations* (two volumes), Princeton University Press, Princeton.

McCoy W (1972) *The Politics of Heroin in South East Asia*, Harper and Row, London.

McMichael J K (1976) *Health in the Third World: Studies from Vietnam*, Spokesman Books, London.

Mounge C (1982) 'The social and economic correlates of demographic change in a Northern Thai community', unpublished Ph.D. thesis, University of London.

Naval Intelligence Division (1943) *Indo-China*, The Admiralty, London.

Neville W (1979) 'Population', in Hill, R D (ed.) *South East Asia: a Systematic Geography*, Oxford University Press, Kuala Lumpur, 52–77.

Ng R C Y (1979) 'The geographical habitat of historical settlement in mainland South East Asia', in Smith, R B and Watson, W (eds.), *Early South East Asia*, Oxford University Press, Kuala Lumpur, pp. 262–72.

Nguyên duc Nhuân (1984) 'Do urban and regional management policies of socialist Vietnam reflect the patterns of the ancient Mandarin bureaucracy?', *International Journal of Urban and Regional Research* **8**, 73–89.

Pardthaisong T (1974) 'The epidemiology of the acceptance and use of depo provera as an injectable contraceptive in Chiang Mai, Northern Thailand', unpublished M.Sc. thesis, University of London.

Prachubmoh V, Knodel J and **Pitaktepsombati P** (1973) 'The longitudinal study of social, economic and demographic change in Thailand: the second rounds', *Institute of Population Paper No. 3*, Chulalongkorn University, Bangkok.

Prasithrathsin S (1973) 'Some factors affecting fertility and knowledge, attitude and practice of family planning amongst rural Thai women', *Institute of Population Studies*, Work Paper No. 2, Bangkok.

Purcell V (1951) *The Chinese in South East Asia*, Oxford University Press.

Salmon C (1981) 'The contribution of the Chinese to the

development of South East Asia: a new appraisal', *Journal of South East Asian Studies* **12**, 260–75.

Santos M (1979) *The Shared Space: the Two Circuits of the Urban Economy in Underdeveloped Countries*, Methuen, London.

Scott J C (1976) *The Moral Economy of the Peasant: Rebellion and Subsistence in South East Asia*, Yale University Press, New Haven.

Silcock T H (ed.) (1967) *Thailand: Social and Economic Studies in Development*, ANU Press, Canberra.

Smith P C and **Ng Shui-Meng** (1982) 'The components of population change in nineteenth-century South East Asia: village data from the Philippines', *Population Studies* **36**, 237–55.

United Nations (1973) *Demographic Year Book*, New York.

United Nations (1976) *Demographic Year Book*, Geneva.

United Nations (1979) *Demographic Year Book: Special issue: Historical Supplement*, New York.

United Nations (1982) *Demographic Year Book*, New York.

United Nations (1982) *World Population Prospects*, Geneva.

United Nations (1986) *Demographic Year Book*, Geneva.

Da Vanzo J and **Haaga J** (1982) 'Anatomy of fertility decline in peninsular Malaysia', *Population Studies* **36**, 373–93.

Whitney V H (1976) 'Population planning in Asia in the 1970s', *Population Studies* **36**, 337–51.

World Bank (1983) *World Development Report 1983*, Oxford University Press, New York.

World Bank (1984) *World Tables*, Washington.

World Bank (1987) *World Development Report*, Washington.

World Bank (1988) *World Development Report*, Washington.

CHAPTER 6

Energy resources

John Soussan

As is true throughout the Third World, much of South East Asia is experiencing a dual energy crisis. On the one hand, the cost of commercial fuels is constraining economic growth and adversely affecting the balance of payments of energy-importing countries. On the other hand, problems associated with the production and utilization of non-commercial fuels derived from biomass materials are jeopardizing both the maintenance of household energy supplies and the viability of the environment in many areas. This dual energy crisis must inevitably form the context within which any discussion of energy resources in South East Asia takes place.

The relationship between the expansion of consumption of commercial fuels and the rate of economic growth is well established (Gordon 1981, O'Keefe *et al.* 1984, Soussan 1988). Energy can be considered as a vital factor of production which, literally, fuels the process of economic growth. This is particularly true for developing countries, where the growth of consumption of commercial fuels has typically been more rapid than the growth of the economy as a whole. The relationship is expressed in the 'energy coefficient', which is a ratio of the growth of energy consumption to the rate of growth of gross national product. The UN (1984) demonstrates that the countries in South East Asia have an energy coefficient of between 1.2 and 1.9, a pattern which has persisted despite the traumas of changing oil prices of recent years. The energy coefficient reflects the pattern of economic growth and structural change which characterizes most Asian 'developing countries', with the more energy-intensive sectors of the economy such as transport, industry and urban households growing more rapidly than other sectors. This pattern is particularly characteristic of a number of South East Asian countries. Singapore has experienced an economic transformation based on export-oriented industrialization, a transformation reflected in the rapid development of infrastructure and the urban fabric and the emergence of new patterns of consumption. By 1986, Singapore's per capita energy consumption, at 1851 kilograms of oil equivalent (kgOE) (World Bank 1988), had quadrupled during her emergence as a newly industrialized country (NIC), and is now as high as most of the developed countries of western Europe. Malaysia, Indonesia, Thailand and the Philippines are all attempting to emulate, at least in part, the NIC model of industrial development. This is reflected in the growth of consumption of commercial energy, much of which is concentrated in the urban sector.

A lack of data and the protracted military conflicts concerning Vietnam, Laos and Kampuchea combine to make any assessment of economic growth and patterns of energy consumption extremely difficult for those countries. As with Burma, consumption of commercial fuels is considerably lower than in the market-oriented economies of South East Asia, but as such countries rebuild and develop the significance of commercial energy will increase.

Table 6.1 Production and consumption of commercial fuels in South East Asia in 1986

Fuel type	Production (thousands of tonnes of oil equivalent)
Solids	6 600
Liquids	91 489
Gas	35 978
Primary Electricity	1 303
Total	135 370
Consumption (total)	79 413

Source: World Resources Institute (1988).

South East Asia is well endowed with energy resources and, as Table 6.1 shows, produces far more commercial energy than it consumes. These resources are unevenly spread, however, with Indonesia, Brunei and, to a lesser extent, Malaysia being major fuel exporters and Thailand, Singapore and the Philippines importing large quantities of commercial energy. Vietnam also imports commercial energy (mainly oil), but exports some coal and is not a major net importer. Consumption of commercial energy in Laos and Kampuchea is at a very low level, and is of little importance when compared to the other countries of the region. This distinction between energy exporters and energy-importing nations is of great significance in South East Asia. The cost of energy imports is a major constraint upon the development prospects of the importing nations. Typically, they were the equivalent to over 40 per cent of total merchandise export earnings for Thailand, the Philippines and Singapore in the first half of the 1980s. In contrast, for Indonesia,

Malaysia and Brunei, income from energy exports and the absence of costly imports provides capital (and in particular scarce foreign exchange) which creates the opportunity for economic development in other sectors of the economy. The fall in oil prices (in 1986) went some way to mitigating this pattern. The oil importers found fuel costs a less onerous burden, with Thailand and the Philippines both saving between $500 and $700 million per year, or about 40 per cent of their total bill in 1986 and 1987 when compared to 1985 (UN 1987). Against these savings, Indonesia, Malaysia and Brunei between them lost $9200 million in revenue in 1986 when compared to 1985 (over 40 per cent of total oil revenue). As a result, South East Asia as a region is a major loser from the oil price fall, but again its impact varies from country to country.

Patterns of energy consumption

Primary energy consumption in South East Asia is split fairly evenly between commercial sources (principally fossil fuels) and biomass fuels (which in South East Asia consist mainly of wood and charcoal, but which also include crop and animal residues in many places). As Table 6.2 demonstrates, there is some variation in this pattern, particularly since the significance of non-commercial fuels decreases with increasing per capita income. The importance of wood and other biomass materials as a fuel source in Asia has became increasingly recognized (Soussan and O'Keefe 1985), and it is significant that even a major energy exporter such as Indonesia still relies on non-commercialized, biomass fuels for a major part of her energy supplies. In Vietnam, Laos and Kampuchea biomass fuels provide most of the energy consumed, a pattern which parallels those found in Africa (Simoes 1984, Barnes *et al.* 1984), South East Asia (Soussan *et al.* 1985, Leach *et al.* 1985) and elsewhere in the Third World (Eckholm *et al.* 1984). The use of biomass fuels is closely related to the level and form of economic development, but for the low-income countries of South East Asia the dominance of these fuels reflects the overwhelmingly rural nature of their economies and societies as much as their impoverishment.

In Laos and Kampuchea intense users of commercial energy such as industry and modern transport are restricted in their development and over 80% of the population live and work in the agrarian sector. For these communities, biomass fuels are the dominant (and frequently the only) source of energy. Throughout South East Asia biomass fuels are used predominantly in the household sector, mainly for cooking and food processing. In the centrally planned South East Asian economies there is only very limited use of any fuel in agricultural production. The main source of energy is human and animal labour, and if these sources could be satisfactorily measured the significance of commercial fuels in total national energy demand would be even more limited than the

Table 6.2 South East Asia: primary energy balances 1982 (thousands of tonnes of coal equivalent)

Production	Brunei	Kampuchea	Indonesia	Laos	Malaysia	Philippines	Singapore	Thailand	Vietnam
Wood[1]	32	182	44 935	1 434	2 856	11 106	–	14 168	25 296
Liquids	19 850	–	95 758	–	19 629	673	–	12	–
Solids[2]	–	–	481	–	–	379	–	650	6 100
Gas	13 617	–	22 054	–	1 192	–	–	1 708	–
Electricity[3]	–	–	192	117	208	899	–	391	197
Total	33 499	182	163 420	1 551	23 885	13 057	–	16 929	31 593
Imports	–	26	12 269	226	10 368	16 573	10 961	15 165	1 841
Exports	30 437	–	102 133	90	15 202	76	–	14	1 000
Consumption total	3 062	208	80 776	1 686	17 151	27 817	10 961	31 839	32 434
Wood[1]	32	182	44 935	1 434	2 856	11 106	–	14 168	25 296
Liquids	264	26	29 753	224	11 727	15 181	10 959	14 773	1 828
Solids[2]	–	–	298	–	178	613	2	711	5 113
Gas	2 766	–	5 598	–	2 182	–	–	1 708	–
Electricity[3]	–	–	192	28	208	899	–	479	197

Notes:
1. Includes wood for charcoal, but excludes agricultural residues.
2. Coal and lignite.
3. Hydro and other primary sources.
Source: United Nations (1982a).

143

Table 6.3 Primary energy demand by sector in Indonesia and Thailand (thousands tonnes of oil equivalent)

	Households	Industry	Transport	Power generation	Others	Total	
Indonesia							
1978							
Biomass	35 471	259	–	–	98	35 828	(68%)
Oil	4 380	2 930	4 535	1 150	1 683	14 678	(28%)
Others	83	1 496	21	532	–	2 132	(4%)
Total	39 934	4 685	4 556	1 682	1 781	52 638	
	(76%)	(9%)	(9%)	(3%)	(3%)		
Thailand							
1982							
Biomass	4 813	2 009	–	–	–	6 822 (42%)	
Oil	1 027	1 945	4 221	–	1 744	7 937 (49%)	
Others	686	798	–	–	–	1 484 (9%)	
Total	6 526	4 752	4 221	–	(5%)	16 243	
	(40%)	(29%)	(26%)				

Source: Soussan *et al.* (1985).

figures in Table 6.2 suggest. Vietnam uses indigenous coal resources for both electricity generation and to fuel her industrialization, which has historically been orientated towards heavy capital goods production (White 1983). This is reflected in the untypical dominance of solid fuels in commercial energy consumption. Economic constraints and government policy have restricted the expansion of use of imported petroleum products, enabling Vietnam to avoid the reliance upon costly energy imports which characterize many Third World countries of a similar level of income and development.

Table 6.3 gives details of the sectoral demand for different fuels for Thailand and Indonesia, patterns of consumption which can be taken as characteristic of the middle-income market economies of South East Asia. Although Thailand relies on imports for most of her commercial energy and Indonesia is a major exporter, the patterns are essentially similar. In both cases energy consumption is dominated by wood and oil, with other fuels contributing no more than a few per cent each to total demand. A similar picture is found in the Philippines and, to a lesser extent, Malaysia, where biomass fuels are of less importance. Households, industry and transport are the most important users of energy, with the relative role of each varying according to the level of industrial and infrastructural development. In general, commercial fuels are used in roughly equal proportions in these three sectors, with other sectors such as agriculture, services and commercial activities being far less significant. In contrast, biomass fuels are used overwhelmingly in the household sector, with industrial use at times being a significant secondary user.

Transport is almost wholly dependent upon petroleum products, and is the most important user of oil throughout South East Asia. In most countries of the region energy use in transport, and in particular in road transport, is growing rapidly. This reflects in part the pace of urbanization, as urban dwellers typically have greater access to and need for motorized transport than their rural counterparts. The rapid growth of fuel use in transport is also attributable to the increased integration of rural areas into the national economy in South East Asia. This is reflected in greatly increased flows of goods and people to and from rural areas – movement which occurs predominantly by road.

The growing use of oil for transport is consequently closely related to the pattern of development occurring in South East Asia, development which expresses itself in growing cities and increasing rural-urban interaction. For example, the number of motorized vehicles in Indonesia grew from less than 600 000 in 1967 to 3 300 000 in 1979, and these vehicles used 72% of the fuel consumed in the transport sector (World Bank 1983b). Trucks and buses use about two-thirds of the fuel consumed in the road sub-sector, reflecting both the importance of the long-distance movement of goods and people and a conscious policy of 'dieselization' by the Indonesian government. Diesel is considerably cheaper than petrol in Indonesia, and its use has grown extremely rapidly as a consequence. Shipping (both inter-island and international, and including oil tankers) accounts for most of the remaining fuel use in the transport sector, with air transport accounting for the residue. Over 99% of energy used in Indonesia's transport sector comes from petroleum products.

Much the same is true of Thailand, where diesel is the largest single component of the total, closely followed by petrol. Much of the energy used in transport goes in moving goods and people to and from Bangkok, whilst fuel demand for transport within the capital is also of major importance. Aviation fuel is also a significant component of demand in the transport sector, reflecting the emergence of Bangkok airport as a focus of air routes in the South East Asian region.

Throughout South East Asia, energy demand for transport will continue to grow rapidly, and will inevitably be based on petroleum products. Transport is a sector in which fuel substitution possibilities are extremely limited. This is particularly true where, as in South East Asia, road transport predominates. Countries such as Thailand, Malaysia and, in particular, the Philippines (see below) are experimenting with alternatives to oil such as gasifiers (which use wood and charcoal) and fuel alcohol (from sugar cane), but as yet these alternatives have made little impact beyond the experimental stage.

Industrial use of energy varies greatly in the South East Asian region, inevitably reflecting differing levels of industrialization. Laos and Kampuchea have little or no industry of significance, whilst countries such as Thailand and the Philippines are experiencing increasing industrialization and for them manufacturing now comprises between 20% and 25% of GNP. Singapore has, of course, emerged in the last 20 years as a sophisticated industrial economy, and this is reflected in the pattern of energy consumption. Singapore imports all of her energy, and relies almost entirely on oil. In industry, energy is provided either by directly using oil products or by electricity, which is in turn generated in oil-fired stations. Although the oil price rises of 1973/74 and 1979/80 inevitably affected Singapore, and the economy as a whole went into a severe but temporary recession in the mid-1980s, the general buoyancy of her export economy, diversification into sectors such as construction and financial services and the growth of domestic consumer demand have permitted her to weather oil shocks and recession more easily than most South East Asian economies. Consumption continues to grow. That of oil rose from 4.6 million tonnes of coal equivalent (TCE) in 1973 to 10.8 million TCE in 1986 for example. Electricity generation has grown at a similar rate, from 3.7 billion kilowatt hours (kWh) in 1973 to 7.4 billion kWh in 1981 (UN 1983). As part of her industrial growth, Singapore has developed a significant oil-refining capability, and indeed exports refined petroleum products to Malaysia, Indonesia, Thailand and elsewhere.

In the middle-income market economies of South East Asia, industry is concentrated in major cities such as Manila, Bangkok and Jakarta, reflecting their high level of primacy and reinforcing the dominance of these centres in the consumption of commercial energy in these countries. In Thailand, industry in and around Bangkok uses a wide variety of petroleum products, with fuel oil and, to a lesser extent, diesel being particularly important. Similarly, electricity is an important source of energy in a number of industrial processes such as textiles, food processing and light engineering (National Energy Administration 1984a). Both Thailand and the Philippines have expanded their consumption of solid fuels in recent years with the intention of switching from oil in electricity generation. Demand for solid fuels in the Philippines rose from 35 000 TCE in 1973 to 590 000 TCE, or 3.4% of total energy demand, in 1981. Similarly, demand for coal and lignite rose from 136 000 to 608 000 TCE in Thailand in the 1973–81 period (UN 1982a). Thailand and the Philippines increased rapidly the use of coal in electricity generation during the 1980s. Thailand also uses indigenous natural gas in electricity generation. Similarly, in both countries the use of bagasse in the sugar industry (wastes from sugar cane processing) has become an important power source. In Thailand, bagasse use in the rural-based sugar industry now constitutes over 30% of total industrial primary energy consumption.

Industrial use of energy in Indonesia is more varied, with oil, coal, natural gas, electricity and fuelwood all contributing. Growth of demand for energy in the industrial sector has been rapid, averaging 12.9% in the 1970s (Soussan *et al.* 1985). There has been a trend towards the more intensive use of energy in industry in Indonesia, a pattern which will continue as industrial growth continues and, in particular, the large-scale manufacturing sector develops. Eighty-five per cent of manufacturing enterprises in Indonesia are located in Java (World Bank 1983b), principally in and around Jakarta. The range of industrial energy consumers is broad, but textiles, food processing, paper products and construction materials are all major energy users. For industries such as glass, cement and paper production energy costs are a major component of total costs (for all of these in Indonesia, over 40% of value added). The viability of these industries is consequently strongly related to the cost of energy, an issue which is true of other industries in South East Asia. Throughout the region, industrialization has increased dramatically in recent decades, but in a number of cases industrial growth has been seriously affected by energy costs and any future sharp rise in these costs can be expected to have an impact upon industrial growth prospects.

This point is particularly salient for the oil-importing nations of South East Asia, all of whom have embarked upon major industrial drives as the basis of their development strategies. The Philippines is attempting to create a broad industrial base. Energy-intensive industries such as chemicals and iron and steel production consume one-third of all the primary energy used in industry (Leach *et al.* 1985). Although attempts have been made to diversify the fuels used, oil is still the dominant energy source for Philippine industry (for example, providing 63% of energy used in the manufacture of iron and steel in 1981 and 97% of fuel used in cement production in 1978). Energy consumption, and in particular demand for oil products, will continue to grow as the Philippines experiences further industrialization. The impact of increased energy costs means that a high growth momentum must be maintained if industrial drives such as that of the Philippines are not to run into increasing difficulties. This problem is of less urgency for countries such as Indonesia and Malaysia, which have plentiful indigenous energy resources, but even for them is one which cannot be ignored.

Transport and industry between them account for a significant proportion of the commercial energy consumed in South East Asia. By far the largest sector of total energy consumption, and a major user of commercial energy, is domestic energy: the use of fuel in the household for cooking, lighting and other purposes. A distinction can be drawn between urban households, who as a group use both commercial and non-commercial fuels, and rural households, who overwhelmingly rely on non-commercialized, biomass fuels for the bulk of their energy needs. Similarly, within each of these sectors there is a clear income effect, with the quality and quantity of commercial fuel use increasing with increasing incomes and the use of biomass fuels commensurately declining. This effect is noticeable both within and between the countries of the region. Amongst urban households the use of biomass fuels for, in particular, cooking is characteristic in the towns and cities of Laos, Kampuchea and Vietnam, and is widespread among lower-income urban households in Malaysia, Thailand, Indonesia and the Philippines.

There is some variation from country to country in the actual fuels used by urban families, reflecting both tradition and economic considerations. In Thailand, charcoal has traditionally been the main cooking fuel in towns and cities, and is still widely used. Boats and trucks laden with sacks of charcoal for Bangkok's market are a common sight on the canals and roads leading to the city, bringing supplies from all parts of Thailand. In Thailand, as elsewhere, urban households rely on a variety of fuels to meet their energy requirements. Electricity is available to most households in the major urban areas except for families in unauthorized housing. Bottled gas (LPG) and kerosene are both widely used, with the latter used as a cooking fuel and for lighting by poorer families.

Patterns of fuel consumption in the cities of South East Asia are strongly influenced by government policies. Investment policy in electricity generation capacity is the most important factor determining the use of this fuel. The growth of demand electricity in urban areas is frequently supply constrained; there is just not enough generating capacity or installed distribution grid to meet the growing urban demand. New residential developments frequently experience considerable delay in receiving electrical connections, whilst power sharing and blackouts are far from unknown in the region. Levels of consumption of LPG and kerosene are strongly affected by pricing and subsidy policies. For example, the use of kerosene grew rapidly in Indonesian cities in the 1970s when it was very heavily subsidized by the government, but has declined (on a per capita basis) in recent years following the removal of much of the subsidy and consequent heavy price rises. In the Philippines, LPG is roughly one-third more expensive as a cooking fuel than is kerosene, and three or four times the price of fuelwood in urban areas. This is reflected in the pattern of household fuel use, with 75% of higher-income families in Manila using LPG, whereas only 45% of lower-income families do so, the remainder using kerosene (35%) and wood (20%). The use of the more expensive fuels declines in urban areas outside the capital, where wood becomes more frequently used by higher income families and is the dominant source of cooking fuel for low-income households (Leach 1984). The use of LPG, which is heavily subsidized by the Thai government, is growing rapidly in Bangkok, as this fuel is now both cheaper and more convenient than most alternative fuels. Islam *et al.* (1984) demonstrate that the use of LPG in Bangkok is strongly related to income, a factor which probably reflects the cost of cooking appliances as much as the cost of the fuel.

Biomass fuels

There is far less variation in patterns of fuel consumption among rural households. Throughout the region, rural domestic fuel consumption is dominated by biomass fuels, with commercial energy being used only for lighting except by a small, privileged minority which can afford the fuel and the appliances they power.

The importance of biomass fuels for rural communities is just beginning to be understood. Until the very recent past their role has consistently been under-estimated and the complex problems associated with their use (discussed below) have remained largely hidden. This position is now changing, and national governments and international agencies are beginning to express grave concern over the ability of the rural population to provide for their energy needs, the environmental consequences of this pattern of resource use and the elusiveness of effective policy solutions to these problems.

Many of these difficulties reflect the lack of reliable data concerning biomass fuel use in rural areas. For most countries we do not even possess accurate data on the quantities of fuel used (with most figures available being extrapolated estimates), and much less is known about the types of biomass fuels used, the sources of these fuels, etc. Despite these data problems, a clearer picture of the rural energy system in South East Asia is beginning to emerge, if only on a selective basis.

As we have seen, biomass fuels constitute the largest component of Indonesia's energy balance, and provide nearly 90% of all the energy used in the household sector. Soesastro (1983) estimated that 95% of biomass fuel demand in Indonesia is found in rural areas. This constitutes about 130 million cubic metres wood and agricultural wastes, an enormous quantity of material even for a biomass resource-rich country such as Indonesia. Within Indonesia, a sharp distinction can be drawn between densely populated islands (principally Java), where the rural population is faced with serious problems of access to energy resources, and more sparsely populated regions such as much of Kalimantan and Irian Jaya, where there is no significant energy problem. In contrast to the lush and extensive forests of these regions, Java has very few areas of forest cover and, with a population of over 100 million, is one of the most crowded agricultural regions in the World. Atje (1979) suggests that per capita biomass fuel consumption (at 0.79 cubic metres per annum) in Java is significantly lower than that for the rest of the country (where it is estimated to be 0.96 cubic metres per annum). Such differences in patterns of biomass fuel consumption reflect variations in the resource base of different areas. In densely settled regions the availability of biomass fuels is restricted by the conflicting demands and lower levels of tree cover, meaning that more time is needed to collect fuel supplies and consumption levels are constrained. Despite the greater efficiency of use, however, the local environment in many places is unable to sustain the conflicting demands placed upon it, with existing levels of extraction of material leading to the problems of tree loss and environmental deterioration which characterize wide areas of the regions. These internal variations are dramatically highlighted by the island nature of Indonesia, but may be similarly found in the other countries of South East Asia. Within the Philippines, biomass fuel scarcities are far more acute in densely-settled areas such as Luzon and Samar than in more sparsely populated regions. Islam *et al.* (1984) identify a similar pattern in Thailand.

In rural areas where most consumption occurs, biomass fuels are overwhelmingly gathered freely from the local environment. These fuels are produced and consumed at a local level, with rural people (almost invariably women) collecting biomass from the vicinity of their settlements. They are a 'free' good, in that they do not have a commodity value. The costs associated with their use are the time which it takes to gather the fuel and the opportunity cost of the biomass materials (that is, the alternative uses to which the wood or residues could be put, uses such as construction, fodder or fertilizers). In consequence, any discussion of biomass fuel issues must concentrate on the problems at a local rather than a national level. This is less true for wood fuel which serves the urban/industrial market, as this fuel is commercialized and may be transported considerable distances.

In South East Asia, as elsewhere, biomass fuels are generally derived from agricultural land. This is as true for tree fuels as it is for crop residues. Except in districts where forested areas are situated very close to agricultural settlements, the main source of biomass fuels in South East Asia is trees outside the forests. Trees scattered around the agricultural landscape, along field margins, roads and canals, and around the homestead are the source of fuel, not designated forest areas which are in general remote from concentrations of rural population.

A widely believed myth is that fuelwood collection causes deforestation. There is increasing evidence that this is not the case. Authors such as Eckholm *et al.* (1984), Plumwood and Routley (1982), Foley *et al.* (1984) and Soussan (1988) demonstrate that factors such as clearances for agricultural land and commercial logging lie behind the rapid deforestation which threatens South East Asia (and the forests of the rest of the Third World). Compared to these factors, fuelwood collection is of minimal importance. Where a relationship exists between the two, it is the reverse: deforestation may threaten the energy supplies of rural population. The main adverse environmental consequences of biomass fuel use impinge upon agricultural land rather than on forested areas. The evidence available is almost non-existent, but what there is suggests that fuel scarcity is becoming a major factor in the devegetation of the countryside. Many agricultural areas of South East Asia are suffering from loss of tree cover, increased erosion and declining soil fertility, all of which reflect the pressures upon the biomass resource of which fuel demands are a major, but not the only, component.

The third point concerning biomass fuel use stems from this. Vegetation, whether from trees or agricultural residues, has a variety of uses for rural people (Soussan 1988). Trees are valued for shade, fruit and fodder, as a source of construction materials and, where there is cash market for wood, as a potential source of income, as well as for their role in providing fuel. Similarly, residues provide fodder, fertilizer and construction materials as well as fuel. Where there exists pressure upon local resources, these alternative demands come into conflict, in general with immediate needs necessarily taking precedence over longer-term environmental maintenance.

Such pressures impinge upon the ability of the rural population to meet their basic survival needs. This may be true even in places where enough materials are produced locally to meet everyone's needs, for the key issue is not the production of biomass materials, but access to those materials for all sectors of the population. Access to biomass materials depends upon who controls land within a rural community. In consequence, questions of access to biomass fuels are embedded in the structure of social relationships within rural communities.

The biomass fuel crisis in South East Asia is a consequence not only of increased pressures upon environmental productivity, it also reflects changes which are occurring in the agrarian economy and society of the region. Of particular importance in many areas is the increased commercialization of the agrarian sector and the consequent erosion of rights and obligations within rural communities. During this process, traditional rights of the rural poor (and in particular the landless) of access to biomass materials for fuel from the land of others are increasingly breaking down, so that even where enough fuel exists the poorest sections of the community are unable to collect it. This process is particularly strong where pressures exist upon the biosphere; it is invariably the poor who feel these pressures first and most severely. These pressures are at their strongest where the commercial fuelwood market, which caters for urban demand, creates a conflict between local needs and the potential cash value of the wood. It is in these circumstances that groups which control local resources will quickly deny traditional rights of access in order to cash in on the new commodity value of biomass materials. Such characteristics of the system in which biomass fuels are produced and consumed are crucial if the crisis is to be understood and addressed.

As Fig. 6.1 shows, the biomass fuel situation varies greatly within South East Asia. Over considerable areas population densities are low and little pressure exists upon biomass fuel resources. This appears to be true of Laos, Kampuchea, southern Thailand, Brunei, Sarawak, Mindanao in the Philippines and large areas of Indonesia such as Kalimantan, Irian Jaya and northern Sulawesi. In southern

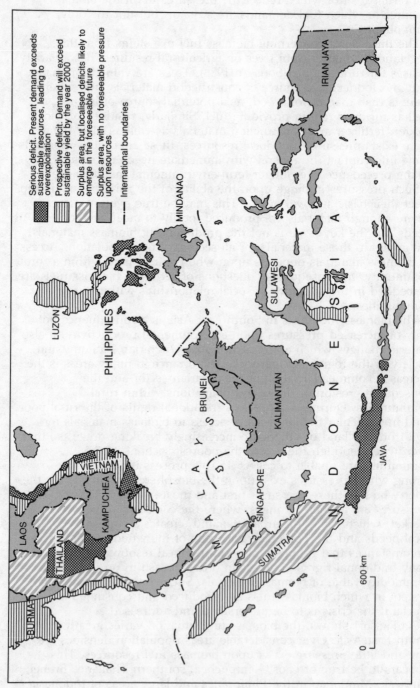

Fig. 6.1 The biomass fuel situation in South East Asia

Serious deficit. Present demand exceeds sustainable resources, leading to overexploitation

Prospective deficit. Demand will exceed sustainable yield by the year 2000

Surplus area, but localised deficits likely to emerge in the foreseeable future

Surplus area, with no foreseeable pressure upon resources

International boundaries

IRIAN JAYA

MINDANAO

LUZON

PHILIPPINES

SULAWESI

INDONESIA

BRUNEI

KALIMANTAN

MALAYSIA

SINGAPORE

JAVA

VIETNAM

SUMATRA

LAOS

KAMPUCHEA

THAILAND

BURMA

600 km

0

Sumatra, peninsular Malaya and much of Thailand the position also appears satisfactory at present, but may in the near future deteriorate to a situation where serious deficits emerge at least at a local level. Throughout the rest of South East Asia, including areas which contain the bulk of the region's population, there already exist problems of biomass fuel supply of a greater or lesser severity. The problem is particularly acute in central Thailand, Java, many areas of the Philippines and the Mekong Delta and other densely populated areas in Vietnam. In all these districts demand for biomass fuels exceed sustainable supply, an assessment which takes no account even of alternative demands for such materials.

Biomass fuel problems characterize countries of differing energy resource bases and political complexions. They are acute in communist Vietnam and unashamedly capitalist Thailand; in energy-resource rich Indonesia and in the Philippines, which contains few commercially exploitable energy resources. This is because the biomass fuel crisis is, as we have seen, a product of population pressure and local-level supply and demand imbalance, not only of national-level economic and political forces or resource endowments. This is illustrated classically by Indonesia, which contrives to contain the region's most extensive forest resources and one of the most serious deficit areas in Java.

In the less-densely inhabited areas, wood is used almost exclusively and is freely gathered from, predominantly, forest areas. The abundance of the resource is such that difficulties of access exist only in very localized areas such as, for example, around some agricultural resettlement schemes. Much of the forest in these areas is high-density rainforest, which typically has high sustainable yield. Considerable quantities of biomass material are removed from many of these forest areas, but this is principally for commercial timber extraction, and indeed such operations leave vast quantities of potential woodfuel behind as residues. Most of this material is not utilized, but either rots or is burnt. In the production of round and sawn wood in Indonesia, for example, a conservative estimate would be that a similar quantity of tree mass is discarded as waste. There is little doubt that in Indonesia as a whole the total forest area could easily sustain the fuelwood requirements of the total population for the foreseeable future, but the different locations of demand and supply and the prohibitive cost of transporting the fuel to the people mean that this theoretical potential has little practical relevance.

For the population of densely settled areas of Indonesia, the main source of biomass fuels is not forest areas but is rather materials extracted from agricultural land, in the form of tree fuels from homestead plots, field margins, etc. and of agricultural residues. Atje (1983) presents results of a survey of rural energy consumption in West Java. Of the households surveyed, 76 per cent relied entirely upon their homesteads, gardens and agricultural land for woodfuel, with 13 per cent purchasing the fuel and only 8 per cent collecting part or all of their fuel from forest or brushwood areas. Atje cites similar studies by Sumarna and Sudiono (1973) in East Java which showed 65 per cent of fuelwood to be gathered from agricultural land, 7 per cent to be gathered in forests and 28 per cent to be purchased, and by Wiersum (1976) in central Java which showed that 95 per cent of households rely principally on fuels from agricultural land. Atje (1983) further demonstrates that the bulk of the tree fuel used is in the form of twigs gathered from the ground or small branches lopped from standing trees. Trees in agricultural areas are rarely, if ever, felled specifically for fuel purposes unless they have outlived their productive usefulness.

There is evidence that the increasing woodfuel crisis in Java is leading to the progressive commercialization of supplies. Soesastro (1983) cites a study which shows that households in the Solo River catchment (an area of acute fuel shortage) sell 47 per cent of wood produced from their homesteads and land, principally for the urban market. This provides a welcome additional source of revenue, but highlights the conflicts facing communities struggling to sustain their basic needs with insufficient resources.

The contribution of crop residues to fuel supplies in densely settled areas of Indonesia is not known but generally does not appear to be great. The total national crop residue output used as fuel is possibly around 7 per cent of current demand for biomass fuel. Much of this residue is produced in densely settled areas where the fuel crisis is at its most acute, however, and the local impact of these residues could consequently be significant. It appears probable that, in the future, significantly greater quantities of crop residues will be used as fuel. The consequences of this for the local environment are difficult to predict, but are unlikely to be desirable.

The picture presented for Indonesia has parallels throughout South East Asia, and reveals the close relationship which exists between the local environment and biomass fuel problems. An assessment of biomass fuel resources must consequently look at both forest areas and non-forest areas, with the latter perhaps the crucial factor in the debate. Having said this, such an assessment is a difficult, if not impossible, task. The data does not exist to permit any meaningful discussion of potential biomass fuel resources of non-forest areas.

Only limited information concerning the forest resources of South East Asia is available. Estimates from an FAO study (FAO 1981b) are summarized in Table 6.4. These results show that, whilst forest cover is generally high in the region, there is considerable variation in this, with Thailand, Vietnam and the Philippines having a noticeably lower level of coverage at a national level than the other countries of the region. Deforestation is a major problem throughout the region, with 140 000 square kilometres per annum being lost in the 1980s. Such a rate of loss cannot be sustained in the long term, and the threat of continued deforestation jeopardizes the

Table 6.4 South East Asia: forest cover and deforestation trends

	1980 Total forest area (thousand of hectares)	1980 Percentage of country	1985 Total forest area (thousands of hectares)	1985 Percentage of country	Average annual deforestation (thousand of hectares) 1976–80	Average annual deforestation (thousand of hectares) 1981–85
Brunei	323	56	298	52	7	5
Kampuchea	12 648	70	12 498	69	15	25
Indonesia	116 895	61	113 800	59	550	600
Laos	13 625	58	12 950	55	125	100
Malaysia	20 995	63	19 721	60	230	255
Philippines	9 510	32	9 050	30	101	91
Thailand	15 675	30	13 780	27	333	252
Vietnam	10 110	30	9 785	29	65	65

Source: FAO (1981b).

environmental stability of the whole region. The main causes of deforestation are clearances for agriculture and clear felling associated with the timber industry. Clearances for the commercial fuelwood and charcoal market also have a localized impact in some areas, but non-commercial collection of fuel from forest areas has little, if any, effect upon the rate of deforestation. This point can be illustrated by looking at the picture in Thailand in more detail.

Thailand's forest resources are disappearing at an alarming rate. The forest area declined from over 27 million hectares in 1961 to about 15 million hectares in 1982, a deforestation rate of 2.6% per annum. The quality of many of the remaining forest areas also gives cause for concern. The FAO (1981b) estimates that in 1980 closed forest (over 75% crown cover) covered only 92 350 km^2, with a further 64 400 km^2 consisting of 'open forest' with a crown cover of between 45% and 75%. This report and other sources suggest that much of Thailand's remaining forest area is seriously degraded, with commercial logging, shifting cultivation and illegal removals for charcoal-making all significant contributors to this process. In addition, many designated forest areas have been encroached upon by small farmers who clear patches on the edge of forests for agriculture. Some efforts at forest management and replanting have been made, but their impact has been negligible when compared to the extremely rapid rate of deforestation. At current rates, Thailand's forests will have disappeared within two to three decades.

Policies towards alleviating the biomass fuel crisis have been introduced in a number of South East Asian countries. Most policies centre upon some form of supply enhancement, with some countries also attempting to induce conservation of biomass fuels by the introduction of improved cooking stoves. Supply enhancement policies inevitably mean attempts to grow more trees. The localized nature of biomass fuel and the problems of tree loss in rural areas are increasingly recognized, and are reflected in programmes of 'community forestry'. Frequently, for example in the 'greening' programme in Java, a large number of trees are planted, but follow-up care is poor and the survival rate is very low. In addition, the concentration on growing trees for fuelwood alone does not meet the needs and perceptions of rural communities, who value trees for many other purposes.

The Philippines leads the way in attempts at the 'high-tech' use of biomass fuels. Whilst most of these fuels are burnt in simple stoves or open fires in the home, their potential as substitutes for commercial fuels is being actively exploited there. By 1984, 34 dendrothermal (wood-fired) projects had been established and 17 power stations were under development. Gasifiers (burners which use wood or charcoal to power internal combustion engines) have a long history in the Philippines, and are now used to power irrigation pumps, small trucks and fishing boats (Foley et al. 1983). The Philippines also has the most advanced biogas programme in South East Asia, including one of the world's largest commercial integrated biogas plants. Similarly, there is an active 'gasohol' programme, which uses alcohol produced from sugar cane to blend with gasoline in order to lower oil imports. Originally, it was intended to replace 30% of oil imports by 1988, but in 1981 the programme was scaled down considerably (Kovarik 1983). The experience of the Philippines in the range of technologies is unique in the region, however, and it is unlikely that this range of new technologies will make a significant contribution to South East Asia's energy in the near or distant future.

Fossil fuels

As Table 6.2 shows, there is considerable variation in the scale of production of fossil fuels in South East Asia. Indonesia, Malaysia and Brunei all contain extensive deposits of fossil fuels and have major oil and gas industries. Vietnam contains significant coal deposits and Thailand limited reserves of lignite and natural gas. Elsewhere, reserves of fossil fuels in South East Asia are confined to small deposits of marginal significance.

The income from oil has permitted Malaysia to diversify her export economy, with dependence upon agricultural products diminishing rapidly. This has provided an important buffer against the rapid decline in the international price of many primary commodities such as tin and rubber in the 1980s. The collapse of commodity prices, in conjunction with the price of oil, has had a devastating impact upon countries such as Malaysia in the past. Malaysia has been able to minimize such impacts through the development of her energy industries. Oil production is now several times that in the 1970s (24.9 million tonnes in 1987), reflecting the rapid increase in exploration efforts and the greater economic attractiveness of previously marginal fields which followed from the 1973–4 oil price rises. As is true for all oil exporters, output fell in the mid-1980s.

As Table 6.3 shows, Malaysia's proven oil reserves (that is, inventoried reserves which are economically and technically recoverable) are the equivalent to about thirty years' annual production at the 1982 rate of extraction. The proven reserves are likely to increase with further exploration and development efforts. Since 1980 the Malaysian government has followed a policy of careful production control, restricting future expansion of output in order to stretch the life of the Malaysian oil fields. Most of Malaysia's oil production (which is primarily of superior light crude) is exported, and local demand is catered for with cheaper Middle Eastern imports. This pattern will continue as long as the relative prices make this unusual pattern of trade advantageous. Malaysia's natural gas resources are also considerable, and offer the prospect of significant expansion of production over the next few years. By 1983 both the Bintulu and the Trenggana fields had been developed and were producing a thirty million cubic metres of natural gas per day each. There are plans to substantially expand electricity generation capacity and reduce dependency on oil as a primary fuel from 85% to 38% by 1990 (UN 1982b), with the rapid expansion of gas-fired thermal stations forming the basis of this policy.

Oil has changed Brunei from a small, isolated enclave on the northern margins of Borneo to a rapidly changing nation which has one of the highest per capita incomes in the world. The income from this resource has been used to finance many forms of social, economic and infrastructural development. At current levels of production (6.9 million tonnes in 1987), proven reserves will last about 15 years, but there is the possibility of further discoveries which, along with increased extraction rates from known reservoirs, may increase the life of Brunei's oil industry. There have been reservations expressed concerning the profligate use of a short-term, non-renewable source of income, but potential tensions are unlikely to emerge in the near future, whilst the flow of oil money lasts.

Brunei's gas deposits have a similar potential life-span. Production has increased rapidly in the 1980s, and the bulk is exported. A significant proportion is used for domestic consumption, however, and somewhat unusually, energy demand in Brunei is dominated by natural gas. Much of this is used to generate electricity, a use which makes sense but which requires more careful management of resources if it is to be sustained.

Table 6.5 Fossil fuel resources of South East Asia (1986–7)

	Oil (million tonnes)	Natural gas (billion cubic metres)	Coal Geological resources	Proven reserves
Brunei	200	182	–	–
Indonesia	1 200	1 870	23 232	3 000
Malaysia	407	1 501	75	7
Philippines	3	–	82	170
Thailand	12	105	78	15
Vietnam	–	–	3 000	472

Sources: World Resources Institute (1988), British Petroleum (1988).

Indonesia contains by far the largest fossil fuel resources in South East Asia, and indeed is one of the world's leading oil exporters and a key member of OPEC. As Table 6.5 shows, Indonesia has considerable deposits of oil, coal and natural gas, resources which form the basis of her major and vigorous energy industries. These resources are located in many parts of Indonesia (Fig. 6.2), but the eastern areas of Kalimantan and Sumatra are particularly important areas of current production. The exploitation of fossil fuels in

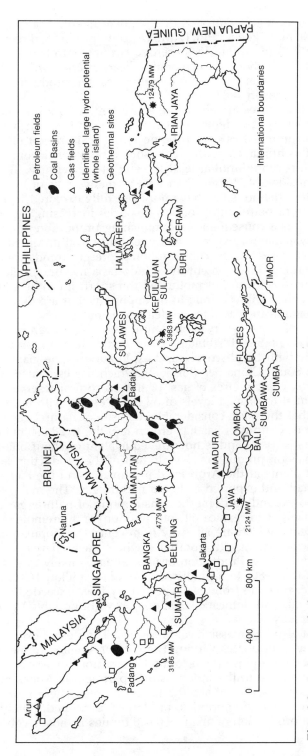

Fig. 6.2 Indonesia: fuel resources

Indonesia is strongly controlled by the government. The government's oil agency, PERTAMINA, only produces directly very small quantities of oil but does exercise strong controls over the granting of production and exploration licences and owns and operates all of Indonesia's main oil refineries. Oil production was 6.4 million tonnes in 1987.

Proven oil reserves are about the equivalent of 20 years' output. Production levels have remained fairly stable (and indeed have declined in the mid-1980s with the declining real price of oil), so that oil will continue to be the main source of export income and government revenue until at least the end of the century. The oil may last considerably longer than this, as many areas of Indonesia (in particular offshore sites) have not been fully explored. Domestic consumption of petroleum products has been increasing by 12% per annum, and as a consequence a diminishing proportion of oil output is available for export. Over one-third of oil production will soon be used for domestic consumption if present trends continue. Given the strong development drive Indonesia has undertaken, and the rapid increase in use of motorized transport, it can be expected that oil revenues will continue to be important, but are likely to decline in significance as internal demand is met.

Indonesia has vast reserves of natural gas. The proven reserves shown in Table 6.5 of 1870 billion cubic metres have expanded considerably with the discovery and exploration of the giant Natuna field in the South China Sea (Fig. 6.2), which has estimated reserves of one trillion cubic metres of gas, or half of current reserves. This is almost equal to proven levels of oil reserves. As with oil, it is estimated that there are considerable deposits of natural gas as yet undiscovered. Output of gas has grown rapidly, to 28 000 million cubic meters in 1986, and is now the equivalent of about one-third of Indonesia's oil production. Nearly half of the production is liquefied at plants at the Arun Field in Sumatra and the Badak Field in Kalimantan and exported as liquid natural gas. The main drawback to expanding domestic consumption of natural gas (and in particular using it to replace oil consumption) is the remoteness of the main fields, which are all away from centres of population and are in particular far distant from Java, where most energy consumption occurs. Production of natural gas is likely to expand throughout the 1980s to a predicted level of 70 million TCE by 1990. By this time, revenue from gas exports may have exceeded those from crude oil, and domestic consumption will undoubtedly have increased greatly.

Indonesia also has massive coal reserves. Their exact extent is unknown, with estimates varying from about 3 billion tonnes to as high as 15 billion tonnes (Directorate General for Power 1981). Recent estimates from the Indonesian Department of Mines put 'mineable' coal reserves at about 3 billion tonnes. These represent an energy resource of enormous, but largely untapped, potential. Present coal production of about 300 000 tonnes is mostly confined to two small, state-owned mines at Bakit Asam and Ombilin in southern Sumatra.

Fossil fuel deposits in South East Asia outside of Malaysia, Brunei and Indonesia are limited in their extent. The Philippines has a little oil and coal. Thailand some lignite and offshore gas deposits (the exploitation of which is expanding rapidly) and Vietnam coal. The exact extent of Vietnam's coal resource is uncertain. Geological resources are put at 3 billion tonnes, but the proportion of this which is technically and economically recoverable is not known. Coal forms the basis of Vietnam's commercial energy economy, and the exploitation of these resources can be expected to continue apace. Current levels of production do not appear to jeopardize Vietnam's medium-term future, but the continued expansion of production may lead to rapid resource depletion unless new deposits are discovered.

There can be little doubt that fossil fuels will continue to dominate commercial fuel consumption in South East Asia, and will be central to the development prospects of the region's major energy exporters. There are, however, efforts being made to develop alternative commercial energy resources in South East Asia.

'Alternative' energy resources

The so-called 'alternative' forms of energy resource, such as wind, water and solar power, have received a great deal of publicity since the 1973–4 oil crisis, with grandiose claims being advanced concerning their potential as the fuels of the future. Such energy sources have been seen as particularly appropriate for developing countries, as these countries were hit most severely by the 1970s oil crisis, are believed to contain the best potential resources of these energy forms and frequently do not already have well-established energy infrastructures. The grand designs for these alternative energy sources have rarely been fulfilled, however, and such forms of energy make little more than a marginal contribution to the energy economies of South East Asia (apart, that is, from hydro-electric power (HEP), which can hardly be considered an 'alternative' form of electrical generation).

It is very difficult to identify and measure in any meaningful way energy sources such as the sun and wind. This is because that these 'resources' do not exist as a physical material such as coal or oil, but are rather a latent potential which may or may not be harnessed. The theoretical potential of solar, wind, wave and geothermal energy in South East Asia is enormous, almost limitless, but this has no practical relevance. Such potential cannot really be considered a resource until it can be meaningfully measured, technically harnessed and be shown to be economically viable.

To an extent, these comments do not really apply to HEP, as this power source has a long history of exploitation and the technical

Table 6.6 Hydro-electric power in South East Asia

	Estimated potential (gigawatts)		Installed capacity (gigawatts) 1985	HEP generation 1981 (million kWh)	Percentage of total electricity generation
	World Bank(a)	World Resources(b)			
Kampuchea	–	10	0.02	–	–
Indonesia	30.0	81	0.5	2800	36
Laos	–	28	0.2	n/a	n/a
Malaysia	1.3	68	1.2	1300	14
Philippines	7.5	38	1.8	3800	20
Thailand	6.2	20	1.8	3659	19
Vietnam	53.6	18	0.3	700	18

Sources: (a) World Bank (1980, Table II.5), (b) World Resources Institute (1988), (c) United Nations (1987, pp. 75–6).

and economic issues involved are, in general, well understood. Table 6.6 gives figures for the gross theoretical HEP potential for the main countries in the South East Asian region (and vividly illustrates the degree of uncertainty surrounding such estimates). For all countries in the region, HEP is a major potential resource. Inevitably, much of this potential is theoretical. Economic or technical considerations preclude the development of much of the resource base, and in particular the very high capital costs involved in HEP schemes are a major constraint. There are also a number of wider social and environmental costs involved with developing HEP, costs such as deforestation and the large-scale resettlement of displaced peoples, which render many HEP schemes highly undesirable. In any case, most South East Asian countries do not possess sufficient installed distribution capacity, or even sufficient demand for electricity, to make the full development of HEP potential worthwhile.

In most countries of the South East Asia there are already a number of HEP schemes in production. As Table 6.6 shows, HEP currently contributes a significant proportion of total electricity production, and there can be little doubt that in most countries the role of HEP will grow in the future. Countries such as the Philippines, Thailand and Indonesia are currently developing new HEP schemes and have plans to significantly extend generating capacity in the next decade. Indonesia had plans to develop 20 HEP schemes in the 1985–8 period. The combined installed generating capacity of these schemes would be 2.2 gigawatts, or over twice the HEP capacity of the early 1980s, and the energy generation potential of these new schemes was estimated at over 1000 gigawatt hours per annum. The bulk of existing HEP capacity in Indonesia is located on Java, with much of the rest on Sulawesi. The new capacity is similarly concentrated in areas of high population density. Six of the 20 new schemes are on Java, six on Sumatra and five on Sulawesi, whilst just two are on Kalimantan and one on Irian Jaya. The new schemes account for most of Java's remaining significant HEP potential. The water resource position in Indonesia closely resembles that of biomass fuels. Resources exist in abundance but are located far from centres of demand and are consequently of little practical relevance. Nearly half of HEP potential is located on Irian Jaya, which contains only 1 per cent of Indonesia's population and has little potential for industrial development.

Power from water can consequently be expected to expand in most of the countries of South East Asia, and the role of HEP in meeting urban-based commercial energy needs will grow. If the widely advocated small-scale HEP projects become widely adopted, HEP may also play a significant role in rural development. At present, such schemes are rarely economically viable, but the success of small-scale HEP in China provides a model some South East Asian countries may follow in the near future.

Geothermal energy systems exploit heat from 'hotspots' within the earth's crust, either by tapping naturally occurring hot water or by injecting water into hot rocks. A number of countries in the region contain such potential, but it is only in Indonesia and the Philippines that any action has been taken to tap this energy source. The Philippines is second only to the United States in the exploitation of geothermal energy. The first experimental 10 kW generation plant opened at Tiwi in south-western Luzon in the 1960s and there was an active development programme in the 1970s. By 1980, the Philippines had developed a geothermal generation capacity of 446 MW, some 18 per cent of the world's total (Earthscan 1981); Taylor (1983) predicted that this capacity would triple to 1500 MW by the mid-1980s. Indonesia is not confronted with the same lack of conventional energy resources, and in consequence the exploitation of her widespread geothermal potential has been far less active. Taylor (1983) suggests, however, that Indonesia may have as much as 1500 MW of installed capacity by the end of the century, a scale of development which would make her one of the world's leaders in geothermal development. The World Bank (1981) shows that over half of Indonesia's geothermal resources are located on Java, a factor which may make this energy form highly attractive as it would be able to supply power directly to the area of greatest need.

Sun and wind are, of course, energy sources of great antiquity, but recently the harnessing of sun and wind via more sophisticated technology (generally to generate electricity) has been widely discussed, and indeed most countries in the region have some form of research programme into the potential of these technologies. Although solar collectors and photovoltaic cells are used in a few places, this potential is largely latent, and is likely to remain so for the foreseeable future.

Conclusion

It consequently seems clear that, apart from HEP and in some places geothermal energy, new and renewable sources of energy are of little significance in South East Asia, however fervently they are advocated by their proponents. Discussion of energy resources in South East Asia centres upon the dual crises associated with biomass fuels and fossil fuels. These two sources of energy provide 99% of all energy produced and consumed in the region, a proportion which will only marginally diminish in the forseeable future. The basic issues are, first, how can the supplies of these fuels be sustained in the face of deteriorating environments and increasing costs? Second, what are the interactions between biomass and fossil fuels, the 'energy transition' discussed above. Third, what are the implications of these trends for the provision of the energy needs of the bulk of the population (energy for survival) and the energy needs associated with economic growth and change (energy for development)? It is clear from the discussion above that the extent and nature of these two energy crises varies significantly within South East Asia. Substantial areas are faced with serious biomass fuel deficits, whilst others have these resources in abundance. Similarly, some countries have extensive fossil fuel resources whilst others have a dearth of resources and have been seriously affected by moves in international energy prices.

Such national comparisons are of major importance, but the case of Indonesia illustrates graphically that a national-level analysis is not enough, for major energy problems exist within resource-rich as well as resource-poor countries. This is particularly true for the use of biomass fuels in rural areas, as the system of production and consumption of these fuels is essentially a local one, with many rural communities facing acute problems of providing for their energy needs despite the existence of extensive resources in other regions of the same country. This issue is not confined to biomass fuels, however, as poorly developed supply infrastructures frequently constrain access to commercial fuels as well.

Above all, the main lesson to be learnt from South East Asia is that energy problems cannot be understood by looking at energy issues and resources alone. In all cases, they are essentially a reflection of broader economic, social and environmental trends. Energy crises, like food crises, environmental crises, political crises and others, are inseparable from these broader forces. They are a facet of the general crisis of development which confronts the contemporary Third World, and are intimately entwined in the complex structures at a local, national and international level which combine to create this crisis of development. It is from this perspective, and this perspective alone, that one can understand questions of energy resources and energy problems, whether in South East Asia or elsewhere in the Third World.

Further reading

Atje R (1979) 'Konsumi energi di sektor rumah tangga desa' Mimeograph, Centre for Strategic and International Studies, Jakarta.

Atje R (1983) 'Konsumi dan ketersdiaan energi rumah tangga' in Soesastro H et al. (1983) *Energi dan Pemerataan* Centre for Strategic and International Studies, Jakarta.

Barnes C, Ensminger J and **O'Keefe P** (eds.) (1984) *Wood, Energy and Households: Perspectives on Rural Kenya*, Beijer Institute, Stockholm.

BP (1988) *BP Statistical Review of World Energy 1987* BP, London.

Directorate General for Power (1981) *Energy Planning for Development in Indonesia*, Ministry of Energy, Jakarta.

Earthscan (1981) *New and Renewable Energies*, London.

Eckholm E, Foley G, Barnard G and **Timberlake L** (1984) *Fuelwood: the Energy Crisis that won't go away*, Earthscan, London.

El-Hinnawi E, Biswas M and **Biswas A** (1983) *New and Renewable Sources of Energy*, Tycooly, Dublin.

Foley G, Barnard G and **Timberlake L** (1983) *Gasifiers: Fuel for Siege Economics*, Earthscan, London.

Foley G, Moss P and **Timberlake L** (1984) *Stoves and Trees*, Earthscan, London.

FAO (1981a) *Map of the Fuelwood Situation in the Developing Countries*, Rome.

FAO (1981b) *Forest Resources of Tropical Asia*, Rome.

Gordon L (1981) 'Energy and development: crisis and transition' in: Auer P (ed.) *Energy and the Developing Nations*, Pergamon Press, New York, pp. 1–12.

Islam M N et al. (1984) *A Study of Fuelwood, Charcoal and Densified Fuels in Thailand*, Asian Institute of Technology, Bangkok.

Kovarik B (1983) *Fuel Alcohol*, Earthscan, London.

Leach G (1984) *Household Energy Handbook for Developing Countries*, International Institute for Environment and Development, London.

Leach G, Jarass L, Hoffman L and **Obermair G** (1985) *Energy and Development: a Comparison of 13 Countries*, International Institute for Environment and Development, London.

National Energy Administration (1984a) *Thailand: National Energy Balance for 1982*, Ministry of Science, Technology and Energy, Bangkok.

National Energy Administration (1984b) *Thailand Energy Situation 1983*, Ministry of Science, Technology and Energy, Bangkok.

O'Keefe P, Raskin P and **Bernow S** (1984) *Energy and Development in Kenya*, The Beijer Institute, Stockholm.

Plumwood V and **Routley R** (1982) 'World rainforest destruction: the social factors', *The Ecologist* **12**, 4–22.

Simoes J (ed.) (1984) *SADCC: Energy and Development to the Year 2000*, Beijer Institute, Stockholm.

Soesastro H (1983) 'Strali energi pedesaan dan masalah pemerataan' in Soesastro H *et al.* (1983) *Energi dan Pemerataan*, Centre for Strategic and International Studies, Jakarta.

Soussan J (1988) *Primary Resources and Energy in the Third World*, Routledge, London.

Soussan J and **O'Keefe P** (1985) 'Biomass energy problems and Policies in Asia', *Environment and Planning A* **17**, 1293–301.

Soussan J, O'Keefe P and **Ferf A** (1985) *Fuelwood Strategies and Action Programmes in Asia*, Asian Institute of Technology, Bangkok.

Sumarna K and **Sudiono Y** (1973) *Konsumsi Kayu Bakar oleh Rumah Tangga, Industri dan Perusahaan Jawatan Kereta api di Jawa Timur*, Caporan No. 8, LPHH, BOGOR.

Taylor R (1983) *Alternative Energy Sources*, Adam Hilger, Bristol.

United Nations (1982a) *Statistical Yearbook for Asia and the Pacific*, Bangkok.

United Nations (1982b) *Economic and Social Survey of Asia and the Pacific 1981*, Bangkok.

United Nations (1983) *Economic and Social Survey of Asia and the Pacific 1982*, Bangkok.

United Nations (1984) *Economic and Social Survey of Asia and the Pacific 1983*, Bangkok.

United Nations (1987) *Economic and Social Survey of Asia and the Pacific 1986*, Bangkok.

Wiersum K (1976) *The Fullwood Situation in the upper Bengawan Solo River Basin*, Upper Solo Watershed Management and Upland Development Project, Solo.

White C (1983) 'Recent debates in Vietnamese development policy' in White G, Murray R and White C (eds.) *Revolutionary Socialist Development in the Third World*, Wheatsheaf Books, London.

World Bank (1980) *Energy in the Developing Countries*, Washington DC.

World Bank (1981) *Indonesia: Issues and Options in the Energy Sector*, Washington DC.

World Bank (1983a) *World Development Report*, Washington DC.

World Bank (1983b) *Indonesia: Selected Issues of Energy Pricing Policies*, Washington DC.

World Bank (1984) *World Development Report*, Washington DC.

World Bank (1988) *World Development Report*, Washington DC.

World Resources Institute (1988) *World Resources 1988–9*, Basic Books, New York.

CHAPTER 7

Agriculture and fisheries

Donald Fryer

South East Asia's agricultural record over the past quarter century stands in striking contrast to the generally lackluster agricultural performance of the Third World chronicled year by year with such deep concern by the FAO's annual *Report on World Agriculture*. In the ten years after 1972, for example, agricultural output in South East Asia rose nearly 60 per cent, an increase unmatched in any other major world region, more than double that of South Asia and more than triple that of East Asia (United States Department of Agriculture 1983, p. 5). These large gains were achieved in both subsistence and export oriented cash crop agriculture alike. South East Asia remains, as it has been throughout this century except the years during and immediately after World War II, in overall food surplus, a situation shared by no other major region of the Third World, and its rising exports of both food and industrial crops have been a powerful factor in a spritely overall economic performance.

Nevertheless, the record permits of no complacency. Wide variations in agricultural performance are inevitable over a region as large as South East Asia, but the greatest gains have been concentrated in a few countries with pragmatic economic policies, notably Thailand and Malaysia. Even in these, large sections of the national territory have achieved limited or only modest increases in productivity. For the region as a whole, the disparity between agricultural and non-agricultural incomes per head, which tends to be greater the lower the national GNP per capita, appears to be widening. Under the pressure of rapidly increasing populations, rural poverty has increased absolutely, and notwithstanding most government claims to the contrary, perhaps relatively to the total rural population also. Extreme rural poverty coincides with another problem, all the starker in view of the broad success achieved in the food producing sector. Famine and starvation have been rare in South East Asia; only in Kampuchea during and immediately following the murderous regime of Pol Pot, has real famine occurred in modern times, a supreme irony in a country that traditionally is in large food surplus. Nevertheless, many rural people regularly experience real hunger before the harvest, a period called *peceklik* in Indonesia.

In parts of Indonesia and notably in Central and East Java, in parts of Luzon and the Visayan Islands in the Philippines, and widely in Vietnam, Laos and Kampuchea, malnutrition reaches serious proportions throughout much of the year. In rural Java food supplies are kept constantly under lock and key in all but well-to-do households; meal portions are carefully rationed by the mother, a process in which infants and young children, and hence those most at risk, fare poorest. Food is also heavily spiced, a practice which in part is intended further to restrict individual consumption (Chapman 1983, pp. 63, 168). Approximately one-third of all pre-school children in Indonesia suffer from protein-energy malnutrition. Keropthalmia, a vitamin A deficiency disease which in extreme cases produces blindness, and goitre, an iodine deficiency disease, are also serious problems in Central and East Java (Mubyarto 1982, p. 121). Clearly, a large and sustained increase in food output and what is equally important, a substantial change in the composition of that output, are urgently necessary. But Indonesia is not expected to achieve self-sufficiency in food even by 1990 (Mubyarto 1982, p. 33) and it is difficult in view of the faltering economy and the growing evidence of political instability in the Philippines to entertain any sanguine hope of an early solution to the problem of malnutrition in that country.

The wretched condition of agriculture in Vietnam and its Indochinese satellites is in the large part the result of Communist economic policy and these countries are repeating the dreary agricultural experience of other Communist states. Elsewhere however, rural poverty, and malnutrition, while obviously influenced by many environmental factors, have a common and overriding cause: they arise from limited or nonexistent access to land for a substantial, and probably for a growing, section of the rural population. In the 1970s, more than a quarter of total households in Indonesia and more than 10 per cent of those in the Philippines, fell into the landless, or near landless category (Booth and Sundrum 1976, Asian Development Bank 1978). In Malaysia, whose government has consistently claimed both an absolute and relative decline in the number of poor households in recent years, squatting on government land by landless Malays has provoked several clashes with authority. In Thailand, where before World War II landlessness was virtually absent, the number of landless families has also increased rapidly. Moreover, throughout the region, a large proportion of land-owning households own insufficient land to provide an income much above the poverty level; almost half of farm holdings in Indonesia are less than 0.5 hectare in extent, and almost 60 per cent of those in Java (*Census Pertanian* 1973, quoted by Mubyarto 1982, p. 233). Closely associated with landlessness and poverty, is a growing burden of rural indebtedness.

National development plans recognize the seriousness of rural poverty, but investment expenditures to relieve it are mostly indirect. Seldom do they address the question of defective agrarian organization, which is its principal cause. Government's main response to landlessness is through official planned land settlement schemes, or through encouraging voluntary migration to less densely populated parts of the national territory. In relation to the magnitude of the problem, land settlement schemes are a marginal palliative at best, and an expensive one. In short, governments of South East Asia, outside the Communist countries, do not choose to act on the pressing issue of land reform. Most existing land reform legislation, and certainly that of the Philippines, is largely an exercise in window dressing. Governments believe that technological change, in the guise of the 'Green Revolution' will bring about an increase in productivity and living standards that will make grappling with the social and political difficulties of devising and implementing land reform programmes unnecessary. In a magisterial study of Asian poverty two decades ago, Myrdal emphatically rejected this strategy (Myrdal 1968, 1970). There is much of South East Asia's subsequent agricultural experience to confirm the validity of his analysis.

Agricultural systems

For a broad overview, the three-fold division of South East Asia's agricultural systems into (a) shifting, or swidden, agriculture, (b) *sawah*, or wet-rice agriculture, and (c) plantation, or plantation crop, agriculture (Pelzer 1948, Spencer 1952, Geertz 1965, Fryer 1979) is an adequate representation of reality. The three systems are sharply differentiated on the ground, and the contrasts among them are starker than those encountered in the various types of temperate-latitude agriculture. Nevertheless, they interact at many points, and under the pressure of technological and socio-economic change some of their more salient characteristics are becoming blurred; in particular, indigenous systems have over the past two decades shown a marked tendency towards a greater commercialization. Each, moreover, shows wide variations over the region according to differing environmental conditions and cultural attitudes, so that generalized descriptions can sometimes be misleading. The emphasis here is on factors making for change, and on official policy in agriculture for attaining national development goals.

Shifting cultivation

In shifting cultivation discrete plots of land are cleared and cultivated in a relatively short production cycle of two or three years (Spencer 1966). Thereafter, the land is abandoned to regenerating bush and secondary forest, which in time restore to the soil the nutrients removed by cultivation. After an interval which may vary from five to twenty or more years, the land may again be reoccupied in a new cultivation cycle. Termed swidden (a fire field) by anthropologists, it has a rich variety of vernacular names and is generally anathema to national governments, which strive to persuade or to coerce shifting cultivators to adopt sedentary forms of agriculture.

Clearings for plots are made in secondary forest with the onset of the dry, or drier, season, so as to give the felled material the best chance to dry for a good burn, which releases the nutrients pent up in the woody debris. Without further soil preparation, and using no more than a dibble stick, the cultivator sows a range of crops, which can include cereals, roots and tubers, legumes, leafy green vegetables, and herbaceous perennials. The wide range of crops, with varying growing seasons, ensures a continuity of food supply and is security against natural hazards. Though growing in apparent confusion, each particular crop is planted in the environment that the cultivator's long experience has proved best; Geertz (1965) argued that the swidden represents an attempt to reproduce the ecological balance of the original forest with its multiplicity of species. In continental South East Asia it is practised by most hill people who, however, may also have flooded rice fields in valley bottoms. In archipelagic South East Asia, much of which has no marked dry season, it is also a common practice among lowland dwellers, such as the Lampung people of southern Sumatra and the Dayaks of Borneo.

Contrary to the long-held beliefs of colonial administrators, shifting cultivators can attain surprisingly high yields per unit area (up to 1750 kg/ha for dry rice) and with comparatively little labour input. Thus they can enjoy much time for other activities such as hunting and religious observances. Hence shifting cultivators strongly resist attempts to convert them to other forms of land use; in their view any higher return is inadequate compensation for the extra labour involved (Freeman 1955, Boserup 1965, Seavoy 1973). Where population density allows a long enough period of regeneration between successive cycles of cultivation so that the original soil properties are restored, many scientists hold swidden cultivation as an acceptable and rational form of land use.

But with the swidden cultivators rapidly increasing contacts with other cultural groups, and its concomitant, rapid population increase, such desiderata are absent in many parts of the region. For governments, shifting cultivators and their practices constitute serious obstacles to national development and cohesion. The swidden causes deforestation and progressive environmental degradation. Moreover, the traditional rights of shifting cultivators to land may conflict with official plans for land settlement schemes, mining projects, water utilization schemes, and for commercial forestry; and to buy off or relocate such people can cause expensive delays or modifications. The four million or so hectares of forest land affected each year by swidden agriculture is certainly substantial. But such loss is not permanent where true swidden cultivators in ecological balance with their environment are concerned, and is far less than that arising each year from legal and illegal removal of fuel wood by other indigenes, from legal and illegal clearing for other forms of agriculture, and from the depredations of a commercial forestry that in most parts of the region is still unacceptably close to timber mining.

There are other and more compelling reasons for government dislike. Shifting cultivators are often sharply differentiated ethnically or linguistically from the dominant cultural group of the lowlands that in each state has a firm grip on political power. In continental South East Asia tribes of shifting cultivators often extend across national frontiers, for which they have small regard. Wherever tribes of shifting cultivators maintain connections with kinsmen across poorly demarcated borders, they can appear to government as a potentially separatist elements, and the danger is heightened when tribesmen's grievances over land are exploited by agitators. Autonomy for the Miao hill people who inhabit the Thailand-Laos-Burma border country, for example, is perceived by Bangkok as a Communist plot; Vietnam's autonomous areas for the hill peoples of Tonkin are as fraudulent as are the autonomous areas of the USSR. The 'Golden Triangle' of the Thai-Burma borderlands, moreover, produces some 600 – 1000 tonnes of raw opium a year, the principal source of cash income for many hill people. The opium traffic is controlled by private armies of dissident groups such as the Shans of Burma who use the profits of the trade to finance their rebellion against the Rangoon government, by rump elements of the defeated Chinese Nationalist (KMT) army, and by local opium war lords, who seldom find difficulty in reaching accommodation with national police and security forces. Government therefore, seeks to relocate shifting cultivators to areas of greater security, and to encourage them to adopt sedentary agriculture. Yet the problem of an alternative to opium cultivation is a difficult one. The opium poppy needs fresh swiddens and its harvesting is anomalous in demanding a far greater labour input than that of other swidden crops. Nevertheless, its high return continues to make cultivation economically attractive, and the enormous profits of the opium trade exert an insidiously corrupting influence on officials.

In South Sumatra, the indigenous Lampung people have created an important modification of the swidden system by planting rubber in the later stages of the cultivation cycle. As *Hevea* rubber competes successfully with regenerating secondary growth, or *belukar*, large plantings of what are effectively smallholdings have been created at little or no capital cost. Where ignorance of these plantings has led Indonesian officialdom into incorporating such mature rubber into designated land settlement, or transmigration schemes, indigenous reaction has been extreme.

Illegal forest clearing by peoples whose material culture is essentially based on sedentary agriculture is a more pressing problem than that caused by shifting cultivators. Growing landlessness and lack of employment opportunities, in large part the consequence of rapid population growth, have encouraged illegal squatting on any accessible uncultivated land. Fearful of detection and expulsion, squatters grow quickly maturing food crops such as banana and tapioca in an apparent replica of the practices of swidden cultivator, but without the latter's appreciation of ecological reality. This 'false swidden' agriculture quickly destroys or debases the original forest cover and greatly accelerates soil erosion. During the parliamentary interregnum of 1973–6 in Thailand's long series of military administrations, illegal clearing greatly accelerated, particularly in the eastern region, as the population correctly judged that an elected regime would take a more tolerant view of such activities. Palawan, Mindanao and the eastern Visayas in the Philippines, and in Indonesia, Sumatra, have also experienced much new illegal clearing. Smoke from fires set by shifting cultivators and squatters has at times constituted a serious hazard to navigation in the Malacca Strait, and caused the closing of Kuala Lumpur airport.

Sawah agriculture

For all major indigenous ethnic groups, the dominant form of agricultural land use is the cultivation of rice on flooded fields. Such rice cultivation may depend on rainfall alone, on supplementary water provided by irrigation systems, or on natural flooding as river discharges follow the precipitation regime. All of these may conveniently be described as *sawah* agriculture, from the Indonesian word for flooded rice fields. Rain-fed fields still account for about one-quarter of Indonesia's rice areas, and rainfed and naturally flooded fields for some 60 per cent of the rice area of both Thailand and the Philippines. Irrigation and improved water control permit higher productivity, and are essential for the cultivation of high yielding varieties that popular usage has dubbed the 'Green Revolution'.

The scope for extension of irrigation in most South East Asian countries appears considerable. Nevertheless, sites for major new irrigation projects are increasingly difficult to find, and the very heavy capital costs and long construction times of such ventures are daunting. Moreover, several South East Asian countries carry an immense burden of foreign debt. Small-scale pumping projects have the advantage of low initial cost but suffer from high maintenance costs.

Where irrigation water can be made available during the dry, or drier, season for which storage dams are usually necessary, the cultivation of second rice crop may also be possible. If the supply of off-season irrigation water is insufficient for rice, the most profitable crop, other food crops such as maize, cassava, sweet potatoes and pulses (crops collectively called *polowijo* in Indonesia) may be grown. For Java and Bali, the index of multiple cropping (i.e. (harvested area/physical area cropped) x 100) approaches 150, and for the country as a whole about ·120. In Malaysia, where anomalously for the region rice is not the most important crop, the double cropping index was some 156 at end of the Third Malaysian Plan period in 1980, a consequence of the intense government investment in major new irrigation schemes to improve the economic position of the Malay rice farmer.

More widespread adoption of the new technology of high yielding varieties is everywhere a major part of national economic development strategy, although Thailand and Burma, traditionally large rice-surplus generating countries, long doubted its wisdom. Modern high yielding varieties (HYVs) can mature in as short a time as three months, so that theoretically it is possible to produce three crops a year. Such a goal is still distant, but in parts of the Yogyakarta area and in certain districts in Central Java, five crops are taken in twenty-four months. Multiple cropping entails rapid preparation of the harvested fields for the new crop, a task that is increasingly performed either with hand tractors or, as in Malaysia, by conventional tractors operated by contractors. As a result, the draught animal population progressively declines. The once common water buffalo is now rarely seen in the important rice areas of north east and north west peninsular Malaysia.

The capital costs of growing traditional varieties, of which there are more than 3000 in South East Asia, and which are adapted to many microclimatic regimes, are very low. This is decidedly not the case for HYVs, which are still heavily dependent on increasingly expensive imported inputs in the form of oil-derived fertilizers and pesticides, (United States Department of Agriculture 1974). Thus the charge has been made that the enhanced productivity of the Green Revolution has exacerbated an already great socioeconomic inequality in the rural areas, and that the fruits of the new methods have accrued mainly to the larger and wealthier farmers who have been able to acquire the inputs. Most rice-growing areas show a high, sometimes an inordinate, incidence of tenancy as in the Central Plain of Luzon and in the Hong Ha (Red River) Delta of Tonkin before the agrarian changes of the Vietnamese revolution. Rents, typically a share of the harvest, have tended to rise with HYV cultivation, and most landowners demand another rent payment for a second crop. Dwyer (1964) recorded a case of a Filipino farmer who was worse off as a result of adopting double dropping. Moreover, in Java, the wider adoption of HYVs has led to a diminution in traditional employment opportunities in the village. Under the *bawon* system, all who so desired could assist in the harvest of the tall traditional varieties, each head of which is cut individually with the harvest knife, for a one-eleventh share of the harvester's output. The dwarf HYVs must be cut by sickle, and are threshed in the field. Such work is increasingly performed by hired

Table 7.1 South East Asia: rice area and production, annual averages for 1974/6 and 1984/6

	Area Harvested (thousands of hectares)	1974/6 Yield (kg/ha)	Production (thousands of tonnes)	Area Harvested (thousands of hectares)	1984/6 Yield (kg/ha)	Production (thousands of tonnes)
Burma	4 942	1 829	9 037	4 744	3 125	14 825
Indonesia	8 458	2 685	22 705	9 846	3 942	38 815
Kampuchea	1 002	1 309	1 312	1 717	1 172	1 933
Laos	682	1 306	891	616	2 161	1 427
Malaysia	741	2 378	2 029	640	2 799	1 795
Philippines	3 555	1 713	6 092	3 365	2 638	8 882
Thailand	7 952	1 834	14 585	9 811	2 015	19 756
Vietnam	5 134	2 184	11 213	5 682	2 791	15 859
South East Asia	32 466	2 092	67 864	36 466	2 833	103 292

Source: Based on FAO (1984, 1986a).

outside contractors. As much as 80 per cent of Java's rice output was still hand-pounded in 1970 but over the following decade the widespread adoption of small hulling machines of Japanese origin drastically reduced the demand for labor in traditional rice processing.

Every technological change throughout history has produced losers as well as gainers; most texts stress only the latter. The Green Revolution has proved a mixed blessing, and may contain indeed potential for disaster. The cultivation of a multiplicity of traditional rice varieties ensured some protection against widespread loss through diseases and pests. A concentration of effort on but a few and genetically close varieties over a large part of South East Asia's rice area – more than half of the Philippines rice area was under HYVs as long ago as 1971 and more than half that of Malaysia by 1976 – greatly enhances the risk of major crop losses through new disease or pest infestation. There is a danger that traditional varieties, immune to such epidemics, might disappear totally. The International Rice Research Institute (IRRI) at Los Banos in the Philippines is thus building up a seed bank of every known rice variety within the region to guard against such a contingency, and to aid further research.

Adoption of HYVs in the great deltas of continental South East Asia, traditionally generators of large exportable rice surpluses, proceeded slowly but by the early 1980s the wider use of HYVs finally enabled Burma after decades of fruitless effort to surpass its pre-World War II rice area, if not its level of exports. Burma's rice exports are in reality far larger than officially stated as allowance must be made for an unknown but probably very large volume of smuggled rice. Simultaneously, Thailand had raised its rice exports to more than 4 million tons in some years, in part through wider use of HYVs and also through expansion of the cultivated area; and despite a shortage of inputs, Vietnam had by the mid 1980s turned the corner in a grim battle to raise rice output to a level that would permit some relaxation of the harsh rationing that has endured since the North's victory of 1975.

Most countries now grow HYVs developed in their own national research stations rather than those from the International Rice Research Institute, where the first HYVs were produced. But HYVs have still not lost their reputation for unpalatability and for their own consumption many farmers prefer to grow the traditional tall varieties. The adoption of HYVs has thus transformed masses of formerly subsistence farmers, selling merely what was surplus to their own requirements into commercial farmers who sell the greater part of their output. Typically, where multiple cropping has established itself, virtually all of the dry, or off-season, crop is sold. This greater commercialization has heightened rural inequality, and has hastened migration to the cities.

Such changes indicate perhaps an early demise of 'involution', Geertz's term for the process of continuous labour-intensive

Table 7.2 Rice imports and exports

	1974/6	1984/6	1986
	(annual averages)		
Exports			
(thousands of tonnes)			
Burma	365	564	500[1]
Thailand	1 289	4 400	4 131
Total	1 654	4 964	4 631
Imports			
(thousands of tonnes)			
Indonesia	1 042	158	28
Kampuchea	230[2]	85	30[1]
Malaysia	229	354	199
Philippines	119	24	2
Vietnam	127	402	500[3]
Total	1 747	1 023	759

Notes:
1. FAO estimate.
2. 1974 only; unofficial figures place imports for 1976 as high as 800 000 tonnes.
3. Unofficial figure.
Source: FAO (1974, 1977, 1986a).

refinements, but without basic change, of the *sawah* system in Java.
But the model of the *sawah* system in Java developed by Peltzer
(1948), De Terra (1954), Geertz (1965) and others, with its tripartite
division of intensively cultivated house-lots (*pekacaosao*) of fruits and
vegetables, dry fields (*tecalan*) for the cultivation of rice substitutes
and supplements (*polowijo*), and flooded rice fields (*sawah*), does not
accord with reality in most other parts of South East Asia, and even
in Java is most appropriate to the Javanese heartland of Central and
East Asia.

The long-sought goal of national rice sufficiency has been slow in
coming in many countries. Time and again Indonesia approached
the target, only to see it recede, and with a larger deficit than
before. In the early 1980s, rice imports of some 2 million tonnes
annually, double those of the mid-1960s, made Indonesia by far the
world's largest rice importer. Then with dramatic suddenness the
prolonged and largely aid-financed investment in the new
technologies combined with favourable weather to give Indonesia
the largest national rice yields in the region, and the deficit was
transformed into embarrassing oversupply. Meantime, the enhanced
productivity of other rice producers had created a world glut, and
the surplus situation in other grains generated much acrimony

among exporters. The result was a substantial fall in rice prices, and even though per capita consumption increased, thus arresting a decline of more than three quarters of a century in the proportion of rice in overall diet, after years of exhorting farmers to greater efforts the surplus was an unpleasant problem facing government. Shortages of *polowijo*, the rice substitutes and supplements that still constituted the diet of the poor, meanwhile indicated that government had placed too high a priority on raising rice production.

The rice surplus also affected relations between Jakarta and the *Tanah Seberang* (Outer Islands), always sensitive to changes in the relative prices of rice, largely a product of Java, and those of their own export staples such as rubber, coffee and palm oil. The desirability of reducing the large rice deficit of the Outer Islands has been an important factor in the expansion of planned land settlement schemes for mainly Javanese immigrants, which Indonesia terms transmigration (Hardjono 1977, Jones 1979); a substantial national surplus was another powerful argument against this expensive programme, reinforcing the charges of many scientists that poor site selection and unsuitable clearing and cultivation practices were producing major, and perhaps irreversible environmental damage (*The Ecologist*, 1986).

The transmigration schemes result in the replacement of indigenous agricultural activities and landscapes by those of Java, as it is usually many years before irrigation facilities can be provided for the new settlements, and having to grow crops on dry fields is for most migrants an agricultural retrogression. Once viewed largely as a means of relieving, if only in a small way, growing landlessness and population pressure in Java (Pelzer 1948), transmigration is now conceived more in terms of its overall development impact, and in improving the productivity of marginal lands. Long focussed on southern Sumatra, transmigration schemes are increasingly located in other parts of the Outer Islands such as Sulawesi and Kalimantan. Migrants are now drawn from Bali, as well as from Java. Not least, transmigration also serves the military regime's national strategy based on *perang wilayah* (territorial war), born of the Army's experiences of the guerrilla campaigns of the revolutionary struggle of 1945–9. In the densely populated *sawah* landscapes of Java, Indonesian forces moved freely and undetected, like 'fish through water' to use the revolutionary terminology of Mao Zedong. In contrast, in the more sparsely populated and very different environments of the Outer Islands, the performance of nationalist forces was much poorer.

Plantation crop agriculture

This term is a convenient collective for the cultivation of crops grown largely for an export market, whether by large units (plantations, or estates), or by indigenous farmers and other smallholders. It embraces both food crops and agricultural raw materials, some in a double role, and is largely, though not exclusively, concerned with tree crops (Courtenay 1968). Although some lines of production arose initially from European capital and enterprise, and involved exotics from the American or African tropics such as natural rubber and the oil-palm, others such as pepper and spices are native to South East Asia and have largely remained a small grower's prerogative. Moreover, smallholders have often been able to compete successfully with large units in all but the most capital-intensive lines of production, i.e. those requiring elaborate and expensive processing, and the smallholder's response to the price-stimulus of an expanding world market has important lessons for development strategy.

The contribution of foreign-owned estate companies to the development of the natural rubber and other export-oriented cash-crop industries has received much comment. In contrast, the almost equally great contribution of the 'Nanyang' ('South Seas'), i.e. Chinese, immigrants to South East Asia, is less well known. Some large Malaysian estates are Chinese owned, but more typically Chinese are owners and operators of medium-sized holdings (those of approximately 5–50 ha), or are smallholders, and are also dealers and traders, processors and exporters. Government procurement and exporting agencies notwithstanding, the output of the indigenous smallholder is more likely than not to reach the world market through a chain of Chinese dealers and exporters. Desire to avoid government marketing systems may generate a substantial illegal trade, such as that from Indonesia across the Malacca Strait. But Indonesian local officials are apt to turn a blind eye, as also does Singapore, a principal beneficiary. South East Asia's intricate coastline and multiplicity of islands present an environment ideal for the activities of the smuggler and the barter trader, and suppression is virtually impossible. Government policy that prevents growers from realizing the full prices that their produce commands on the world market, and the widespread use of transistor radios, makes most growers well aware of Singapore quotations, and merely ensures that illegal trading will continue.

Perhaps the most salient facts of the geography of plantation crop agriculture are the relatively limited area it occupies – less than one-sixth of the area under rice – and the marked concentration of this area in equatorial (i.e. continuously humid) South East Asia, and particularly in the lands bordering the Malacca Strait. It is the dominant form of agriculture in Malaysia, and a major reason for that country's high level of well-being (Table 7.3). In wet-and-dry South East Asia, its only really important representatives are the sugar and to a lesser extent, the coconut, industries of Luzon and the Visayas in the Philippines. Elsewhere, save for portions of the Burma-Thailand peninsula, of southern Vietnam (the Cochinchina of French colonial times) and adjacent eastern Kampuchea – all areas of continuously humid soils suitable for the cultivation of equatorial crops – it is almost entirely absent.

Table 7.3 Planted area and production of selected plantation crops 1984

	Area (thousand ha)	Production (thousand tonnes)
Indonesia		
Rubber estates	493	357
Rubber smallholdings	2 164	687
Oil palm estates	431	1 212[1]
Oil palm smallholdings	38	
Sugar cane estates	67	1 500
Sugar smallholdings	317	1 390
Coffee estates	49	23
Coffee smallholdings	830	302
Coconuts	3 000	1 790
Malaysia		
Rubber estates	412	548
Rubber smallholdings[2]	1 506	982
Oil palm estates	720	
Oil palm smallholdings[3]	580	3 715[4]
Philippines		
Coconuts	3 217	2 500[5]
Sugar cane	423	2 460[6]

Notes:
1. Fruit.
2. Includes land settlement schemes.
3. Palm oil.
4. Estates and small holdings
5. Copra equivalent.
6. Centrifugal sugar.

Sources: Government of Indonesia 1985; Government of Malaysia 1984; Government of Philippines 1985.

The reasons for this distribution pattern are complex, but the concentration on crops with continuous or protracted harvests, above all on rubber and oil palm which both provide a year-round income and a uniform deployment of labour throughout the year, and thus avoid the expensive discontinuities inevitable in the production of annuals or herbaceous crops, is striking. Malaysia, having by the mid-1950s re-established and consolidated its position as the world's leading rubber producer that its colonial forebear lost to the Netherlands Indies in the interwar period, had also made itself the world's largest producer of palm oil by the 1970s, and planned to achieve the same status with the third great equatorial crop, cacao, by the later 1980s. This goal appeared well within its reach. But while an equatorial regime offers optimal conditions for the cultivation of continuously bearing tree-crops (cacao does not bear as continuously as either rubber or palm oil, but it does have two principal harvest seasons in the year), it is obvious that only a very small part of the area suitable for cultivation is in fact so planted.

Access to land is fundamental, and the present distribution still reflects the heritage of the economic policies of the various colonial administrations. Much of the estate land under natural rubber and other crops was made available by government on extremely favourable terms, which newly independent states would inevitably seek to modify sooner or later. But whenever indigenes had access to land they vigorously took up the planting of the exotic new crop, natural rubber, and responded so promptly to rising world demand for traditional crops such as pepper and other spices, coffee and coconuts, that estate enterprise never found these attractive. Moreover, as noted earlier, certain swidden cultivators were able to respond to the expansion of the world market by incorporating new cash crops within their traditional systems. *Sawah* cultivators, in areas of complex irrigation systems, were clearly not. But in Java they were also constrained by the Dutch designation of lands apparently unused, but in which the population had traditional rights, as *Woestegronden* (wastelands) and state property (Hansen 1981). Thus though Java has a small but significant estate rubber sector, a smallholder industry was never possible. It was not possible either in the Philippines, where the US administration resisted the establishment of an estate industry in Mindanao. In Thailand, where a timorous but still independent government also set its face against foreign agricultural enterprise, a smallholder industry developed in the southern peninsula only through the response of Thai-Muslims (Malays) and Chinese immigrants, originally attracted by tin mining, to the growth of the rubber industry in adjacent Malaya.

The smallholder strikingly demonstrates the essential simplicity of both rubber cultivation and processing. 'Rubber' incomes have always been larger than 'rice' incomes, in boom and slump alike. A hectare of double-cropped riceland under HYVs still generates a lower return than a hectare under modern high-yielding clonal rubber, and a hectare of oil palm yields more than either. The smallholder possesses a greater social and economic independence than that of the estate worker, but even where his holding is under modern material, his productivity per unit area is still lower than that of a well-managed estate, and most smallholder's real incomes are significantly lower than those of estate workers. Governments have found that these facts can conflict with national development plans and aspirations, and have attempted, with mixed success, to effect a reconciliation.

The control of large areas of land by foreign corporations is now politically unacceptable, and the estate sector has thus been progressively brought under national control, or under that of indigenous nationals (*peribumi* in Indonesia, *bumiputera* in Malaysia). Additional inducements to such (nationalization) policies were that technological change both in the field and the factory has swung the balance of advantage heavily towards the large unit in the natural rubber industry, and that for the production of high quality palm oil large units are a technical necessity. Indonesia acquired a substantial government estate sector from the Netherlands Indies on the 1949 transfer of sovereignty, to which it added through the take-overs of Dutch and other foreign-owned properties in the regional revolts of 1957–8, and in the confrontation with Malaysia in the mid-1960s. Some properties were later returned to the former owners by the Suharto administration, but with substantial reductions in rights of land use, both in area and over time. Estates also lost much of their former housing, and when handed back most processing plants were generally in poor or very poor condition. As the performance of government's own estates has always been lacklustre, disincentives to reinvestment in private estate enterprise reduced the entire estate sector to a parlous state. With major World Bank assistance, Indonesia has planned to regain former levels of efficiency, but clearly this will be a lengthy process.

The Malaysian government's approach to 'nationalization', on the other hand, has been through the thoroughly capitalist techniques of the purchase of estate company shares on the stock market. In theory, the government entities involved in this programme are intended to promote the acquisition of a stake in the estate sector (and in other capital intensive sectors of the economy) by the ordinary Malay. But the rapid acquisition of control of several agency houses on behalf of *bumiputera* interests has prompted the Malayan Chinese Association (MCA), a member of the Alliance Government, to create an entity for fulfilling a like function for the Chinese community. In 1984 the Malayan Indian Congress (MIC), another member of the ruling Alliance, established a fund with similar intent for the Indian population. Thus with great rapidity a very large part of the European estate sector had passed to the control of Malaysians of one or other ethnic group during the 1980s, and the early completion of this 'Malayanization' process is assured. Additionally, the federal government through its own land development agencies, above all Felda (Federal Land Development Authority) (Shamsul Bahrin and Perera 1977), and the various regional authorities (DARA, Kejara, etc.) created for the development of large areas in eastern Peninsular Malaysia, are heavily engaged in estate operations, as also are the State Development Corporations of the various states, either on their own account or in joint ventures with private capital (Voon 1981). A large part of the huge increase in the planted area under oil palms since 1970, has been created through such official enterprise.

To improve incomes of rubber smallholders and to cope with growing landlessness, the Malaysian government has established elaborate replanting and new planting schemes. Risda (Rubber Industry Smallholders Development Authority) makes grants to cover a large part of the costs of replanting smallholdings with modern material, and those with holdings of uneconomic size can obtain additional land through Felcra (Federal Land Consolidation and Rehabilitation Authority). Malaysia in fact has a plethora of public land-development agencies and some consolidation would appear desirable.

Felda however, is in a class by itself; it is by far the largest beneficiary of all government development expenditure, and has received high praise from multilateral development aid organizations such as the World Bank. Felda's operation are among the most successful, and the most expensive, land-development schemes in the world. Simply, the authority acquires, clears and plants the land, either to rubber or oil palm, in a phased series of operations, and provides housing and other infrastructural facilities for the settlers. These are thus relieved of the arduous labour of land clearing and preparation, and merely have to plant their house lots and maintain their holdings until the latter come into bearing, that is, five years for rubber and three for oil palm, during which time settlers receive a cash subsistence allowance. The size of holdings, 6 hectares for rubber and 4 for oil palm, is fixed to provide the settler with a predetermined level of income. In theory, after repaying over time the capital costs of establishing his holding, the settler finally receives title. Felda's Gedangsa scheme, on the Selangor-Perak border, some 110 km north of Kuala Lumpur, for example, was begun in 1960. It has a good macadamized road connecting with Tanjong Malim, and has little of the rawness and lack of facilities that make life hard in newly established and inaccessible schemes still in the 'pioneer' stage. Its 3349 ha (8350 acres) were planted to rubber in four phases, and in 1970 an oil-palm area was added. Gedangsa in the mid-1970s supported some 520 settler families, mostly drawn from nearby parts of Selangor, and had a total population of nearly 3200. The settlers' median income was nearly M$220 per month, a little below the national average for all rubber schemes at the time, and substantially below that for oil-palm schemes; nevertheless, some enterprising settlers were making nearly three times this figure.

The Malaysian government insists that Felda schemes are aggregates of smallholdings in the private sector, but legal niceties apart, most are essentially 'quasi-estates', with central processing facilities, and inevitably therefore, central accounting. This is all the more so with oil-palm schemes, where technological considerations make it necessary for settlers to consent to work as required by management on lots other than their own. True, Felda smallholders do not consider themselves as estate workers. But the capital costs of Felda schemes per unit area are substantially larger than those for new estates, and their productivity per hectare is lower. Nor can Felda, or any other government agency for that matter, cope with more than a very small portion of the problem of landlessness arising from population increase, despite the fact that in more recent years the Malaysian government has dramatically increased the scale of land development schemes, notably with the commencement of the massive Pahang Tenggara (971 000 ha) and Johore Tenggara (283 000 ha) integrated projects (Fig. 7.1). The selection of settlers has

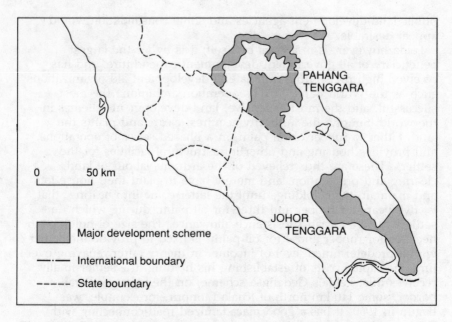

Fig. 7.1 The Pahang Tenggara and Johore Tenggara integrated development areas. (Ooi 1976)

become highly politicized. Non-Malays now stand scant chance of being accepted; although this was not so in the early days of the authority, and states providing the land for Felda schemes insist on a large quota of settlers from their own citizens. As so often with land reform, Felda schemes, too, replace one kind of inequality with another.

Perhaps the most striking object lesson for the Third World in Malaysia's success with export crop agriculture is the supreme importance of a flourishing estate sector. Notwithstanding government's own research efforts, without the private sector the development of vastly more productive new material, more efficient processing and the development of new markets, would have been impossible. Both the personnel and the planting material for government's own replanting and new planting programmes came originally from the private estate sector. In trying to replicate Malaysia's success with natural rubber and oil palm, and for which Malaysia itself is providing assistance, Indonesia and Thailand face major problems of personnel and material; Indonesia because it actively promoted the demise of estate industry, Thailand because it could never countenance any such development.

Although Mindanao has witnessed a great expansion of a largely smallholder coconut growing industry since World War II, it was not until the 1970s that the Philippines launched a serious effort to encourage the cultivation of the equatorial tree crops for which the island appears well suited, and to supplement the long established, but very restricted, area under rubber with large plantings of oil palm. A major scheme in the Agusan valley with the assistance both of Britain's Commonwealth Development Corporation and the now largely 'Malayanized' agency house of Guthrie on land made available by presidential decree, ran into conflict with existing occupants. The use of force to expel the original cultivators caused the aid principals to reconsider, and provided further encouragement to the Communist New People's Army in its operations against the regime in eastern Mindanao. As western Mindanao and Sulu are racked by another guerilla struggle – as the Moro National Liberation Organization strives for a greater autonomy for the Muslims (Moros) of the south – Mindanao appears far from propitious for further large-scale capital intensive agricultural enterprise. The two large American pineapple canners, Dole and Del Monte, are phasing out their Hawaiian operations not by expansion in Mindanao but rather in South East Thailand and in Central America.

The 1980s, moreover, brought serious trouble for the Philippine coconut and sugar industries. Prices for these traditionally first and second Philippine exports by value, are as volatile as for any commodity, but the depressed levels of the 1980s reflected more serious and long-term difficulties than the market collapses of the past (Sacerdoti 1982, Sacerdoti and Galang 1985). Both industries are largely aggregations of smallholders. The once largely US-owned sugar industry is now entirely in Filipino hands. Smallholder tenants of estate land produce cane under contract for processing at the *central* (mill), but over the 1970s some of the larger estates resumed interest in direct estate cultivation, as was the practice in the former US-owned industry. Northwest Negros, the principal sugar area, is notorious for its malnutrition problem, a consequence of the seasonal labour peaks punctuated by lengthy periods of under-employment inherent in cane cultivation, and the population's lack of access to land on which food production is permissible, or possible. Sugar and food-crop production are far better integrated in well-fed Thailand, which after a remarkable increase over the 1970s in the area under sugar, had surpassed the Philippines in sugar output by 1983.

Both major Filipino export crop industries have a dismally low productivity and face increasingly severe competition in world markets: sugar from the continued high level of subsidized beet exports from the EEC and from corn syrup and synthetic sweeteners such as *Aspaciams*; coconuts from more price competitive vegetable oils such as palm oil, sunflower seed, cotton seed and soybean oil. Over 70 per cent of all Philippine coconut farms are less than 2.5 ha in extent and some 90 per cent are less than 10 ha. With a productivity of scarcely more than 400 kg of crops per hectare per year, as compared with a national average of nearly 1000 kg in 1985, most small growers are at or below the margin of subsistence

(Sacerdoti and Galang 1985). Productivity is lowest in the oldest established coconut areas of central South East Luzon where a large proportion of all trees are elderly; it is highest in the new plantings in Mindanao, particularly in Zamboanga whose large estates achieve yields of over 1600 kg per hectare per year. Nevertheless, even this is only half that attained by Malaysian or Sri Lankan estates, and the Philippines has an ambitious replanting programme based on high-yielding material from Malaysia.

With the imposition of martial law in 1972, President Marcos rewarded close friends and political allies with control of large sections of the national economy in what became known as 'crony capitalism' (Kerkvliet 1974). Sugar and coconuts, the dominant activities in commercial agriculture, each became a virtual monopoly of a particularly close associate, who created and manipulated marketing and processing organizations, domestic and foreign allocations, and prices, largely for private gain. In 1985 the World Bank and the International Monetary Fund (IMF) insisted on the dismantling of these structures and a return to free market operations in both industries as a condition for further aid and the renegotiation of the country's enormous foreign debt. Following the accession of the Aquino government in early 1986, the controversial and cumbersome marketing machinery of the two industries was swept away, and the country is in the process of establishing a new economic policy for national rehabilitation, or in view of its still formidable Communist opponents, for national survival.

Fishing and fisheries

The importance of fishing in South East Asia is far greater than its meagre imputed contribution of some 2–5 per cent to respective GNPs would suggest. There are probably more than three million professional fishermen in the region, about one-third of whom are Indonesians, but the number of part-time fishermen is far larger. Every rice field, at least before extended use of the pesticides associated with the Green Revolution began to affect wildlife, also produced a crop of fish. Fish is extremely important in diet, and accounts for some 40–80 per cent of animal protein intake in South East Asian nations, though the poor development of storage and marketing facilities outside major urban centres necessitates that much of this consumption is in the form of dried, salted or preserved fish and fish sauces (Marr 1981, pp. 76, 80). The Philippines and Singapore, whose per capita fish consumption of more than 30 kg per year is more than twice the regional average, are unique in that fish is mostly consumed fresh.

Until the mid-1960s, it remained broadly true that the freshwater fisheries of the region were overexploited and marine resources were underexploited (Abdullah 1983). Since that time, marine output has greatly expanded through the introduction of new fishing

Table 7.4 Fishing catch 1984 (thousands of tonnes)

	Inland waters	All areas
Burma	144	613
Kampuchea	59[1]	65[1]
Laos	20	20
Malaysia	4	665
Indonesia	538	2 217
Philippines	600	1 935
Thailand	150	2 250
Vietnam	100	1 650
Total	1 615	10 001

Note: 1. FAO estimate
Source: FAO (1986b).

technologies, and from the early 1970s the regional catch increased dramatically at a time when the level of world output stagnated (Table 7.4). Indonesia, Thailand and the Philippines have thus emerged as major world fishing nations. The scope for further growth is contentious, 'though historically speaking, estimates of fishery potential tend to be conservative' (Marr 1981, p. 84). Nevertheless, evidence of serious overfishing is already clear in some parts of the region, notably the Gulf of Thailand and the Malacca Strait.

The new technologies have embraced both trawling and purse-seining, capital intensive methods which need larger boats than those of traditional fishermen, and powerful inboard engines. Thus they have only been available to operators with access to capital, which in practice has largely meant the Nanyang Chinese, or those of indigenous and Chinese blood, as in Thailand. Through state fishing enterprises, fishing cooperatives, and joint ventures with foreign, notably Japanese, enterprise, governments and especially that of Indonesia, have endeavoured to secure a place for indigenes in capital-intensive fisheries. Occasionally, combining imported modern, with traditional techniques has proved very productive, as with the Philippine *payaw*, or bamboo raft tuna fishing, in which bamboo rafts are used as lures and shelters to attract the live bait for tuna, which are then caught by large purse-seiners (Aprieto 1980, p. 62). In Thailand too, where the rapid development of trawling was largely the fruit of a West German aid programme, only gear capable of being used in indigenous craft was introduced (Marr 1981, pp. 82–3).

The most valuable component of the demersal trawl catch is shrimp. In eastern Indonesia state enterprises and joint ventures with Japanese capital fish the Banda and Timor seas with large modern vessels to catch shrimp for export to Japan from processing plants at Ambon and Sorong. Trawling, however, produces a large haul of 'trash-fish', species which are inedible or for which there is no convenient market. In Thailand, what were formerly 'trash-fish' now support a substantial fish-meal industry, but elsewhere in the region such catches are largely discarded. Indonesia has endeavoured with limited success to compel Japanese operators in eastern Indonesia to utilize a larger proportion of their total catch.

A more pressing problem resulting from the introduction of trawling and purse-seining, whose most valuable product is tuna, is a growing conflict with traditional fishermen. These find their livelihood menaced by the new technologies, and claim with good reason that these are accelerating the depletion of fish stocks. The problem, moreover, has been greatly compounded through the application of the UN Convention on the Law of the Sea, which grants a suite of marine jurisdictions to littoral states, including an exclusive economic zone (EEZ) some 200 miles wide, and under certain conditions, even beyond (Lee 1982). Under the Convention Indonesia and the Philippines are recognized as archipelagic states, and *de facto*, if not *de jure*, all the waters separating their many islands have become national property. In 1957, however, Indonesia declared sovereignty over all the waters of its archipelago, claiming them as national territorial waters, and began either to deny access to former international waters to foreign fishermen, or in eastern Indonesian waters, to compel Japanese operators to buy fishing licences.

In the important fishery of the Malacca Strait, Malaysian fishermen work their traditional fishing grounds, now within Indonesia's EEZ, at their peril. The failure of the two governments to reach agreement over traditional fishing rights, leads to harassment and extortion, or the imprisonment of Malaysian fishermen and confiscation of their boats and gear. Corrupt Indonesian officials levy what the traffic will bear as fishermen attempt to make their own accommodation to the change. The problem of poaching by foreign fishermen whether by Thais, who also make their own arrangements with corrupt Vietnamese officials to fish Vietnamese waters, by Japanese, Taiwanese or by South Koreans, has reached serious proportions. Poaching shades insensibly into smuggling, the 'barter-trade' and piracy, other activities rife in the Malacca Strait. Disputes between fishermen, whether 'modern' or traditional, or of differing nationalities, have occasionally erupted in violence and have caused diplomatic embarrassment.

Governments have thus felt compelled to take action to regulate new fishing techniques, partly to protect the well-being of their traditional fishing nationals, and partly to conserve stocks. Most prohibit trawling within three, sometimes six miles or more, from the nearest shore, and determine what type of gear (length of purse seines, mesh of drift nets, or type of trawl) and length of craft may fish, and where, within their respective jurisdictional waters. In 1970 Indonesia established a system of regulated fishing zones for different classes of vessel and types of gear off the coasts of the most important and populous part of the country, including the Malacca Strait, the Java Sea and Macassar Strait, and divided Indonesian waters into four trawl regions. Trawlers could operate only within the region in which they were registered. Early in 1980, however, a presidential decree banned all trawling operations throughout Indonesia in order to minimize social conflict (Comitini and Hardjolukito, 1983, pp. 25–8).

In Malaysia, where trawling is very largely a Chinese activity, similar action would be interpreted as yet another government concession to *bumiputera* (Malay) interests, and would provoke a storm of protest. It would conflict moreover, with government policy designed to enlarge and accelerate *bumiputera* participation in all 'modern' and capital-intensive economic activities. Harmonization of policies for the Malacca Strait region is thus very difficult. At the same time, governments have also severely restricted the activities of traditional fishermen in certain areas in order to promote other uses of the marine environment, notably the offshore production of, or exploration for, oil and gas, activities which grew very rapidly in South East Asian waters from the mid-1960s. Indonesian fishermen are thus excluded from large areas of the Malacca Strait and Java Sea. Singapore, which has adopted a policy of restricting each section of its coast, has closed its entire south coast to fishing in the interests of shipping using the important but narrow and congested Singapore Strait.

Fishing is thus only one use of the marine environment, and fishing policy has to be harmonized with other uses of the sea, especially shipping, mineral production, as a means of waste disposal, or for defence purposes. The new Convention on the Law of the Sea greatly complicates this task, and enlarges the possibility of conflict.

Such conflicting usages, and the apparent stagnation in world marine fishery production, have suggested that an important role in future alimentation and in enlarging peasant incomes in many Third World countries might come from freshwater fisheries, and from aquaculture, which can embrace both saltwater and freshwater species. Little is known of the extent of freshwater fisheries in South East Asia, but such few statistics that exist may well substantially understate the magnitude of its contribution to total fishery production (Marr 1981, p. 85). The largest freshwater fishery in the region is the Tonle Sap, or Great Lake, of Kampuchea, which produces at least 50 000 tonnes of fish a year, mainly carp species. The lake owes it great productivity to the seasonal inundation of its basin caused by the rising floodwaters of the Mekong and Bassac, which cannot all be discharged into the South China Sea by their lower channels. Before the communist victory of 1975, this fishery was largely in the hands of specialist Chinese firms operating under concession from the government. Elsewhere, however, the use of riverine flood plain areas for freshwater fish production is likely to be reduced through new irrigation works and pollution from pesticides and urban and industrial wastes.

At certain selected spots in the region, the Laguna de Bay in Luzon, the western coast of peninsular Malaysia, the northwest coast of Singapore Island, and at the head of the Gulf of Thailand, acquaculture is an important activity. Marine species such as sea bass, grouper, prawns and mussels are reared in excavated ponds. Such ponds, inundated at high tide, are also used to trap shrimp and other species, as over an extensive area along the coast of the west Java plain. Inland, excavated ponds are used for the cultivation of carp, milkfish or *tilapia*, a business that is typically the preserve of Chinese. The techniques, largely imported from southern China are frequently capital intensive. Fish culture in fact, produces a quality product priced substantially above that of most marine species, and normally beyond the reach of most indigenes. Research and development of aquaculture in the region has so far not succeeded in reducing high production costs, and the prospects of a greatly enlarged contribution to total fisheries output from this source appear slender.

Further reading

Abdullah A H (1983) 'The Marine Fish Resources of Southeast Asia', in Ooi Jin Bee (ed.), *Natural Resources in Tropical Countries*, Singapore University Press, Singapore, pp. 199–288.

Aprieto V L (1980) 'Philippine tuna fisheries: Resource and industry', *Fisheries Research Journal of the Philippines* **5**, 53–67.

Asian Development Bank (1978) *Rural Asia*, Praeger, New York.

Booth A and **Sundrum R M** (1976) 'The 1973 Agricultural Census', *Bulletine of Indonesian Economic Studies* **12**, 90–105.

Boserup E (1965) *The Conditions of Agricultural Growth*, Allen and Unwin, London.

Chapman B A (1983) *A Medical Geography of Endemic Goitre in Central Java*, University of Hawaii, Honolulu, Ph.D. dissertation.

Comitini S and **Hardjolukito S** (1983) *Indonesian Marine Fisheries*, Research Report 13, East-West Environment and Policy Institute, Honolulu.

Courtenay P P (1968) *Plantation Agriculture*, Bell, London.

Dwyer D J (1964) 'Irrigation and land problems in the central plain of Luzon', *Geography* **49**, 236–46.

FAO (1984) *Production Yearbook*, **38**, Rome.

FAO (1986a) *Production Yearbook*, **40**, Rome.

FAO (1986b) *Yearbook of Fishery Statistics*, **58**, New York.

FAO (1974) *Production Yearbook*, **28**, Rome.

FAO (1977) *Production Yearbook*, **31**, Rome.

Freeman J D (1955) *Iban Agriculture*, HMSO, London.

Fryer, D W (1979) *Emerging Southeast Asia*, Philip, London.

Geertz C (1965) *Agricultural Involution*, University of California Press, Berkeley.

Government of Indonesia (1985) *Statistik Indonesia 1985*, Djakarta.

Government of Malaysia (1984) *Yearbook of Statistics 1984*, Kuala Lumpur.

Government of Philippines (1985) *Philippine Statistical Yearbook 1985*, Manila.

Hansen G E (ed.) (1981) *Agriculture and Rural Development in Indonesia*, Westview Press, Boulder, Colorado.

Hardjono J M (1977) *Transmigration in Indonesia*, Oxford University Press, Kuala Lumpur.

Jones G (1979) 'The Transmigration Programme and Development Planning' in Pryor R J (ed.), *Migration and Development in South-East Asia*, Oxford University Press, Kuala Lumpur, pp. 212–22.

Kerkvliet B (1974) 'Land reform in the Philippines since the Marcos coup', *Pacific Affairs* **47**, 286–304.

Lee Y L (1982) *Southeast Asia: Essays in Political Geography*, Singapore University Press, Singapore.

Marr J (1981) 'Southeast Asian marine fishery resources and fisheries', in Chia Lin Sien and MacAndrews C (eds.), *Southeast Asian Sea: Frontiers for Development*, McGraw-Hill International, Singapore, pp. 75–109.

Mubyarto (1982) *Growth and Equity in Indonesian Agricultural Development*, Yayasan Agro-Ekonomika, Jakarta.

Myrdal G (1968) *Asian Drama: An Inquiry into the Poverty of Nations* (3 vols.), Twentieth Century Fund and Random House, New York.

Myrdal G (1970) *The Challenge of World Poverty*, Random House, New York.

Nations R *et al.* (1982) 'Sugar's sweet and sour', *Far Eastern Economic Review* **118**, Nov. 5, 62–76.

Ooi J B (1976) *Peninsular Malaysia*, Longman, London.

Pelzer K (1948) *Pioneer Settlement in the Asiatic Tropics*, American Geographical Society, New York.

Sacerdoti G (1982) 'Cracks in the coconut shell', *Far Eastern Economic Review* **118**, Jan. 8, 42–8.

Sacerdoti G and **Galang S** (1985) 'The seeds of change', *Far Eastern Economic Review* **120**, Oct. 31, 103–10.

Seavoy R E (1973) 'The transition to continuous rice cultivation in Kalimantan', *Annals of the Association of American Geographers* **63**, 218–25.

Shamsul Bahrin and **Perera D P A** (1977) *FELDA: 21 Years of Development*, Ministry of Land and Regional Development, Kuala Lumpur.

Spencer J E (1952) *Land and People in the Philippines*, University of California Press, Berkeley.

Spencer J E (1966) *Shifting Cultivation in Southeast Asia*, University of California Press, Berkeley.

De Terra G J A (1954) 'Mixed garden horticulture in Java', *Malayan Journal of Tropical Geography* **4**, 33–43.

The Ecologist (1986) Indonesia's Transmigration Programme: A Special Report, **16**, 2/3, 58–118.

United States Department of Agriculture (1974) *Development and Spread of High Yielding Varieties of Wheat and Rice in the Less Developed Nations*, Foreign Agricultural Economic Report 95, Washington DC.

United States Department of Agriculture (1983) *Southeast Asia: Review of 1982 and Outlook for 1983*, Supplement 10 to WK5-31, Washington DC.

Voon P K (1981) 'The rural development programme in Sabah', *Malaysian Journal of Tropical Geography* **3**, 51–67.

CHAPTER 8

Mining and manufacturing

Peter Dicken

Industrialization on any significant scale is a relatively recent phenomenon in South East Asia. Historically, the countries of the region have performed the classic role of peripheral economies of the world economic system (Wallerstein 1979). They were, overwhelmingly, producers of primary commodities – both agricultural and mineral-based – for the metropolitan markets of the core economies of the world system and also markets for some of the manufacturered goods produced in the factories of Europe and North America. They played a distinctive role in the kind of international division of labour which had been evolving since the 15th and 16th centuries but which developed so dramatically in the 19th century as the industrialization of the Atlantic Basin countries spread its global reach through trade and investment. In South East Asia, all except Thailand were integrated into this larger system of economic activity directly through their colonial ties. But, as the experience of this single exception demonstrates, integration into an international division of labour was not necessarily conditional upon colonial status. Economically, Thailand became a part of the larger world system even though, politically, it was independent.

To describe the South East Asian economies as primary producers/exporters for most of the past few hundred years is not to deny the presence of any kind of manufacturing activity. Manufacturing indeed existed but it was, for the most part, small-scale and fragmented in organization and predominantly local in orientation (Chi 1986). As Singh (1986, p. 287) points out, the production of manufactured goods

'was limited to a very narrow range of consumer goods, handicrafts and ancillary industries which sprang up with, and were dependent upon, the primary industries of agriculture and mining. Cottage industries were devoted to the production of hand-made consumption goods in everyday use. They also satisfied meagre needs of the rural population but, all in all, manufacturing activity was not an important sector in the economy' (present author's emphasis). This situation prevailed virtually unchanged until roughly thirty or so years ago and, indeed, still prevails in countries such as Burma, Kampuchea and Laos. Such larger-scale industry as did exist before World War Two was mostly that of the processing of agricultural and mineral products by European firms.

Today, although primary production and exports still dominate the regional economy, the position has changed dramatically in a number of South East Asian economies. Though admittedly from a low base level, manufacturing growth has been spectacular. Indeed, some of the South East Asian economies experienced manufacturing growth rates in the 1960s and 1970s which were among the fastest in the world. Consequently, their role is now no longer simply that of primary exporters and manufactured importers; they have become part of a new international division of labour in which manufacturing activities are no longer confined predominantly to the core economies. More than this, they are beginning to play an important role in what many observers see as a fundamental shift in the global geography of economic activity – a shift in the centre of gravity – away from the Atlantic Basin towards the Pacific Basin (Benjamin and Kudrle 1984, Conkling and McConnell 1985, Hofheinz and Calder 1982). Many see these countries of the 'East Asia Edge' as being crucial elements in the 'Pacific Century' of the year 2000 and beyond.

Such industrialization as has occurred so far is, however, regionally uneven. But for the market economies in particular – the ASEAN countries of Indonesia, Malaysia, the Philippines, Singapore and Thailand – deliberate policies of industrialization have been a crucial, and often dominant, national goal in recent decades. The geographical unevenness of industrial development within South East Asia arises from a number of factors. Most of all, it arises from the particular way in which external, global forces of demand, new technology and the reorganization of production intersect with the specific internal characteristics of individual economies – their resource endowment (both physical and human) and the direction and effectiveness of national industrial policies.

In this chapter, the focus is on industrial developments in South East Asia during the last thirty years or so; that is, from around 1960. Most of the emphasis is on developments in the manufacturing sector because it is here that the greatest changes have occurred. It is in their emergence as increasingly important producers and exporters of manufactured goods that the South East Asian economies of today differ so markedly from their earlier history. It is upon the development of the manufacturing sector that many of the governments of the region have focussed their economic policies. It is in the manufacturing sector, too, that the recent and rapid growth of these, and the East Asian, economies have impinged most directly on the economies and labour forces of the older industrialized countries of Europe and North America.

Industrial growth and change in South East Asia

Most of the major industrial developments in South East Asia have occurred since the early 1960s. Indeed, this is also true of East Asia. In 1960, Japan was still relatively undeveloped industrially compared with the USA and Western Europe and only on the threshold of its subsequent spectacular industrial take off. Such comments apply equally to three of today's leading Newly Industrializing Countries (also known as NICs) in East Asia: Hong Kong, South Korea and Taiwan. For a detailed description of the South East Asian economies at that time the reader is referred to the report by the Asian Development Bank (Asian Development Bank 1971, Myint 1972). In many ways, that report serves as a useful benchmark from which to examine subsequent developments. By 1960, according to Myint, South East Asia was already characterized by an internal division between 'two groups of countries with contrasting rates of export expansion and economic development' (Myint 1972, p. 26). The first group of countries with high rates of growth included Singapore, Malaysia, the Philippines and Thailand. The second group of low growth countries included Indonesia, Vietnam, Cambodia (Kampuchea), Laos and Burma.

A closer examination of the South East Asian economies in the early 1960s shows just how far their production and trade was dominated by the primary sector and by non-manufacturing industries. Table 8.1 shows that industry (including manufacturing) typically accounted for less than one-fifth of each country's GDP. The manufacturing sector alone was substantially less important than this. Except for Singapore, where manufactured goods represented 26 per cent of merchandise exports, manufactures were an insignificant component of South East Asian exports in 1960. The structure of the labour force reflected this primary orientation. Even when 'industry' is taken to include mining, manufacturing, construction, electricity and gas, its relative significance as a source of employment was extremely small with the sole – not surprising – exception of Singapore.

By the mid-1980s, however, the picture was very different in a number of countries in the region. Even so, there had been, apparently little overall change in the economic structures of Burma and Laos and probably also in Kampuchea and Vietnam. Such apparent stability, of course, masks cataclysmic political and military conflicts in the latter three countries. In none of them has industrialization – particularly manufacturing industry – made significant progress. They remain low-income, primary producers with a predominantly agricultural orientation. But in the other five main countries of the region (five of the six members of ASEAN) the extent of change between the early 1960s and the 1980s was substantial in all cases and spectacular in some, as comparison of Table 8.2 with Table 8.1 reveals. Industry's share of the GDP of each of the ASEAN countries grew by between 4 and 20 percentage

Table 8.1 South East Asia: the industrial position in 1960

Country	Structure of GDP (%)				Structure of merchandise exports (%)			Structure of employment (%)		
	Agriculture	Industry[1]	Manufacturing	Services	Fuels, minerals, metals	Other primary commodities	Manufactured goods	Agriculture	Industry[1]	Services
Burma	33	12	8	55	4	95	1	68	11	21
Kampuchea	–	–	–	–	0	100	0	–	–	–
Indonesia	54	14	8	32	33	67	0	75	8	17
Laos	–	–	–	–	–	–	–	83	4	13
Malaysia	37	18	9	45	20	74	6	63	12	25
Philippines	26	28	20	46	10	86	4	61	15	24
Singapore	4	18	12	78	1	73	26	8	23	69
Thailand	40	19	13	41	7	91	2	84	4	12
Vietnam	–	–	–	–	–	–	–	81	5	14
Brunei	–	–	–	–	–	–	–	–	–	–

Notes:
1 Includes mining, manufacturing, construction, electricity, gas.
–Data unavailable.
Source: Based on World Bank (1980, Tables 3, 9, 19).

Table 8.2 South East Asia: the industrial position in 1986

Country	Structure of GDP (%)				Structure of merchandise exports (%)			Structure of employment (%)		
	Agriculture	Industry[1]	Manufacturing	Services	Fuels, minerals, metals	Other primary commodities	Manufactured goods	Agriculture	Industry[1]	Services
Burma	48	13	10	39	3	84	13	53	19	28
Kampuchea	–	–	–	–	●	99	●	–	–	–
Indonesia	26	32	14	42	58	21	21	57	13	30
Laos	–	–	–	–	–	–	–	76	7	17
Malaysia	21	35	19	44	26	38	36	42	19	39
Philippines	26	32	25	42	14	26	60	52	16	33
Singapore	1	38	27	62	21	12	67	2	38	60
Thailand	17	30	21	53	4	54	42	71	10	19
Vietnam	–	–	–	–	–	–	–	68	12	20
Brunei	–	–	–	–	100	0	0	–	–	–

Notes:
1. Includes mining, manufacturing, construction, electricity, gas.
– Data unavailable.
● Less than 1 per cent.
Source: Based on World Bank (1988, Tables 3, 12, 31).

points, although the corresponding employment increase was a good deal lower. Even so, by the 1980s, a much higher proportion of the labour force was employed in industry than in 1960. By far the largest relative increase occurred in Singapore; by 1986, 38 per cent of its labour force was employed in industry compared with 23 per cent in 1960. Industrial employment in Malaysia and the Philippines was 19 and 16 per cent, respectively, followed, some way behind, by Indonesia and Thailand.

The most significant developments, however, occurred in manufacturing industry. In 1960, as we have seen, manufacturing industry was of relatively little importance in most South East Asian countries. By the early 1980s, this situation had been transformed to a greater or lesser degree. In each one of the ASEAN economies, manufacturing had become a much more significant element in the economy. Most important of all was the vastly increased importance of manufactured exports in each country's trade position. In Singapore in 1960, manufactured exports represented 26 per cent of total merchandise exports; by 1986 they accounted for almost 70 per cent. There were even more spectacular relative gains in the Philippines, Thailand and Malaysia. In the Philippines, manufactured exports increased from 4 per cent of the total to 60 per cent; in Thailand from 2 to 42 per cent; in Malaysia from 6 to 36 per cent.

Table 8.3 shows the distribution of value added in manufacturing in South East Asia together with growth rates of individual countries for two time periods, 1965–1980 and 1980–6. To put these figures in a broader, and possibly more meaningful, perspective comparable figures are also given for one of the leading East Asian NICs (South Korea) and also for Japan and some older industrialized countries. The first point to make is that the manufacturing sector in South East Asia is still very small compared with that of the major industrial countries as well as in comparison with South Korea. In absolute terms, Indonesia has the largest manufacturing sector in South East Asia, followed by the Philippines, Thailand, Malaysia and Singapore. But in terms of value added per capita of the population it is clear that Singapore is the dominant manufacturing country. Indeed, in per capita terms Singapore's manufacturing economy is comparable with that of the United Kingdom. But the most striking feature of Table 8.3 is the comparative growth rates of manufacturing production between the South East Asian economies on the one hand and the older industrialized countries of Europe and North America on the other. Not only did the South East Asian economies experience much faster manufacturing growth during the 1960s (though from a low base) they also continued to do so at least up to the mid-1980s, during the period when growth in Western Europe plummeted to extremely low levels. Since then, a number of the fast-growing South East Asian economies have experienced considerable problems, notably Singapore which, in 1985, experienced a decline in GDP of −1.8 per cent compared with

Table 8.3 South East Asia: industrial performance 1965–86

Country	Value added in manufacturing 1985		Growth of manufacturing production (average annual percentage change)	
	US$ million	**Per capita**	**1965–80**	**1980–6**
Burma	680	17.9	3.9	5.8
Kampuchea	–	–	–	–
Indonesia	11 447	68.8	12.0	7.7
Laos	–	–	–	–
Malaysia	3 287[1]	220.6[1]	–	6.0
Philippines	8 048	140.5	7.5	–1.7
Singapore	4 311	1 658.1	13.3	2.2
Thailand	7 696	146.3	10.9	5.2
Vietnam	–	–	–	–
Brunei	–	–	–	–
Comparative performance of selected countries:				
South Korea	24 466	589.5	16.5	10.2
Japan	395 148	3 252.2	9.4	7.8
United States	803 391	3 325.3	2.7	4.0
West Germany	201 640	3 311.0	3.3	0.8
United Kingdom	101 470	1 789.6	1.1	1.2

Notes:
– Data unavailable.
1. 1983.
Source: Based on World Bank (1988, Tables 2, 8).

growth in 1984 of 8.2 per cent (Dicken 1986a). However, by 1987 the Singapore economy had bounced back with an annual growth rate of 8.7 per cent. In 1988, 11 per cent growth was achieved.

Overall, therefore, each of the ASEAN countries has experienced varying, but very substantial, manufacturing growth rates since 1960. Figure 8.1 shows the variation between the individual countries in terms of their concentration on particular broad types of industry. For all except Singapore, industries based on food and agriculture remain especially important. For Thailand, Burma and the Philippines, textiles and clothing are far more important than machinery and transport equipment, which is the dominant manufacturing sector in Singapore. The chemicals industry makes a larger contribution to value added in manufacturing in Thailand and Indonesia than elsewhere. However, each of the countries now produces a considerable diversity of 'other manufacturing' products. Most of these, as well as textiles and clothing and the increasingly important electrical and electronic products, are light, labour-intensive consumer goods. In employment terms it has been the electrical and electronics industries, together with textiles and

199

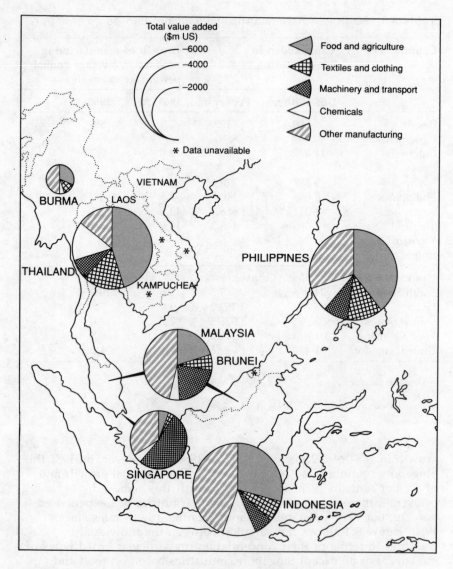

Fig. 8.1 Level and composition of manufacturing value added in South East Asia 1982. (Source: Data from World Bank (1985)

clothing manufacture, which have shown the most significant growth in South East Asia (Lee 1986).

It is important to re-emphasize, however, that the production and processing of primary products, based on both renewable and non-renewable natural resources, continue to be, in value terms, the dominant industrial sectors in all South East Asian countries, apart from Singapore. Even in Singapore, a large-scale primary processing industry has been established on the basis of imported materials, notably crude oil. In the agriculture and land-resource based sectors, the dominant regional activities are the milling and processing of commodities such as rice, sugar, tobacco, vegetable oils, timber and rubber. In the mining of non-renewable resources, the most important commodities produced in South East Asia are: oil and natural gas, tin, nickel, copper, chromite, tungsten, antimony, lead, bauxite, manganese and some iron ore. Figure 8.2 summarizes the major geographical features of the production of these commodities in South East Asia in 1982.

By far the most important mined mineral in the region is tin. Three countries produce almost 50 per cent of the world's tin: Malaysia (23 per cent), Indonesia (15 per cent) and Thailand (11 per cent). In no other single mineral does the region's share of total world production exceed 10 per cent. Nevertheless, individual countries are of world rank in specific minerals. For example, 9 per cent of the world's nickel is produced by Indonesia (5.4 per cent) and the Philippines (3.5 per cent). The Philippines is also responsible for the production of 5 per cent of the world's copper and 4 per cent of the world's chromite. Thailand (2.1 per cent) and Burma (1.3 per cent) together account for a significant proportion of world production of tungsten, while Thailand produces just over 2 per cent of the world's antimony.

The region's major producer of crude oil and natural gas is Indonesia, with about 2.5 per cent of world crude oil output and 1.3 per cent of natural gas production. Crude oil is also a most important element in the Malaysian economy whilst in Brunei, the newest member of ASEAN, oil and gas are of overwhelming economic importance. So far, the refining of oil and gas into petrochemical products is less well developed in the region, although some important refining complexes have been built. By far the most important refining centre is Singapore. Although it has no indigenous oil or gas resources, it has built itself into the major refining location in South East Asia. However, its position is being challenged, particularly by Indonesia which has been developing large petrochemical complexes to refine its own indigenous oil and gas.

Thus, by the 1980s the industrial economy of South East Asia had changed substantially from the situation in 1960. Although industries based upon renewable and non-renewable natural resources remained extremely important, the most dramatic change has been the growth of manufacturing industry. In terms of manufacturing output per head of population a clear manufacturing hierarchy has emerged within South East Asia:

1. Singapore,
2. Malaysia,
3. Philippines, Thailand,
4. Indonesia,
5. Burma, Kampuchea, Laos, Vietnam.

The forces underlying industrialization in South East Asia

Much has been written in recent years about the rapid industrialization of parts of the global periphery, primarily within the conceptual framework of a new international division of labour (Fröbel *et al.* 1980). But, as is so often the case, a very complex and dynamic situation often has been oversimplified and overgeneralized (Jenkins 1984, Corbridge 1986). There has been a particular tendency to focus primarily on the role of external forces in the industrialization of developing countries and to neglect, or at least to underplay, the role of specific internal forces – political, cultural, social, as well as economic – in the industrialization of individual countries. But 'it is impossible to ignore the part played by specific national conditions and the role of the state in the emerging pattern of differentiation' (Jenkins 1984, p. 47). Industrialization in any particular country is the specific outcome of the combination of both

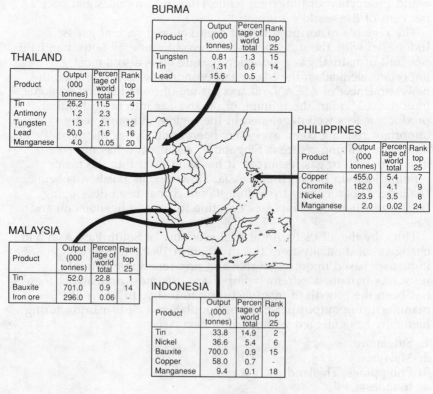

BURMA

Product	Output (000 tonnes)	Percentage of world total	Rank top 25
Tungsten	0.81	1.3	15
Tin	1.31	0.6	14
Lead	15.6	0.5	-

THAILAND

Product	Output (000 tonnes)	Percentage of world total	Rank top 25
Tin	26.2	11.5	4
Antimony	1.2	2.3	-
Tungsten	1.3	2.1	12
Lead	50.0	1.6	16
Manganese	4.0	0.05	20

PHILIPPINES

Product	Output (000 tonnes)	Percentage of world total	Rank top 25
Copper	455.0	5.4	7
Chromite	182.0	4.1	9
Nickel	23.9	3.5	8
Manganese	2.0	0.02	24

MALAYSIA

Product	Output (000 tonnes)	Percentage of world total	Rank top 25
Tin	52.0	22.8	1
Bauxite	701.0	0.9	14
Iron ore	296.0	0.06	-

INDONESIA

Product	Output (000 tonnes)	Percentage of world total	Rank top 25
Tin	33.8	14.9	2
Nickel	36.6	5.4	6
Bauxite	700.0	0.9	15
Copper	58.0	0.7	-
Manganese	9.4	0.1	18

(a)

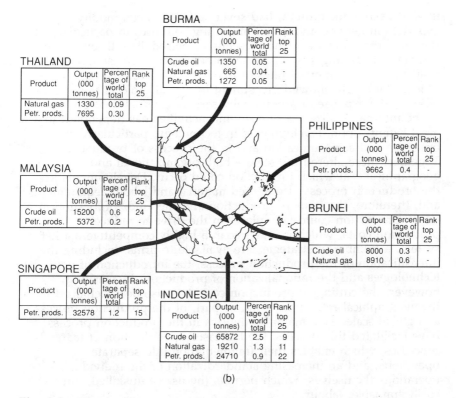

BURMA

Product	Output (000 tonnes)	Percentage of world total	Rank top 25
Crude oil	1350	0.05	-
Natural gas	665	0.04	-
Petr. prods.	1272	0.05	-

THAILAND

Product	Output (000 tonnes)	Percentage of world total	Rank top 25
Natural gas	1330	0.09	-
Petr. prods.	7695	0.30	-

PHILIPPINES

Product	Output (000 tonnes)	Percentage of world total	Rank top 25
Petr. prods.	9662	0.4	-

MALAYSIA

Product	Output (000 tonnes)	Percentage of world total	Rank top 25
Crude oil	15200	0.6	24
Petr. prods.	5372	0.2	-

BRUNEI

Product	Output (000 tonnes)	Percentage of world total	Rank top 25
Crude oil	8000	0.3	-
Natural gas	8910	0.6	-

SINGAPORE

Product	Output (000 tonnes)	Percentage of world total	Rank top 25
Petr. prods.	32578	1.2	15

INDONESIA

Product	Output (000 tonnes)	Percentage of world total	Rank top 25
Crude oil	65872	2.5	9
Natural gas	19210	1.3	11
Petr. prods.	24710	0.9	22

(b)

Fig. 8.2 Distribution of (a) mining, and (b) oil and petroleum production in South East Asia, 1982. (Source: data from World Resources Institute (1986), *The Economist* (1984)

external and internal forces. In this chapter, particular attention is devoted to the role of transnational corporations (TNCs) and national governments in the differential industrialization of South East Asia.

External forces in the industrialization of South East Asia

Without doubt, a major factor in the post-war industrialization of South East Asia was the unprecedented growth of world manufacturing production and trade which occurred between the 1950s and the early 1970s. Such growth was associated with an increased internationalization of economic activities (Dicken 1986b). The very high rates of economic growth created vastly increased demands for the basic raw materials of industrial production, demands which were especially significant for the resource-rich nations of South East Asia. Conversely, the oil crises of 1973, 1979 and 1983 though not solely responsible for the dramatic slowdown

203

in world economic growth, had severe effects on commodity exporters in general and on oil importing countries in particular. At the same time, however, the recycling of petrodollars through the commercial banking system led to an enormous increase in credit available – at commercial rates of interest – to developing countries. This undoubtedly allowed investment in capital projects to continue, albeit at the expense of a potentially serious external debt problem.

The internationalization of economic activities has depended, to a great extent, upon developments in technology, particularly in transport and communication and in the process of production. Together, these altered the scale – both organizational and geographical – at which production could occur. Developments in the production process also altered material and labour requirements and, therefore, the costs of production and the kinds of location in which production could take place. In the drive to reduce production costs, which is endemic in a highly competitive market system, a number of strategies are available to firms, including the intensification of the production process, the introduction of new technologies and the rationalization of production. Increasingly, however, the cutting of production costs has also been achieved by the geographical relocation of all, or part, of the production process at a global scale. Two major tendencies in the production process have facilitated this process: the increased specialization of many processes, which enables their fragmentation into separate operations, and an increasing standardization of the individual operations themselves, which permits the use of unskilled, but easily trainable, labour.

Technologies, in themselves, do not cause the internationalization of production; rather, they make possible an enlarged and more complex spatial division of labour. But the decisions to employ such technologies rest with the major business organizations which have the necessary resources; increasingly, these are the transnational corporations. Of course, business enterprises with world-wide horizons are by no means new. Indeed, their influence has long been extremely important in South East Asia. The earliest ancestor of today's TNC to make its mark in the region was probably the British East India Company whose representative, Sir Stamford Raffles, established a trading base in 1819 on the Malacca Straits which eventually became Singapore. Subsequently, TNCs became especially significant in the resource-based extractive industries of the region. During, and after, the colonial era, the plantation agriculture and mining activities of most countries in the region were dominated by foreign-controlled companies. Such dominance is still apparent in some cases, although a good many of the transnational resource corporations' activities have passed into domestic control since national independence. As Manners (1986) has pointed out, national governments in Third World countries have rightly regarded the ownership and exploitation of their indigenous natural resources with some sensitivity. Thoburn's (1981) study of TNCs in the tin mining industry illustrates the situation very clearly.

However, the large-scale geographical spread of TNCs in the manufacturing sector is relatively recent compared with their long-established involvement in resource-based ventures. The real take-off occurred between the 1950s and the 1960s. There is no doubt that TNCs have been an exceptionally important catalyst to the industrialization of South East Asia and particularly to the growth of manufacturing industry in the region. Within manufacturing industry at a global scale, TNCs tend to be most heavily engaged in three broad types of manufacturing industry: the technologically advanced sectors, such as pharmaceuticals and electronics; large-volume medium-technology consumer goods industries, such as motor vehicles and televisions; mass-production consumer goods industries supplying branded products, such as processed foods, drinks and cigarettes. Within these industries, TNC operations – other than their corporate headquarters and research and development activities – can be divided into two broad categories: plants which produce for the local geographical market (host-market plants) and those which are part of a rationalized product or process strategy (Dicken 1986b, p. 202).

South East Asia has become a major destination for the activities of TNCs. Indeed, Asia – primarily East and South East Asia – experienced the fastest rate of growth of foreign direct investment (FDI) in developing countries between 1967 and 1978, its share almost doubling during this period from 14 per cent to 24 per cent. But the importance of TNCs varies considerably between individual South East Asian countries (Hill and Johns 1985, Lee 1981, Lee 1984). Table 8.4 shows the changing distribution of FDI within South

Table 8.4 Distribution of the stock of foreign direct investment in South East Asia 1967–78

	1967		1978	
	Value (US$m)	Percentage	Value (US$m)	Percentage
Brunei	85	3.5	300	2.3
Burma	10	0.4	65	0.5
Kampuchea	84	3.4	–	–
Indonesia	254	10.3	5 760	44.4
Laos	83	3.4	–	–
Malaysia	679	27.6	2 880	22.2
Philippines	723	29.3	1 820	14.0
Singapore	183	7.4	1 700	13.1
Thailand	211	8.6	445	3.4
Vietnam	152	6.2	–	–

Note: – Data not available
Source: UNCTC (1983, Table 22).

Table 8.5 The relative importance of foreign direct investment to ASEAN countries 1981/2

	FDI as a percentage of gross domestic investment	Relative significance (Singapore = 100)
Singapore	9.1	100.0
Indonesia	8.0	87.9
Malaysia	3.1	34.1
Thailand	2.1	23.1
Philippines	1.0	11.0

Source: Hill and Johns (1985, Table 2).

East Asia between 1967 and 1978, the latest date for which comprehensive figures are available. Both Indonesia and Singapore increased their share of the regional total – Indonesia markedly so – while the other major host nations, Malaysia, the Philippines and Thailand had a reduced share. Table 8.5 shows the relative importance of FDI to gross domestic investment in the ASEAN countries in 1981/2 expressed as a percentage of its importance in Singapore, the country in which the foreign-controlled sector is most dominant. In 1981, in fact, foreign firms accounted for 59 per cent of Singapore's total manufacturing employment, 76 per cent of its manufacturing output and 87 per cent of the country's exports.

Despite some recent exceptions, notably in Singapore, the operations of most TNCs in South East Asia are, overwhelmingly, branch plants which use relatively low-skill, often female, labour. This is especially the case in the electronics, textiles and clothing and other light consumer goods sectors. These are the highly export-oriented activities which have a strong assembly type character. But not all TNC manufacturing operations in South East Asia are of this kind. A substantial proportion of the foreign investment is in plants which serve the domestic market, many of which were established under the import-substitution policies which were adopted by most governments in the region (see below).

The geographical origins of TNCs operating in South East Asia show some diversity although there is a predominance of United States and Japanese firms (Table 8.6). As the Table shows, Japanese investment has grown especially rapidly, a trend reinforced by the massive revaluation of the yen after 1985. An important emerging trend, which is only faintly reflected in Table 8.6, is the increasing significance of TNCs originating from other newly industrializing countries within the region and from East Asia, notably Singapore, Hong Kong, South Korea and Taiwan (Lall 1983). For example, a number of Hong Kong textile and clothing firms have established manufacturing plants in other Asian countries to get around the

Table 8.6 Sources of foreign direct investment in ASEAN countries, 1972–8 (percentage of country total)

	Indonesia 1972	Indonesia 1978	Malaysia 1972	Malaysia 1978	Philippines 1972	Philippines 1978	Singapore 1972	Singapore 1978	Thailand 1972	Thailand 1978
Developed market economies	60.8	53.6	53.8	63.3	67.6	81.8	77.7	79.8	62.8	65.0
United States	15.6	6.0	15.3	18.6	57.7	34.4	33.0	32.9	13.8	13.9
Western Europe	13.7	8.9	23.4	17.8	6.3	16.8	38.7	31.6	11.5	14.2
United Kingdom	1.5	0.9	21.4	12.1	2.1	8.5	18.3	13.7	5.3	5.4
Australia	3.8	2.7	2.0	2.6	–	4.1	–	–	–	–
Japan	22.9	35.8	11.5	23.4	1.3	24.5	6.0	15.3	37.5	36.9
Hong Kong	9.0	11.4	9.1	10.4	–	1.4	5.7	5.6	1.4	5.1
Malaysia	1.2	0.5	–	–	–	–	3.5	3.5	3.6	2.6
Philippines	8.4	4.1	–	0.1	–	–	–	–	–	–
South Korea	2.3	1.0	–	–	–	3.0	–	–	–	–
Singapore	2.8	1.4	28.6	22.1	–	0.2	–	–	–	1.3

Source: UNCTC (1983, Table 24).

quota constrains of the Multi Fibre arrangement and, more recently, because of uncertainty over the political future after 1997. In the case of Singapore, the government has actually begun to encourage local manufacturers who cannot pay rising local wages to relocate overseas (Lim and Fong 1986). Thus, with the increased importance of Japanese investment and the growth of investment from other Asian countries, TNCs within South East Asia have become more Asian in character. Nevertheless, US and, to a much lesser extent, European, investment remains very significant.

Internal forces in the industrialization of South East Asia: The role of national governments

Despite their undoubted importance, external forces alone do not fully explain the recent spectacular industrialization of South East Asia and, especially, its uneven distribution within the region. Too often, explanation has been confined to the level of TNCs seeking to increase or sustain their profits by utilizing low-cost labour. Certainly this is an important factor, though by no means in all types of industrial activity. But such narrow explanations 'deny any independent dynamic within the Third World. . .(and lead to). . a general underestimation of the role of the state in Third World industrialization' (Jenkins 1984, p. 34). In fact, the state has played an exceptionally important role in industrialization throughout South East Asia (Lee 1981) though, of course, its precise role has varied according to individual political, cultural and social circumstances. Here, our attention is focussed upon the five major ASEAN countries – Indonesia, Malaysia, the Philippines, Singapore and Thailand – which have been most active in promoting industrialization within the framework of a world market economy. The other states in the region – Burma, Kampuchea, Laos and Vietnam – have all adopted centrally planned and relatively inward-looking economic policies in which industrialization plays a relatively minor role.

> 'Given their common pre-war economic experience, all the South East Asian countries started the postwar period with two commonly shared national goals: to promote industrialization both as a symbol of modernization and as a means of reducing their dependence on the export of a few primary products, and to transfer a growing share of their national incomes and economic activities to the indigenous section of their population.' (Myint, 1972, p. 28)

It became conventional wisdom in both the theory and the practice of economic development 'that industrialization, and in particular manufacturing industry, had a key role to play in the process of long-run growth' (Weiss 1984, p. 39).

Within this general policy thrust towards increased industrialization there has been substantial variation in national practice and in the degree of direct government involvement in the structure and operation of national economies. Of the ASEAN countries, Indonesia and Malaysia have probably the greatest degree of direct government involvement. In both cases, important segments of the economy are controlled by the government. In Indonesia the government controls the major components of the modern sector, including most large-scale manufacturing, such as cement, fertilizer, steel and paper. Policies which discriminate in favour of the *Pribumi* – indigenous Indonesian – sector are practised, most notably in the award of government contracts. Indigenization policies are pursued even more strongly in Malaysia where, under the New Economic Policy of 1970, the aim is for 30 per cent of total corporate equity to be in the hands of the *Bumiputra* (indigenous Malays) by 1990, and in Malaysia too the government owns important units of manufacturing.

Direct government economic involvement in the Philippines was relatively limited until the early 1970s; however, in 1980 it began a major restructuring programme for manufacturing industry. In Thailand, the military-backed government has adopted a policy of giving the private sector the major economic role but, as Warn points out (Wawn 1982), in practice the Thai government's involvement is substantial in several manufacturing sectors, including paper manufacture, sugar milling, as well as in banking. The South East Asian government with apparently the least involvement in the economy when measured in terms of direct control of productive activities is that of Singapore. But this is an illusion. The Singapore economy is a strongly guided market economy with very marked government influence; an influence which has a very significant social, as well as economic, dimension. Thus, although the Singapore government has equity shareholding in shipbuilding and steel companies, its major influence is exerted through its selective investment promotion policies, its influence on wages and trade union activities and its channelling of savings through the state social security system (Lim and Fong 1986 Dicken 1986a, Grice and Drakakis-Smith 1985).

Broadly speaking, there are three kinds of industrialization strategy which can be adopted by national governments:

1. the *local processing* of indigenous raw materials;
2. the manufacture of products which would otherwise be imported – *import-substituting industrialization*; and
3. The manufacture of products for export – *export-oriented industrialization*.

Each of these strategies has been adopted in South East Asia to a greater or lesser degree and with varying degrees of success.

For the resource-rich countries of the region, industrialization via the local processing of indigenous raw materials was an obvious strategy to pursue. Such processing industries have, indeed, often been established by state-owned enterprises as well as by foreign companies and by privately-owned domestic enterprises. However, there are two major obstacles to such development which pose problems. One is where the extractive operations are controlled by foreign-owned companies whose strategy may be to locate their processing activities elsewhere. Undoubtedly, one of the reasons for several nation-states acquiring greater control of their natural resource ventures has been to avoid this kind of problem. But even where this obstacle does not exist a second inhibiting factor to the successful establishment of local materials processing is the tendency for industrialized countries to impose higher import tariffs on processed, compared with unprocessed, raw materials. Such tariff escalation, together with the generally higher average freight rates charged on processed materials, may well prevent the full development of materials processing industries in their source countries. Nevertheless, in each of the major South East Asian economies the local processing of materials has formed an important strand in national industrialization strategies.

In the case of manufacturing industries, the initial strategy of all the major economies of the region (other than Singapore) was one of import substitution (Kirkpatrick and Nixson 1983). The aim of this policy was to protect domestic industries from external competition and, therefore, to allow the development of infant industries. To this end, most countries employed a battery of industry incentives and trade restrictions. In particular, very high import tariffs were imposed on those sectors chosen for domestic development. Import quotas and other devices were also used, as were financial and fiscal incentives to encourage domestic production. An import-substitution strategy is, in theory, a long-term sequential process involving the progressive domestic development of industrial sectors through a combination of protection and incentives. The sequence begins, generally, with an emphasis on the domestic production of consumer goods; subsequently, when this sector is satisfactorily developed, the policy emphasis shifts to the domestic production of intermediate goods and, eventually, to that of capital goods.

Import-substitution policies undoubtedly achieved some success. Domestic production of consumer goods, such as processed food, textiles, leather goods, pharmaceuticals and chemicals, grew substantially in most of the South East Asian economies (Chi 1986, Singh 1986). However, although dependence on imported consumer goods certainly declined, dependence on the import of intermediate and capital goods – and, therefore, on foreign technology and capital – increased. Indeed, substantial direct foreign investment occurred as foreign TNCs set up local plants to serve the protected domestic market. In most cases, the predicted progression through the import-substitution sequence, the hoped for domestic multiplier effects and the stimulus of a broader industrial structure, did not materialize. Where the domestic market was small – as was the case throughout South East Asia – local production of consumer goods could not achieve appropriate economies of scale so that domestic prices remained high. The necessarily high level of imports of intermediate and capital goods imposed balance of payments constraints. Yet there were strong pressures from internal vested interests, especially the protected consumer goods manufacturers, against reducing the protection of those industries.

The realization that an import-substituting strategy could not, on its own, lead to the desired level of industrialization became apparent in South East Asia during the 1960s. In particular, even though industrial growth had been achieved it had not been particularly successful in creating sufficient employment to absorb the huge volume of surplus labour. Consequently, the conventional wisdom shifted very strongly towards an emphasis on an export-oriented industrialization strategy but one which, in particular, would be more labour intensive. Such a shift was strongly encouraged by the Asian Development Bank: 'South East Asian countries' it asserted, 'should move away from import-substitution policies towards a radically different approach to industrialization' (Asian Development Bank 1971, p. 19). Singapore's unique characteristics on its separation from the Federation of Malaysia in 1965 meant that, from the beginning, it had pursued an export-oriented strategy. Other ASEAN economies successively adopted an explicit export strategy based on the manufacturing sector during the 1970s.

Export-oriented industrialization policies in South East Asia have consisted of a number of related measures. Common to all of them – though with differences in degree – has been the luring of foreign firms to set up export-oriented plants. It so happened that in industries such as electronics the policy switch by South East Asian countries coincided with 'the timely search for third world sourcing by large global electronics firms' (Chi 1986, p. 37). Each country, as Table 8.7 shows, began to offer a package of investment incentives to export-serving plants. The most common incentive for firms in defined priority manufacturing sector was the tax holiday. Each of the ASEAN countries also operated very liberal regulations on the remittances of dividends or capital by foreign firms to their parent operation. Overall, South East Asian governments have been far more liberal towards foreign investors than the industrializing countries of Latin America and of the other Asian countries such as India. This is especially true of Singapore.

But the incentive package being offered consists of more than merely fiscal and other financial inducements. Most of all, the ASEAN countries offered the prospect to inward investors of a plentiful supply of low-cost, easily trained labour which, on the whole, could be guaranteed to be relatively docile and malleable. In some cases, such guarantees had the backing of legal (and sometimes extra-legal) sanctions. In general, the activities of labour unions are very closely regulated in South East Asia where 'increasingly stringent labour laws either ban strikes or make it difficult to call them in a region where political leaders are inclined

Table 8.7 Policies adopted towards inward foreign direct investment by ASEAN countries

Country	Government agency responsible for promoting/regulating foreign investment	Ownership requirements	Incentives offered to priority industries	Export processing zones	Policy on dividend and capital remittances
Indonesia	Capital Investment Co-ordinating Board (BKPM)	Joint ventures preferred with 51% Indonesian equity in most cases	Tax holiday of 2–6 years. Investment allowance. Import duties waived	1 established since 1978	No restriction on dividend remittances. Limited restrictions on capital remittances
Malaysia	Malaysian Industrial Development Authority (MIDA)	Objective of 70% Malaysian equity by 1990. Exceptions for key export and high technology sectors	Tax holiday in pioneer industries of 2–10 years. Level varies by location, technology and export involvement	Very important: 10 established since 1971	No restrictions
Philippines	Board of Investments (BOI)	No restriction up to 30% foreign ownership, then BOI approval. Up to 100% foreign ownership allowed in pioneer sectors and where 70% output exported. Filipino control within 40 years	Accelerated depreciation allowances. Partial tax holidays. Export incentives	Important: 3 established since 1972	No restrictions

Singapore	Economic Development Board (EDB)	None	Tax holiday (5–10 years) for firms granted pioneer status	Singapore is a free port	No restrictions
Thailand	Board of Investment (BOI)	Joint ventures preferred with aim of 51% Thai equity. Some exceptions in priority industries	Tax holiday (3–8 years). Additional tax concessions if in specified area	1 established since 1981	No restrictions on dividend remittances. Limited restrictions on capital remittances

Source: Individual country sources.

to believe that a meek workforce is needed to promote economic growth and lure foreign investment' (*The Economist*, 10 December 1984, p. 66).

Amongst all the measures used by South East Asian countries to stimulate their export industries and to attract foreign investment, one device in particular – the export processing zone (EPZ) – has received most attention (Spinanger, 1984; Wong & Chu, 1984). The UN Industrial Development Organization defines an EPZ as

'a relatively small, geographically separated area within a country, the purpose of which is to attract export-oriented industries, by offering them especially favourable investment and trade conditions as compared with the remainder of the host country. In particular, the EPZs provide for the importation of goods to be used in the production of exports on a bonded duty free basis'. (UNIDO 1980, p. 6)

EPZs are, in effect export enclaves within which special concessions apply including, very often, exemption from certain kinds of legislation. Within an EPZ, all the physical infrastructure and services necessary for manufacturing activity are provided. In a number of cases, the restrictions on foreign ownership which apply in the country as a whole are waived for foreign firms locating in the zones, allowing 100 per cent foreign ownership of export-processing ventures.

Figure 8.3 shows the location and size of EPZs in South East Asia. Altogether in the early 1980s there were at least 15 EPZs in the region, excluding the large number of industrial estates in the free port of Singapore. By far the largest concentration was in Malaysia, all of them located on the west coast. In total the South East Asian EPZs, including those in Singapore, employed more than 200 000 workers. Overwhelmingly, the labour force in EPZs is female. For example, 83 per cent of the workers in the Bayan Lepas zone in Penang, Malaysia is female, as is 74 per cent of the labour force of the Bataan zone in the Philippines. This predominantly female workforce is also very young: 85 per cent of the Bayan Lepas workers are aged between 18 and 24 years.

Not all of the region's export activity is located in EPZs but a very large proportion of the production of manufactured exports is. In fact, most of the EPZs contain a narrow range of export industries: production of electronics, textiles and clothing dominates in most cases. In Malaysia, for example, as Chi (1986) demonstrates, several of the zones have become highly specialized. The Sungai Way/Subang zones specialize in electronics while the Prai zone in Penang is the largest textile complex in South East Asia. Overall, almost half of the total labour force in the South East Asian EPZs is engaged in the electronics industry. Overwhelmingly, these EPZ operations are carried on in the branch plants of foreign firms, mostly US and Japanese. Use of such offshore processing sites by US firms has been further encouraged by the Offshore Assembly

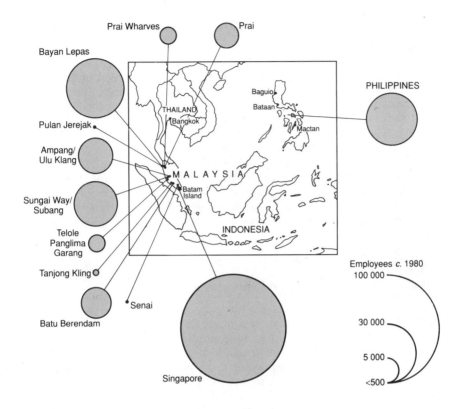

Fig. 8.3 Estimated employment in export processing zones in South East Asia 1980

Provisions of the US tariff regulations which permit a US firm to export domestic materials and components for processing overseas and then re-import the processed or assembled product paying duty only on the value of the foreign processing.

During the late 1970s, the region's most successful manufacturing exporter, Singapore, began to make deliberate efforts to move away from the cheap labour type of production and to upgrade the technological and skill level of its economy (Dicken 1986a; Lim and Fong 1986). The government embarked on a 'Second Industrial Revolution' to attract the knowledge-intensive service industries, more sophisticated manufacturing operations and the research and development operations and regional headquarters of international companies. A major drive to upgrade technical education was launched and a Skills Development Fund established. Most significantly, in some respects, the government embarked on a three-year wage correction policy to raise wage levels and discourage the cheap labour image.

Structure and organization of industry in South East Asia

As a result of both external forces, particularly the spread of TNCs, and the industrialization policies of national governments, the structure and organization of mining and manufacturing industry in South East Asia has become far more complex. The structure of industry, even within the formal sector of the economy, is strongly dualistic, while the informal sector operates on the fringes and in the interstices and remains a highly significant component, particularly in employment terms. Within the formal sector, three major types of business organization constitute the industrial structure of the South East Asian economies, although there is considerable variation in the precise mix present in individual countries:

1. branch plants of foreign TNCs,
2. state-owned enterprises,
3. private, domestically owned enterprises.

The units in categories 1 and 2 tend to be relatively large while those in the third category cover a range of sizes but tend predominantly to be small-scale. In general, there is a huge technological gap between the small domestic firms and the branch plants of TNCs. It is also important to emphasize that the small domestic firm is numerically the most important type of business organization throughout the South East Asian economies. These three categories of business organization do not exist in total isolation from each other; indeed, a complex network of inter-relationships exists though this is poorly documented in the literature. Not surprisingly, perhaps, most attention has been focussed on the role and relationships of TNCs in the South East Asian countries and their impact on the host economies.

The branch plant of a TNC operating in South East Asia is locked into two major sets of relationships. The first of these is within its parent organization (intra-firm relationships); the second is with other, independent, firms in the host economy or elsewhere (inter-firm relationships). The general consensus view is that the kind of branch plant operating in South East Asia has very little independence or autonomy from its parent company. Its function, its technology, its freedom to purchase its material inputs or to sell its output all tend to be tightly controlled from its headquarters. Thus, the branch plant's position in an intra-firm sense also tends to define its external, inter-firm relationships. For example,

'the market structure of the electrical and electronic group of industries tends to be rigid. This is especially true of the components which are mainly sent to parent companies, sister companies and other affiliated firms overseas. . . . In general, the organizational structure of the global firms provides the major pattern of trade linkage.' (Ch 1986, p. 39)

A good many empirical studies have demonstrated the relatively limited and shallow extent of the linkages formed by TNCs with domestic firms in South East Asia, particularly, those branches located in Export Processing Zones (see, for example, Chi 1981, 1986). In some cases, the foreign TNCs themselves establish their own input supply operations or encourage other TNCs with which they have links to do so, a practice especially common among Japanese firms and also in the electronics and textiles industries. In many cases, of course, it may be that no suitable local supplies of the necessary inputs exist, at least initially. But although the general level of local/domestic linkage formation by TNCs is relatively low this is not invariably the case. For example, Lim and Fong's study of electronics TNCs in Singapore found that, under certain conditions, local vertical linkages are created by TNCs (Lim and Fong, 1982). Indeed, in the more advanced of the South East Asian countries, considerable subcontracting networks have begun to develop whereby domestic firms are connected into export markets through the established marketing channels and brand names of foreign firms. In consumer goods industries, the role of the foreign buying group or trading company has been very significant in this respect (Hone 1974). The international subcontracting process (Germidis 1980) is undoubtedly one of the most important mechanisms through which small- and medium-sized domestic firms in South East Asia gain access to international markets, albeit at the expense of some loss of independence. Of course, this question of the linkages which may, or may not, be forged between foreign and domestic firms is but one part of the much larger debate over the costs and benefits of TNCs to developing countries, a debate beyond the scope of the present chapter (see Dicken 1986b, Lee 1984), although we shall return to some of the issues in the final section.

A second dimension of the structure and organization of industry in South East Asia is its particular geography. Two aspects of this will be discussed very briefly here. The first relates to the emergence of an intra-regional division of labour in the electronics industry. We referred earlier to the development of a general manufacturing hierarchy within South East Asia. In the semiconductor industry, because of the specific nature of the production process, this hierarchy is taking on a very precise functional character. The growth of semiconductor assembly operations in South East Asia by United States, Japanese and, to a lesser extent, European firms is one of the most dramatic developments of the last two decades (Dicken 1986b, Henderson 1986, Lee 1986, Scott 1986). For the most part, however, it is only one stage of the semiconductor production sequence – the assembly stage – which has been transferred 'offshore' to South East Asia. Following the pioneer decision by the American manufacturer, Fairchild, to establish a semiconductor assembly plant in Hong Kong in 1962, all the major US firms in the industry, and many Japanese and some European firms followed a similar path.

Initially, virtually all the offshore plants in Asia were located in the four leading NICs: Hong Kong, South Korea, Taiwan and Singapore. However, as these countries rapidly developed industrially and as the semiconductor industry itself evolved so, too, the geographical structure changed. As wage levels rose in the leading Asian NICs, the TNCs began to shift some of the more standardized production to the lower-wage countries of, first, Malaysia, and subsequently to the Philippines, Indonesia and Thailand. Also, as the skill level of countries such as Singapore increased, TNCs began to locate some of the most sophisticated testing processes there. The outcome is an intra-regional division of labour – functional specialization – in the semiconductor industry which is set within the broader global division of labour in the industry. Within the area covered by this book, Singapore has emerged as the key centre, with a growing sophistication of electronics production as well as of marketing and regional co-ordination functions. Less sophisticated assembly is increasingly being located in the other ASEAN countries. One result is an increasingly complex geographical flow of components between the peripheral assemblers and the testing/marketing operations in Singapore (and in the East Asian NICs). This flow is organized by, and predominantly within, the foreign TNCs.

The second geographical dimension of the structure and organization of industry in South East Asia is the concentration within individual countries. Modern industrialization in the region, other than that based directly on geographically localized raw materials, is predominantly urban in its orientation. Cities, as Armstrong and McGee (1985) point out, are the focus of capitalist expansion and growth – the 'theatres of accumulation' – in the developing world. The vast majority of modern industrial activity in South East Asia is located in the major cities because it is there that the multifarious needs of business firms, both physical and human, are most readily met. In several Asian countries, for reasons of historical development, the urban structure is strongly primate. But even where this is not the case, a disproportionate share of each country's industry (and employment) is found in a very small number of urban centres. In Malaysia, for example, Kuala Lumpur, Penang and Johore Bahru contain most of the country's manufacturing growth. This process of geographical concentration has been strongly reinforced by the construction of EPZs in or near these three urban centres (Figure 8.3). In the Philippines, Manila totally dominates, with some two-thirds of all the country's manufacturing establishments located there (Armstrong and McGee 1985, p. 96). However, the Bataan EPZ was deliberately located in an economically depressed part of the country. In Thailand, the degree of urban primacy is particularly great, with 70 per cent of the urban population located in Bangkok. Hence, the kinds of industrialization processes outlined in this chapter, particularly the strong participation of TNCs, contribute a major push to continued urban concentration in the industrializing countries of South East Asia.

Industrialization in South East Asia: Progress, problems and the future

Without doubt, the pace of industrialization in the market economies of South East Asia has been dramatic during the past two decades. Although most of the region's economies remain predominantly primary producers, several have become significant producers and exporters of manufactured goods and have begun to make a global impact in certain sectors. Such industrialization has, indeed, brought substantial benefits in the form of increased national and personal incomes and a greater range of opportunities for the labour force. In one particular case, Singapore, the transformation of both economy and society has been little short of spectacular, despite the doubts voiced by some critics of the social ramifications of national policy (Grice and Drakakis-Smith 1985). But such industrialization as has occurred within South East Asia is not without its problems; neither can it be assumed that those countries in the lower tiers of the region's manufacturing hierachy will inevitably be able to emulate the leaders.

A first point to make is that, despite its considerable growth in the region, manufacturing industry has made barely a dent in the unemployment and underemployment problem of most countries. Its labour absorption capacity has been limited (Hirono 1986). Only in atypical Singapore, with its lack of an agricultural population of any size, has manufacturing industry absorbed large numbers of people. Much of the increased employment that has been generated by the export-oriented manufacturing activities has been for females. As Lee points out,

'these industries have selectively tapped one particularly vulnerable segment of the labour force. . .that of young female entrants to the labour force, who are in a disadvantaged position in the local labour market. Their wages are generally lower than equivalent categories of male workers.' (Lee 1986, p. 56)

As we have seen, female workers dominate in the EPZs and, especially, in textiles, clothing and electronics. The employment of females for very long hours in tedious and sometimes unhealthy conditions has led to charges of exploitation. This may well be true but, as Jenkins observes,

'there is only one thing worse than being exploited by capital and that is not being exploited at all. Given the alternatives which at present exist in most Third World countries it is by no means clear that workers employed in export production would be better off if these factories did not exist.' (Jenkins 1984, p. 51)

This problem of 'what if?' – the counterfactual problem – complicates the entire debate over the costs and benefits of TNCs to host countries (Dicken 1986b, ch. 11). For developing countries a key issue concerns the transfer of technology. Clearly, by setting up branch plants in such countries TNCs do transfer technology. But is it the kind of technology which will contribute towards more general growth and development? Is it appropriate to local needs and resources? Apart from some recent developments in Singapore, very few research and development activities have been located in South East Asia. Thus, TNCs tend not to 'transfer the capability to generate new technology to affiliates in the Third World. They transfer 'know how' (production engineering) and not 'know why' (basic research and development)' (Lall 1984, p. 10). There has been much criticism, too, that the kinds of investment made by TNCs have been capital- rather than labour-intensive (Hirono 1986), a tendency encouraged by the granting of financial subsidies and incentives to capital investment by host country governments. It has been suggested, however, that the TNCs which originate from other developing countries do operate in ways more consistent with host country conditions. But it is dangerous to generalize beyond specific circumstances. In some cases, appropriate technology is transferred, in others it is not; in some cases, domestic entrepreneurship is stimulated, in others it is inhibited. What is clear, however, is that through the involvement of TNCs the South East Asian economies have become much more tightly and extensively connected into the global economy.

After the heady growth of the 1960s and 1970s most of the region's economies experienced a major slowdown in 1985 (Hirono 1986). This, and the earlier slowdown in 1981–2, reflects both the region's continuing heavy dependence on its external markets, notably the United States and Western Europe, and also the tighter monetary and fiscal policies adopted by most South East Asian countries. A particular problem for some countries, notably the Philippines, Burma, Indonesia and Thailand, is that of rising external indebtedness. More generally, a still heavy dependence on the production and export of primary commodities has been a particular problem as the demand for, and the prices of, such commodities has declined. The collapse of such institutions as the International Tin Council (in 1986) was salutary reminder of the volatility of marketing arrangements in primary commodities. The industrial structure of most South East Asian economies remains quite narrow.

Apart from these particular events, what of the future? Can Singapore resume its growth trajectory and succeed in restructuring its economy to one at a higher technological and organizational level? The high growth rates of 1987 and 1988 suggest that a recovery has been achieved. Will the other industrializing countries of the region – Malaysia, the Philippines, Indonesia and Thailand – emerge as a 'second tier' of Asian NICs to follow the 'Gang of Four'

(Hong Kong, Singapore, South Korea and Taiwan)? It is by no means certain that they will be able to grow at anything like the rate of the earlier generation of Asian NICs. The world is now a different place in terms both of the potential for export-led growth and also technologically. In particular, the increasing automation of labour-intensive processes may be reducing the incentive for TNCs to relocate some of these activities to developing countries. If this tendency should develop more widely, then it would greatly alter the prospects for export-based industrialization through the TNC route. It is also necessary to appreciate the specific internal conditions in the leading NICs which, in conjuction with a favourable global environment in the 1960s, contributed so substantially to their growth.

For tomorrow's industrializing countries in South East Asia more emphasis will have to be placed on enlarging both the domestic and the regional market. So far, ASEAN has been of relatively minor significance in this respect although, at the time of writing, some attempts are being made, notably by the new Philippines government, to create a regional common market by the year 2000. A major problem, apart from the distrust which still exists in some quarters, is that the economic structures of the ASEAN economies tend to be competitive rather than complementary. More broadly, of course, ASEAN is but a part of the much larger East Asia edge of the Pacific Basin, a region of enormous, and much publicized, economic potential. Whether, and how such potential will be realized is a question beyond the bounds of this chapter but it is one of major significance not just for the countries of South East Asia themselves but also for the global economy as a whole.

Further reading

Armstrong W and **McGee T G** (1985) *Theatres of Accumulation: Studies in Asian and Latin American Urbanisation*, Methuen, London.

Asian Development Bank (1971) *South East Asia's Economy in the 1970s*, Longman, London.

Benjamin R and **Kudrle R T** (eds.) (1984) *The Industrial Future of the Pacific Basin*, Westview Press, Boulder.

Chi S C (1981) 'Industrial organizations and industrial estates: a case study', *Tijdschrift voor Economische en Sociale Geografie* **72**, 235–41.

Chi S C (1986) 'The structure and impact of export-oriented industries in Malaysia: a review', in Tornquist, G., Gyllstrom, B., Nilsson, J-E. and Svensson, L. (eds.) *Division of Labour, Specialization and Technical Change*, Liber, Lund pp. 33–50.

Conkling E C and **McConnell J E** (1985) 'The world's new economic powerhouse', *Focus* January, 2–7.

Corbridge S (1986) 'Capitalism, industrialization and development', *Progress in Human Geography*, **10**, 48–67.

Dicken P (1986a) 'A tale of two NICs: Hong Kong and Singapore at the crossroads', *Geoforum* **17**, 151–64.

Dicken P (1986b) *Global Shift: Industrial Change in a Turbulent World*, Paul Chapman Publishing, London.

The Economist (1984) *The World in Figures*, London.

Fröbel F, Heinrichs J and **Kreye O** (1980) *The New International Division of Labour*, Cambridge University Press, Cambridge.

Germidis D (ed.) (1980) *International Subcontracting: A New Form of Investment*, OECD, Paris.

Grice K and **Drakakis-Smith D** (1985) 'The role of the state in shaping development: two decades of growth in Singapore', *Transactions of the Institute of British Geographers* **10**, 347–59

Henderson J W (1986) 'The new international division of labour and American semiconductor production in South East Asia' in Dixon, C. J., Drakakis-Smith, D. and Watts, H. D. (eds.) *Multinational Corporations and the Third World*, Croom Helm, London.

Hill H and **Johns S B** (1985) 'The role of direct foreign investment in developing Asian countries', *Weltwirtschaftliches Archiv* **121**, 355–81.

Hirono R (1986) 'Asian and Pacific developing economies: performance and issues', *Asian Development Review* **4**, 1–16.

Hofheinz R Jr and **Calder K E** (1982) *The East Asia Edge*, Basic Books New York.

Hone A (1974) 'Multinational corporations and multinational buying groups: their impact on the growth of Asia's exports of manufacture – myths and realities', *World Development* **2**, 145–9.

Jenkins R (1984) 'Divisions over the international division of labour', *Capital and Class* **22**, 28–57.

Kirkpatrick C H and **Nixson F I** (eds.) (1983) *The Industrialization of the Less Developed Countries*, Manchester University Press, Manchester.

Lall S (1983) *The New Multinationals: The Spread of Third World Enterprises*, Wiley, Chichester.

Lall S (1984) 'Transnationals and the Third World: changing perceptions', *National Westminster Bank Quarterly Review* May, 2–16.

Lee C H (1984) 'Direct foreign investment and industrial development in the Pacific Basin', in Benjamin, R. and Kudrle, R.T. (eds.), *The Industrial Future of the Pacific Basin*, Westview Press, Boulder.

Lee E (ed.) (1981) *Export-Led Industrialization and Development*, International Labour Office, Geneva.

Lee E (1986) 'Export-oriented industrialization and employment in South East Asia', in Tornquist, G., Gyllstrom, B., Nilsson, J-E. and Svensson, L. (eds.) *Division of Labour, Specialization and Technical Change*, Liber, Lund, pp. 51–69

Lim L Y C and **Fong P E** (1982) 'Vertical linkages and multinational enterprises in developing countries', *World Development* **10**, 585–95.

Lim L Y C and **Fong P E** (1986) *Trade, Employment and Industrialization in Singapore*, International Labour Office, Geneva.

Manners G (1986) 'Multinationals and the exploitation of non-renewable resources', in Dixon, C. J., Drakakis-Smith, D. and Watts, H. D. (eds.) *Multinational Corporations and Third World*, Croom Helm, London, pp. 25–38.

Myint H (1972) *South East Asia's Economy: Development Policies in the 1970s*, Penguin, Harmondsworth.

Scott A J (1986) 'The semiconductor industry in South East Asia: organizations, location and the international division of labour', Department of Geography, University of California at Los Angeles *Working Paper* 101.

Singh M S (1986) 'Industrial dynamics of newly industrialising countries in East and South East Asia: a real path towards a New International Economic Order', in Hamilton, F. E. I. (ed.) *Industrialization in Developing and Peripheral Regions*, Croom Helm, London, pp. 283–310.

Spinanger D (1984) 'Objectives and impact of Economic Activity Zones – some evidence from Asia', *Weltwirtshaftliches Archiv* **120**, 64–89.

Thoburn J (1981) *Multinationals, Mining and Development*, Gower, Farnborough.

UNCTC (1983) *Salient Features and Trends in Foreign Direct Investment*, United Nations, New York.

UNIDO (1980) 'Export Processing Zones in developing countries', *UNIDO Working Paper on Structural Changes* 19.

Wallerstein I (1979) *The Capitalist World Economy*, Cambridge University Press, Cambridge.

Wawn B (1982) *The Economics of Asean Countries*, Macmillan, London.

Weiss J (1984) 'Manufacturing as an engine of growth – revisited', *Industry and Development*, **13**, 39–62.

Wong K Y and **Chu D K Y** (1984) 'Export Processing Zones and Special Economic Zones as generators of economic development: the Asian experience', *Geografiska Annaler* **66B**, 1–16.

World Bank (1980) *World Development Report 1980*, Oxford University Press, New York.

World Bank (1985) *World Development Report 1985*, Oxford University Press, New York.

World Bank (1988) *World Development Report 1988*, Oxford University Press, New York.

World Resources Institute (1986) *World Resources 1986*, Basic Books, New York.

CHAPTER 9

Trade, aid and regional integra

George Cho and Stephen Wyn Williams

For centuries international trade has been the principal economic nexus between nations. Furthermore, trade has provided a powerful mechanism for the transmission of economic growth (and, the possibility of 'development') from one country to another. Indeed, according to dependency theorists (see Ch. 3) international trade has been one of the principal factors promoting the emergence of a hierarchically organized world system divided into a core, periphery and semi-periphery (Wallerstein 1974). During the contemporary period, the importance of trade as a keystone for survival and advancement in the international system has not diminished. The significance of this statement can be seen nowhere more clearly than in the context of the Pacific Basin. The most extensive definition of this region includes Japan, Asiatic Russia, China, South East Asia, Australia, New Zealand and the west coast of the Americas. According to some writers, the economic developments occurring within this region are of such importance that the twenty-first century is already being referred to as the 'Pacific Century' (Linder 1986). In the 1980s, the Pacific Basin became the fastest growing and foremost trading region of the world. Between 1960 and 1982, for example, the ratio of its exports to world exports doubled, while the performance of manufactured exports was even more impressive. Furthermore, trade within the region grew at a faster rate than trade with the rest of the world. Although it is unwise to extrapolate present trends too far into the future and to ignore problems within the region which may significantly slow the tempo of development, it would be difficult not to agree with Linder's (1986, p. 10) conclusion that 'the centre of gravity of the world economy is indeed shifting from the Atlantic Basin to the Pacific Basin'.

While the overall dynamics of growth within the Pacific Basin will no doubt significantly influence the pace of development of all individual countries, there are important variations within the region, for example between the more industrial countries and the predominantly commodity exporters of South East Asia. Similarly,

225

within South East Asia several tiers can be identified. The most advanced and diversified economy is the region's only 'official' NIC, Singapore. Also included in the highest tier (at least in terms of per capita income) is oil-rich Brunei. The second tier is composed of a group of 'proto-NIC's' – Indonesia, Malaysia, the Philippines, and Thailand. Finally, there are the socialist economies of Kampuchea, Vietnam, Burma and Laos, the last two being included in the UN's list of least-developed countries.

Within this broad context, the purpose of the present chapter is to examine the links between development, trade and aid and to explore the prospects of regional integration in South East Asia. It first examines the relationship between development, growth and trade particularly in the context of 'trade as an engine of growth' arguments. Attention is then directed towards a discussion of the nature of appropriate trade policies and their influence on the pace of economic growth and development. An analysis of the structure of foreign trade in South East Asia then follows, focussing on the geographical distribution of external trade and the commodity composition of such activities. As trade is perceived to be the magnet drawing developing countries into the international market-place, a case for and against regional intergration is also examined in the context of ASEAN. Finally, the necessity for many South East Asian nations to look elsewhere for development assistance, or aid, is acknowledged.

Development, growth and trade

In a very simplistic way one may think of 'development' as bringing higher standards of living to more people. Of course, this presumes that, on the one hand, there exists some mechanism which 'triggers' development and, on the other, channels through which the benefits of development filter down to the population at large. 'Growth', in contrast, may be interpreted simply as the quantitative increase of some indicative unit over a given time period. Such increase may be measured in monetary or volumetric terms. However, the manner in which gains from increases in economic activity are distributed to the general population, that is, the process of development, is a vexed social problem, depending to a large extent on the political philosophy of nations.

Growth, for any economy, implies an expansion of production of goods and services for both consumption and export. Gilpin (1987, p. 171) notes that the expansion of trade, in particular, produces a number of related outcomes.

'(1) technological diffusion . . . (2) a demand or Keynsian effect on the economy that, through the operation of the "multiplier", stimulates economic growth and the overall efficiency of the economy, (3) benefits for industrial firms as trade increases the size of the market, promotes economics of scale and increases the return on investment while also stimulating the overall level of economic activity in the economy as a whole, (4) increased range of consumer choice, and (5) reduction in the costs of inputs such as raw materials and manufactured components, which then lowers the overall cost of production. Moreover, in the late twentieth century, export-led growth has itself become a major strategy used to acquire needed imports and promote economic growth.'

The putative benefits indicated above derive from what may be referred to as the orthodox theory of international trade. As initially presented in its classical form by Adam Smith in 1776, international trade is seen as essential for the attainment of economic efficiency, which is the key factor in wealth creation. Economic efficiency itself derives from specialization and the division of labour, the extent of which is dependent on the scale of the market. International trade plays a crucial role because it 'is the principal means by which a nation can extend the market beyond its borders thereby allowing greater specialization in production, enhanced efficiency in the use of scarce resources and greater national income . . . (expanding) . . the capacity to accumulate and hence to grow' (Riedel 1988, p. 26). The advocacy of free trade is a natural corollary of these ideas for without it the 'extent of the market' would be restricted and so too would domestic welfare and economic growth.

According to Adam Smith, specialization of production and the associated territorial division of labour is based on *absolute* advantage. In contrast, David Ricardo (the father of international trade theory) argued that mutually beneficial trade between nations is based on *comparative* advantage or cost. That is, even though a country may have an absolute advantage over others in the production of every good, it would be in the interests of itself and other nations to specialize in producing and exporting those commodities in which it has a comparative cost advantage and importing those goods in which it has a comparative cost disadvantage. As Gilpin (1987, p. 174) notes, 'this simple notion of the universal benefit of specialization based on comparative costs remains the lynchpin of liberal trade theory'.

The principle of comparative advantage, however, provides no explanation as to why geographical differences in production and trade actually arise. Neoclassical reformulations (in particular, the so-called Heckscher–Ohlin model of international trade) have attempted to account for comparative advantages in terms of international differences in 'factor endowments': capital (including human capital), labour, resources, management and technology. A nation's comparative advantage will, therefore, be determined by the relative abundance of its several factors of production. As El-Agraa (1983, p. 77) notes, 'a country will export (import) those commodities which are intensive in the use of its abundant (scarce) factor.' For simplicity, the Heckscher–Ohlin theory was developed by focussing on only two factors of production – capital and labour.

As a developing country is virtually defined by its relative abundance of labour and scarcity of capital, it has been suggested that this approach is particularly useful in explaining the trading patterns of developed countries. Thus, 'poor countries are recommended to specialize in labour-intensive products and trade their surplus of such products for imports of rich countries' capital-intensive goods' (Smith and Toye 1979, p. 3). Without trade, the ability of developing countries to save, invest and grow would be severely restricted. In policy terms, the implication of this discussion suggests that it would be beneficial for all nations if tariff and non-tariff barriers were dismantled. In particular, to exploit their comparative advantages in labour-intensive products, developing countries should adopt liberal, outward-looking commercial policies. As Cairncross (1962, p. 364) put it, trade 'is no mere exchange of goods . . . as often as not it is trade that gives birth to the urge to develop, the knowledge and experience that make development possible'.

For a number of years the idea that trade can act as an 'engine of growth' for developing countries has been influential (Nurske 1961). The basic argument is that the economic growth of developing countries is largely determined by external conditions. Specifically, the trade engine is linked to two gears (see Riedel 1988). One links the growth of developing countries to trade (by exports, but also by the import of investment goods), while the other gear links the trade of developing countries to the growth and prosperity of developed countries. In short, the prosperity of developing countries is ultimately linked to industrial growth in the developed world. Historically, the trade engine worked most efficiently in the nineteenth century when Britain and other European powers were highly dependent on imports of raw materials for domestic manufacturing. This allowed the economies of developing countries to expand, thus providing a ready market for manufactured goods. In the twentieth century, however, writers such as Nurkse (1961) and Prebisch (1959) have argued that the engine of growth has failed due to falling demand for tropical products and deteriorating terms of trade.

According to economic nationalists, the solution was for developing countries to industrialize, thereby reducing the necessity to import goods and simultaneously reducing the need to export. This policy of import-substitution industrialization, it was argued, would increase employment, raise incomes and stimulate further industrialization. In this scenario, the notion of comparative advantage becomes irrelevant, 'general protection would stimulate general industrialization, regardless of the costs of local production. The state had the power, it seemed, to determined the shape of the domestic economy' (Harris 1987, p. 16). Lewis (1980, p. 560) in discussing the 'slowing down of the engine of growth', provided an alternative solution, however. He suggested that as the industrial countries entered a prolonged period of economic slowdown an

Plate XIII   Jakarta: the central core.

Plate XIV Commercial sea fishing, southern Thailand.

Plate XV Jakarta: suburban sprawl.

Plate XVI Jakarta: inner city infilling: squatter huts along a storm drain.

Plate XVII Food hawker, Bangkok.

Plate XVIII Suburban shops along the *klongs* of Bangkok.

Plate XIX Small-scale industry, Bangkok. Making lamp stands out of old car springs.

Plate XX Small saw-mill, Kuala Lumpur.

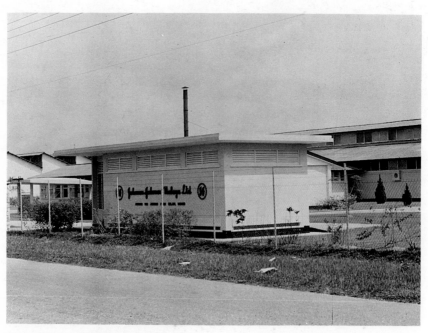

Plate XXI The industrial estate and the multinational company.
Petaling Jaya, Malaysia.

alternative source of fuel for the trade engine was required. This he identified as trade among developing countries, which he argued could 'take up the slack left by MDCs (more developed countries) as MDCs slow down'.

A more radical twist to the above arguments has been provided by writers such as Frank (1969) and Amin (1976) who argue that unequal trade (as exemplified by the deteriorating terms of trade in the periphery) is an outgrowth of the system of unequal economic relationships between rich and poor countries. Trade dependency is, therefore, an inevitable outcome of the process of international capitalism. Such ideas have more recently been subject to considerable debate and the emerging conclusion is that they may have been over-simplified or are even lacking in empirical merit (see for example Peet 1980). The very thought of dependency is, however, objectionable to newly independent countries and there is no doubt that a significant amount of thinking within them, including thinking on trade, is structured around the 'dependence versus independence' dichotomy.

According to Riedel, the 'export pessimism' perspectives associated with the above viewpoints are unfounded. Specifically, Riedel (1988, p. 52) argues that the mechanistic engine of growth analogy rests on an increasingly inappropriate distinction between a primary producing South and an industrialized North. Thus, the 'most important reason why the trade engine no longer works in the old way is the diversification of exports of developing countries, the main exception being the countries of Africa . . . .' Furthermore, the engine of growth model does not allow for the possibility of price competition as a mechanism by which the volume of exports can be expanded. This is a direct consequence of assuming that developing countries completely specialize in 'tropical products', demand for which is inelastic with respect to price – the sole determining factor being the level of expenditure in developed countries. However, in the developing countries of Asia and the Pacific, for example, many countries are now deriving less and less of their GDP from primary producing sectors. In South East Asia, the reduction in primary commodity dependence has been generally lower than in South Asia. Nevertheless, in Thailand the contribution of primary commodities to GDP fell from 32 per cent in 1970 to 23.2 per cent in 1982, while in Malaysia between 1970 and 1984 a fall from 38.9 per cent to 32.4 per cent was recorded (United Nations 1988).

Trade policies for development

The relationship between appropriate trade policies and development is a contentious one. Nevertheless, it would be difficult not to agree with the view that 'the commercial policies that a developing country pursues may have a significant effect on the pattern and pace of economic growth and development' (Milner 1988, p. 55). Broadly speaking, production and trade policies can range from autarchic regimes through to import substitution and finally to export-oriented policies. An inward-oriented strategy such as import substitution is one in which trade and industrial incentives are biased in favour of production for the domestic rather than the export market. In the 1950s and 1960s import substitution was the dominant strategy. A wide range of protectionist policy instruments was deployed in an attempt to encourage new (and expand existing) industries. In the process, both tariff and non-tariff barriers were extensively employed, with many policy-makers preferring the latter, in the form of import licensing, quantitative restrictions and domestic content requirements, for example. In some instances, the effect of these controls was dramatic. The Philippines, for example, reduced its imports of consumer goods from 30.9 per cent to 4.7 per cent of total supplies between 1948 to 1965.

Nevertheless, the failure of import substitution policies to achieve expected levels of industrialization, growth and employment led many developed countries to adopt more outward-oriented, export-promotion strategies. Ideally, these strategies ' do not discriminate between production for the domestic market and exports, or between purchase of domestic goods and foreign goods' (World Bank 1987a, p. 78). In implementing an export promotion strategy, a variety of policy instruments has been employed, including financial and fiscal incentives to export producers, export guarantees and credits, setting export targets, and the establishment of export processing zones and/or free trade zones. The first group of countries to pursue this path consisted of Hong Kong, South Korea, Singapore and Taiwan; they were later followed by other semi-industrialized developing countries. However, as Milner (1988) notes, the precise type of export orientation has varied according to specific national factor endowments. Thus, some countries with favourable primary resource endowments have attempted to move from being mainly primary exporters to manufacturing exporters via local processing. In other countries, the development of manufactured exports has been based on an abundant cheap labour supply combined with investment by multinational corporations.

The success of the four Asian 'tigers' (Hong Kong, Singapore, Republic of Korea and Taiwan) and other NICs in generating exceptional growth rates for manufactured exports and output has provided a model for the principal international development agencies. The World Bank, in particular, is strongly committed to encouraging developing countries to adopt outward- rather than inward-oriented strategies. The benefits of following this strategy are clearly expounded in a survey of 41 developing countries for the period 1963–85 (World Bank 1987a). According to this study, if countries are ranged along a scale from strongly outward oriented at one end to strongly inward-oriented at the other, economic performance, including GNP growth, tends to decline from outward to inward orientation. The evidence on other indicators confirms this picture, and the World Bank authors conclude that 'the important lesson is the strongly-inwardly oriented economies did badly' (World Bank 1987a). Although the evidence appears convincing, Singer has major reservations concerning the validity of the study. Specifically, he points to the fact that the inward-oriented countries are significantly poorer than the outward-oriented group, a characteristic which is more important than relative lack of economic performance. What the World Bank analysis really tells us, therefore, is 'that poor countries find it more difficult to progress than countries already further up the development ladder' (Singer 1988).

The structure of international trade

In attempting a synoptic overview of the structure of their foreign trade, it is useful for comparative purposes to classify the countries of South East Asia into a smaller number of sub-groups. From a strictly geographical perspective the most obvious subdivision would be broadly physiographic; that is between continental and peninsular South East Asia. Clearly, however, this would not provide a rational basis for the comparison of trade regimes. Similarly, a two-fold socio-economic division between a more-developed 'South' and a less-developed 'North' would not be entirely satisfactory as Thailand would be included, anomolously, in the lagging 'north'. In the present context, therefore, the use of purely spatial determinants in creating an appropriate typology is limited. A more meaningful subdivision may be arrived at by considering the political and economic orientation of the constituent countries. On this basis the most significant division is between those countries which are essentially free-market in outlook and those which have embraced socialist philosophies. The socialist states all lie to the north in continental South East Asia, and consist of Kampuchea, Laos, Vietnam and Burma. The majority of the freemarket economies lie to the south and east (that is Brunei, Indonesia, Malaysia, the Philippines and Singapore), the only exception being Thailand.

The socialist states of South East Asia are by no means a uniform grouping. Vietnam, for example, is the only member of the Council for Mutual Economic Assistance (CMEA). While Kampuchea and Laos are Marxist-Leninist in orientation, Burma is a 'marginal' case being a one-party 'socialist' state. In Indochina one of the major factors influencing the overall progress of development has been the existence of various revolutionary conflicts, and also the emergence of Vietnam as the dominating power in Indochina (which was underlined by its invasion of Kampuchea in 1978).

In terms of the overall economic position of the socialist economies, there can be no doubt that they are substantially poorer than the free-market economies. On the basis of per capita income all fall within the World Bank's low income category, while Laos and Burma are included in the UN's category of least developed countries. The structure of production is dominated by agriculture, and until recently all countries have followed a strongly inward-oriented economic policy. In the case of Burma, this has meant that after more than two decades, foreign investment is only now being allowed into the country. However, this general insularity is slowly changing with the prospect that these countries could derive some benefit from the so-called 'fourth wave' of Asian industrial investment arising from the desire of Japan and the Asian 'tigers' to cut their costs even further. Vietnam, for example, is moving towards a more Western-style market doctrine, and as a result of its chronic need for capital now has one of the most liberal investment codes in Asia, allowing one hundred per cent ownership by foreign businesses. Nevertheless, its future growth will depend on political factors, particularly its withdrawal from neighbouring Kampuchea which would lift the trade boycott imposed by the west, ASEAN and international financial institutions. Despite this boycott, however, it has been reported that over twenty foreign companies have opened offices in Hanoi (*South* 1989). In Laos a 'new outlook' policy has been followed since 1986. This has lifted economic controls and encouraged the production and distribution of goods and services while enabling the establishment of private enterprises. Although Laos lacks an investment code, the principles of one hundred per cent foreign ownership and total repatriation of profits in the first five years for export-oriented industries are accepted.

In terms of per capita income the free-market economies of South East Asia are clearly divided between Singapore and Brunei on the one hand and Indonesia, Malaysia, Philippines and Thailand on the other. All countries, except Singapore, rely to a greater or lesser extent on the production and export of primary commodities for foreign exchange earnings, although in recent years all have attempted to diversify their economic base with varying degrees of success. Since the mid-1980s, for example, Indonesia (the poorest of the free-market economies) has been adversely affected by the fall in oil prices. In 1983 oil and gas accounted for 79 per cent of exports; however the oil price collapse of 1986 resulted in revenues from this source falling by more than half. In response a number of reforms were introduced. In the present context the most significant was the reform of trade policy from one of import substitution to a more outward-oriented strategy. In recent years non-oil exports have for the first time exceeded oil and gas revenues. Furthermore, Indonesia has been relatively successful in attracting foreign investment to the non-oil sector. This totalled $4 billion, for example, in the first ten months of 1988 compared with only $800 000 for the whole of 1986 (*South* 1988). In contrast, the model for outward-oriented growth in the region in undoubtedly Singapore; however in 1985 its GDP growth rate declined for the first time in twenty years, due to a fall in external demand and internal constraints. Although growth rates have improved in recent years (GDP grew by 10 per cent in 1988) continued high growth rates cannot be assumed. The basic problem has been identified as Singapore's over-reliance on a productive base established by multinational corporations without a complementary and diversified base of indigenous industries (Singh 1988). Furthermore, Singapore's declining cost competitiveness compared with other Asian NICs has tended to undermine investment promotion in the manufacturing sector.

The trade of South East Asia is about two-thirds of the value of that of the United Kingdom and one-third of the value of the trade of the United States. South East Asian exports by value are more than three times larger and imports more than four times than those of Australia. By any reckoning, therefore, the trade of South East Asia is quite considerable. Table 9.1 provides a broad indication of the importance of international trade for the countries of South East Asia. The most striking feature is the overwhelming dominance of the non-socialist economies in imports and exports, accounting for over 95 per cent of both categories. However, the figures for the four socialist countries are an under-estimate of the volume of trade. For example, in the case of Kampuchea the figures do not include

Table 9.1 South East Asia: imports and exports by value (million US$) and trade balance

| | 1986 | | Exports as a percentage of GNP | Trade Balance 1985–6 | |
	Export	Imports		Value	Percentage of imports
Socialist					
Burma	265	304	3.6	−3	−1.2
Kampuchea	4	35	–	−30	−88.1
Loas	30	70	2.1	−38	−56.3
Vietnam	254[1]	596[1]	1.7	–	–
Non Socialist					
Brunei	2 700	770	75.6	2 144	309.5
Indonesia	14 805	10 718	18.0	6 209	59.2
Malaysia	13 874	10 829	47.0	3 092	26.7
Philippines	4 842	5 394	15.2	−634	−11.9
Singapore	22 495	25 512	117.4	−3 244	−12.5
Thailand	8 753	9 138	20.6	−1 254	−13.6

Note:
1. Kurian (1987) Figures relate to 1984.
Source: UNCTAD (1987, Tables 1.1 and 1.2).

the paired-province trade with Vietnam, nor the private cross-border trade with Vietnam and Thailand, while in the case of Laos barter trade (along the Vietnamese and Chinese borders, for example) amounted to an estimated US$30 million in 1987. In the context of the non-socialist economies, Singapore is pre-eminent, accounting for 33 per cent of exports and 40 per cent of imports, while Indonesia and Malaysia account for a further 43 per cent of exports and 35 per cent of imports. In terms of the balance between exports and imports in the period 1985–6, a positive trade balance was recorded by only the oil exporting countries, most notably Brunei but also Indonesia and Malaysia.

Employing the relatively simple measure of expressing exports as a percentage of GNP and using the World Bank's terminology, Table 9.1 reveals that in terms of trade orientation fairly distinct groupings can be identified. At one extreme are the socialist economies who for largely ideological reasons have followed a strongly inward-oriented trade strategy, with the ratio of exports to GNP not exceeding 5 per cent. In the case of the non-socialist economies, the Philippines and Indonesia can be classified as moderately inward-oriented, while Thailand and, to a greater degree, Malaysia are moderately outward-oriented. Finally, Singapore is a classic example of a small open economy which is strongly outward-oriented, while Brunei's position as the region's major oil exporter places it in this category also.

A comparison of per capita GNP and the degree of openness using the above measure reveals an almost perfect positive relationship. This suggests that those countries which have adopted more outward-oriented strategies have experienced the most significant rates of growth. However, this conclusion must be viewed with a certain degree of circumspection as the relationship could equally imply – to reiterate the point made by Singer (1987) – that the poorer countries find it difficult to progress in general and this would include the establishment of a viable export sector.

Clearly, a critical factor in influencing trade orientation is the commodity composition of exports (which reflects the production structure of an economy) and imports. In particular, successful outward orientation is more likely to be associated with countries who have a 'balanced export' profile (Riedel 1988) in comparison with countries who rely on the production and export of primary commodities.

The strongly inward-oriented economies of Burma, Kampuchea and Laos (there are no data for Vietnam) according to the figures in Table 9.2 are archetypal primary commodity exporters, reflecting the overwhelming importance of agriculture in their economies. However, a note of caution is necessary. The Indochinese states of Kampuchea, Laos and Vietnam have been in a state of flux as a result of political turmoil and internecine wars which have raged since the departure of the United States from Vietnam in the early 1970s and the coming to power of Communist regimes. In addition,

Table 9.2 South East Asia: percentage share of major categories of imports and exports 1986

Exports	Year	All food	Agricultural raw materials	Fuels	Ores/ metals	Manufactured goods	Unallocated	Total (US$ million)
Socialist								
Burma	1976	64.2	28.7	1.0	3.2	2.9	0.0	192.6
Kampuchea	1972	57.4	33.6	0.0	2.9	6.1	0.0	10.0
Laos	1974	2.6	81.5	0.0	11.9	4.1	0.0	11.3
Vietnam	–	–	–	–	–	–	–	–
Non Socialist								
Brunei	1985	0.0	0.0	99.9	0.0	0.0	0.0	2 928.8
Indonesia	1984	7.3	6.7	71.7	3.6	10.0	0.8	21 887.8
Malaysia	1984	19.2	20.2	30.0	4.2	26.4	0.2	16 483.6
Philippines	1986	26.8	4.8	1.3	10.7	30.1	26.3	4 730.0
Singapore	1986	8.0	4.4	20.5	2.7	58.4	6.1	22 427.9
Thailand	1986	44.5	8.2	0.8	3.1	42.7	0.7	8 786.5
Imports								
Socialist								
Burma	1977	9.2	0.9	2.5	10.1	76.8	0.4	299.3
Kampuchea	1972	31.2	2.3	1.7	2.8	62.0	0.0	101.3
Laos	1974	31.7	0.4	11.2	6.5	50.3	0.0	64.8
Vietnam	–	–	–	–	–	–	–	–
Non Socialist								
Brunei	1985	20.6	0.2	1.8	8.5	66.1	2.8	610.5
Indonesia	1985	6.5	3.9	19.5	8.9	60.6	0.6	13 864.5
Malaysia	1984	11.5	1.6	10.2	7.8	68.7	0.3	13 952.9
Philippines	1986	9.7	2.9	17.2	6.7	40.2	23.3	5 041.8
Singapore	1986	9.5	2.8	19.8	4.3	62.0	1.6	25 461.4
Thailand	1986	6.9	4.5	13.5	10.2	59.6	5.4	9 124.0

Source: UNCTAD (1987, Tables 4.2 and 4.4).

as noted earlier in this chapter, significant internal economic changes have occurred and are occurring which will undoubtedly affect the trading positions of these countries in the future. In the case of Vietnam, some liberalization of the economy has already been noted, and a physical expression of this is the creation of free trade zones in Ho Chi Minh City and at Haiphong Airport. In addition, exploration is underway to determine the size of petroleum reserves discovered in the Mekong Delta. Although the Second National Plan (1981–5) envisaged a more balanced export profile, including the export of consumer goods, exports are still dominated by coal and fish. In terms of both exports and imports, an important distinction in the Indochinese context must be made between convertible and non-convertible zone trading*. In Laos, for example, imports from the non-convertible zone (mainly machinery, vehicles and consumer goods) have clearly been more significant than convertible-zone imports (petroleum products, machinery, foodstuffs and raw materials).

Although in terms of their ratio of exports to GNP, the Philippines and Indonesia both fall in the moderately inward-oriented category, their export profiles are markedly different. Indonesia is clearly dependent on petroleum for the majority of its export revenue, while in proportionate terms its manufactured exports are the lowest in ASEAN after Brunei. However, as noted earlier, Indonesia has recently been relatively successful in reorienting and diversifying its export activities towards the non-oil sector (for example, plywood, rattan, coffee, rubber, textiles and handicrafts). In contrast, the Philippines in 1986 displayed a more balanced export structure. In terms of food items, the major exports are coconut oil and sugar, while wood products are also a significant export earner. Although a high proportion of industry is heavily protected and aimed at the domestic market, a number of non-traditional manufacturing export industries have been developed, particularly electronics and clothing. Furthermore, outflows from services and other transfers ('unallocated' in Table 9.2) have become an important source of income in recent years, particularly remittances from Filipinos working abroad and tourists visiting the Philippines. Both countries import similar proportions of fuels; however, in terms of manufactured goods Indonesia's import bill was proportionately higher than the Philippines (which is the lowest in ASEAN) reflecting the higher rates of protection for domestic industry in the latter country.

Malaysia and, to a lesser extent, Thailand are moderately outward-oriented economies, and in recent years trade has functioned as a major engine of growth in both countries. Malaysia

* Non-convertible zone trading is trading within the Communist bloc in non-convertible currencies.

is the most resource-productive nation in South East Asia, being the world's largest exporter of natural rubber and also accounting for a large proportion of world markets in palm oil, tropical hardwoods and tin. However, the majority of its export income is derived from petroleum products. Large investments have been made in developing alternative, mainly industrial, sources of export revenue (with the help of the income from oil production). The manufacturing sector, in particular, has grown relatively rapidly and reached 25 per cent of GDP in 1988. The industrial masterplan for 1986–95 re-emphasizes the importance of manufacturing and export orientation in national development. Specific emphasis has been placed on primary resource-based products over non-resource based industry, but development in the latter category is still envisaged in, for instance, textiles and electronics. In terms of primary resource-based products, high value-added timber products are still not exported in large quantities, for example, although the furniture industry offers potential.

Thailand's economy has grown rapidly in recent years, increasing by 9 per cent in 1988, one of the highest rates in ASEAN. Since the mid-1980s, exports have played a significant role in the overall growth process, increasing by 17 per cent in 1987. In comparison with Malaysia, Thailand exhibited a less diversified export structure in 1986, with two categories (food and manufactured goods) accounting for 87 per cent of the total. During the 1960s primary exports accounted for 90 per cent of total revenues, although still the major export earner this figure, arising largely from rice, vegetables, rubber, fish and maize, had fallen to 42 per cent by 1986. During the mid-1980s, many of Thailand's commodities were affected by decline in prices, as a consequence a major effort has been made to diversify its export structure, specifically in the area of manufactured goods (the principal activities being the processing of primary products, textiles/clothing and electrical machinery). A major stimulus during the latter part of the decade has been the increasing levels of foreign investment, particularly from the newly industrialized economies in the region. Thus in 1986 foreign investment totalled $1 billion, while in the first six months of 1988 $3 billion of new investment had already been recorded. On the basis of its recent performance, predictions are that Thailand will be among the region's next newly industrialized economies (see Hamilton 1987). In terms of imports, manufactured goods, not surprisingly, form the largest category for both countries followed by fuel. Although its staple crop is rice, relatively unsuccessful policies towards food crop production plus a rapid rate of population growth mean that Malaysia has to import a number of foods.

The final, strongly outward-oriented, category consists of Brunei and Singapore. Brunei relies almost entirely on its offshore oilfields for its livelihood. In recent years it has made efforts to attract private industry, and is anticipating developing as a financial centre.

The 1986–90 plan aims to reduce reliance on petroleum and natural gas products. Diversification, therefore, is planned through the expansion of agriculture and industry. In terms of industry, expansion is limited by poor infrastructure and the small domestic market; consequently emphasis is on small-scale industrial projects. With per capita income approaching European levels, it is somewhat of a misnomer to refer to Singapore as a less developed country. Traditionally, Singapore's economy has been based on its position as an entrepôt, both for raw materials from neighbouring countries and also finished goods from around the world. However, during the 1960s stagnating entrepôt trade and increasing unemployment levels provided the stimulant for the diversification and industrialization of the country's economic base specifically directed towards export markets. Emphasis was placed on encouraging foreign investment using a variety of incentives including the creation of export processing zones.

Singapore's industrial success, therefore, is very largely founded on the involvement of multinational corporations. In the period 1960–80 for example, foreign direct investment represented 90 per cent of investment in the manufacturing sector and 20 per cent in the service sector. Even by 1981, foreign firms accounted for 87 per cent of direct exports, while in 1983 38 per cent of the gross fixed assets of foreign companies were in petroleum, 18 per cent in electronics and electrical machinery and 9 per cent in non-electrical machines (Dicken 1987). These sectors were also the most significant in terms of export revenue (see Table 9.2). Although Singapore's growth rate declined in 1985 it emerged from the recession of the mid-1980s relatively unscathed, recording a growth rate of 10 per cent in 1988 (largely as a result of the strong performance of the electronics sector). However, as noted at the beginning of this section, misgivings are now being voiced about the over-reliance on MNC involvement and influence.

The import structure for Brunei is concentrated in two sectors – food and manufacturing. Major items include machinery and transport equipment, basic manufactures, food and live animals and chemicals. Singapore's import structure is similarly dominated by manufactured goods but the second ranking category is 'fuels', consisting largely of imports of oil for subsequent refining. Other specific items include primary commodities, construction-related materials and machinery, many destined for re-export.

The above discussion has provided a picture of the trading patterns of the countries of South East Asia. However, a more generalized and dynamic picture can be obtained by examining the relative change in the number of goods exported. The 'diversification' index used by UNCTAD is computed by measuring the absolute deviation of a country's commodity share from the world structure. The indices range between 0 and 1 with those closer to 0 exhibiting diversity while those nearer 1 complete specialization. This measure discriminates more finely between

Table 9.3 South East Asia: diversification of exports 1970 and 1984

	1970			1984		
	Number of commodities exported	Diversification index	Concentration index	Number of commodities exported	Diversification index	Concentration index
Socialist						
Burma	17	0.92	0.50	49	0.86	0.34
Kampuchea	25	0.92	0.43	–	–	–
Laos	7	0.91	0.45	25	0.83	0.35
Vietnam	–	–	–	–	–	–
Non Socialist						
Brunei	7	0.93	0.98	7	0.86	0.69
Indonesia	48	0.82	0.37	121	0.69	0.50
Malaysia	131	0.79	0.37	161	0.62	0.28
Philippines	78	0.86	0.32	126	0.75	0.30
Singapore	159	0.61	0.30	174	0.50	0.24
Thailand	92	0.83	0.26	138	0.74	0.18

Source: UNCTAD (1987, Table 4.5).

239

countries which are relatively more diversified. To overcome this weakness a 'concentration' index may also be used. This is calculated by taking account of the value of exports of a commodity measured against that of the values of exports of all commodities. This second measure is also known as the Hirschmann index and is normalized to ensure values range between 0 and 1, the latter indicating maximum concentration. It discriminates more finely between countries which are relatively more concentrated in their export structure. Table 9.3 summarizes these indices of functional specialization for 1970 and 1984 to show relative change through time.

In terms of the present-day socialist economies the evidence in Table 9.3 reveals an export structure heavily dependent on a relatively small number of commodities in 1970. A consideration of these indices for the non-socialist economies reveals that while Singapore was tending towards diversification in its exports, the other nations had indices ranging from 0.79 for Malaysia to near complete specialization for Brunei (0.93). While the 'concentration' pattern is not so clear, the primacy of Brunei's oil exports is underscored (0.98). The index of concentration for the remaining nations is rather diffuse, perhaps a weakness in the measure itself in its inability to discriminate between nations whose exports are spread out between two, three or even four products.

Similar comments apply when analysing the values of both indices for 1984. A further observation is that the quantum value of the 'diversification' index had fallen by 1984, indicating a trend towards a diversity of commodities entering the export trade for nearly all South East Asian countries although only moderately so for the socialist economies. The same cannot be said for the measure of concentration because Indonesia is shown to have increased its reliance on some commodities. For all other countries, however, the values are much smaller in 1984 compared with 1970. Finally, except for Brunei, which shows no change in the number of commodities exported between the two dates, all countries had increased the number of commodities they export. Some of these increases are minimal, for example, Singapore from 159 to 174 commodities, whereas others have recorded spectacular increases, for instance the Philippines from 78 to 134 and Indonesia from 48 to 115 goods.

Although increases in the number of goods traded might provide some evidence for economic growth and development, the value of these goods and their relative purchasing powers are equally important. It has been proposed that one way to measure gains from trade is to examine the purchasing power of exports, or the income terms of trade. This is because neither individual values nor volumes are an appropriate measure of trade performance, since these may simply reflect price increases. To capture the impact of changes to both value and volume, export performance may be gauged in terms of a country's income terms of trade. This measure may be derived from the value index of exports deflated by the

Table 9.4 South East Asia: purchasing power of exports[1] for selected years 1970–86

	1970	1975	1980	1985	1986
Socialist					
Burma	78	60	100	74	58
Kampuchea	–	–	–	–	–
Laos	–	–	–	–	–
Vietnam	–	–	–	–	–
Non Socialist					
Brunei	6	37	100	75	63
Indonesia	15	50	100	88	70
Malaysia	47	51	100	134	119
Philippines	83	73	100	87	100
Singapore	–	–	–	–	–
Thailand	44	67	100	120	158

Note:
1. The purchasing power of exports, sometimes known as the income terms of trade, is derived from the value index of exports deflated by the import unit value index, with the base year (1980) showing 100.
Source: UNCTAD (1987, Table 7.2).

import unit value index with the base year shown as 100. Some references exclude oil in calculating this purchasing power (World Bank 1981, p. 20).

The purchasing power of exports for selected years between 1970 and 1986 is shown in Table 9.4. Three patterns may be noted. First, two countries, Brunei and Indonesia, both oil exporters, have benefited from spectacular price increases during this period, increases since substantially reduced, and this is reflected in substantial variations in their purchasing power index. Second, there are two countries, Malaysia and Thailand, which show a steady increase in the purchasing power of their exports, presumably as such exports have become more diversified, in particular away from overwhelming dependence upon primary commodities. Finally, among the non-oil countries Burma stands on its own in showing very poor performance in the purchasing power of its exports since 1970. Falling price levels for rice, lower export volumes and a generally depressed trade picture account for this pattern. In general, the data show only slowly developing trends towards more favourable income terms of trade. A majority of South East Asian countries are still dependent on only a few exports and therefore find it difficult to vary their product mix relative to changes in price levels. There is still relatively little processing of raw materials before trading and as a result the 'value' of the goods sold *vis-a-vis* the international price of the raw material is restricted. One obstacle towards increasing the amount of processing before exchange is the variety of trade barriers and tariffs imposed on 'pre-processed raw materials' by importing countries.

Data exist to show the relative distribution of trade between the individual countries and the major trading blocs – the Pacific, Japan, United States and 'other' for the rest of the world. For South East Asia, no one bloc is dominant but Japan and to a somewhat lesser extent the United States are highly important. Only Malaysia and Singapore have significant exchange relations with the Pacific area compared with their overall activity. Brunei and Indonesia concentrate an important part of their commerce on Japan, while Burma, the Philippines and Thailand have a high volume of interaction with other countries of South East Asia and the rest of the world. Brunei and Indonesia exchange oil and petroleum with Japan for manufactured and industrial products. A large amount of intra-regional dealings is apparent and may be accounted for by rice from Burma and Thailand and sugar from the Philippines going to the other countries of South East Asia, with a concomitant reverse flow of manufactured and consumer products. It is significant that the United States' influence in South East Asian commerce is reduced somewhat by barriers such as tariffs and commodity preference. Only the Philippines sells a fairly large proportion of its products to the United States, due mainly to strong historical ties and the preferred trade status it has obtained.

Until recently Singapore has also benefited under the Generalised System of Preferences (GSP) which allows entry of selected goods duty free (as have the other three East Asian 'tigers'). Since January 1989, however, all four 'tigers' have been deleted from the list of eligible countries. This has raised the possibility that Indonesia, Malaysia, the Philippines and Thailand (who remain beneficiaries under the GSP) could improve their position as a result of increased inward investment. Thailand, in particular, is being viewed as a major growth centre for industries requiring a stable, low-paid labour force. In the context of the socialist economies, trade with other socialist countries, not surprisingly, is the main feature. In the case of Kampuchea and Laos the main trading partners are Vietnam and the USSR. Nevertheless, Kampuchea has a limited amount of undeclared trade with Thailand and Singapore, while Laos has regular trade contacts with Thailand, Japan, Singapore and France. Similarly, Vietnam's major trading partner is the USSR; however, in recent years a major attempt has been made to reduce dependence on CMEA countries. Within the region Singapore, Japan and Hong Kong are significant trading partners of Vietnam, while to a lesser extent France, Sweden and Australia are the major partners outside the region.

In this section, the structure of international trade has been examined in terms of four categories of trade orientation: strongly outward-oriented, moderately outward-oriented, moderately inward-oriented and strongly inward-oriented. On the basis of the ratio of exports to GNP the latter category consisted exclusively of the socialist states, while the non-socialist states were spread throughout the remaining three categories. On this analysis, the implication would appear to be that trade orientation is positively related to levels of per capita income, and that outward-orientation is closely associated with a more balanced export structure (with the exception of Brunei). Furthermore, the weakness of a dependence on a narrow range of primary produce, such as rubber, palm and vegetable oils, rice, wood, sugar, fish products and petroleum and other minerals, has encouraged many South East Asian nations (both socialist and non-socialist) to seek diversification of their external trade.

However, development is a much broader concept and involves a wider range of criteria than simply increases in aggregate growth rates or per capita income (see Ch. 3). Thus, it would be unwarranted to recommend specific trade policies as providing part of a preferred route to development. Trade policy should constitute part of the overall development planning process which should be tailored to fit the conditions and potentialities found in particular countries. As Haq (in Streeten 1972) has commented:

'Trade should not be regarded as a pace setter in any relevant development. . . but as a derivative. The LDCs should first define a strategy for attacking their problems of unemployment and mass poverty. Trade possibilities should be geared to meeting the objectives of such a strategy.'

Nevertheless, most countries in the region are seeking to adopt (if they have not done so already) an export-oriented strategy either by increased local processing of indigenous raw materials or by export-oriented industrialization. Such a policy would also require a parallel restructuring of imports (which appears to be occurring) away from manufactured and consumer goods from the West and Japan, towards the import of capital and intermediate goods. Although this restructuring may have an adverse effect on the balance of payments in the short term, the argument is that these structural readjustments may be beneficial in the long run.

Although the potential benefits of the above strategy are immense it is appropriate to conclude this section with a note of caution which has been clearly expressed by Michael Todaro (1985).

'An export-oriented strategy of growth, particularly when a large proportion of export earnings accrue to foreigners, may not only bias the structures of the economy in the wrong directions (by not catering to the real needs of local people) but may also reinforce the internal and external dualistic and inegalitarian character of that growth. Therefore, the fact that trade may promote expanded export earnings, even increase output levels, does not mean that it is a desirable strategy for economic and social development. It all depends on the nature of the export sector, the distribution of its benefits, and its linkages with the rest of the economy.'

Trade and regional integration: looking both outward and inward

Free trade or protection, nationalism or regional co-operation, are important issues in South East Asia as elsewhere in the Third World. There is the view that greater economic complementarity can be generated through enlarged intra-regional trade. The growth of such trade depends to a large extent on regional co-operation by individual countries. Increasingly, the countries of South East Asia are beginning to realize that the high degree of competition among themselves may be detrimental to their growth and development and that other developing regions in the world may be competing for the same markets. For South East Asian countries in particular, it is as if they were in a zero-sum game in which no group of countries would gain and all would lose from slower growth. The negative effects of competition prompt an examination of trade in the context of regional integration and cooperation. The case study selected is that of the Association of South East Asian Nations (ASEAN).

Regional integration or co-operation here refers to the joining together of two or more countries in economic partnership. The purposes of such a 'union' are to promote more trade among members through economic cooperation, to generate external trade by presenting a 'common front' and generally to raise the standards of living in each country. A successful union should be able to re-orient trade and achieve economic development more rapidly for all member countries. ASEAN is one such organization, though its objectives are also political. It was formally established in Bangkok on 8 August 1967 with the signing of a Joint Declaration by Foreign Ministers from Indonesia, Malaysia, the Philippines, Singapore and Thailand. Brunei joined later. The main objectives are to promote regional cooperation in the economic, social, cultural and technical fields (Broinowski 1982).

Of the five degrees of economic integration that are possible, ASEAN is nearest to the first category, that of a free trade area, since it hopes to remove barriers to the movement of goods between its member countries. The ASEAN agreement initially provided for negotiations leading to the introduction of preferences product by product. In 1982 a system of across-the-board tariff cuts was adopted to replace the earlier approach and by mid-1984 18 000 items had been included. ASEAN cannot be included in the second type, a Customs Union, since this includes the erection of a common tariff wall against non-members. The remaining three forms of integration build on the basic model with increasing degrees of centralized control, from a Common Market, to an Economic Union, and finally to complete Economic Integration. Invariably all forms of integration are designed to bring about both internal as well as external economies of scale.

The merits of establishing a regional grouping, especially by developing countries, have been summarized by Balassa (1961). The first benefit results from the elimination of barriers to trade by partner countries, thereby widening the market for the primary and second industries of integrated economies. Second, as a trade group, integration may increase the bargaining power of participants in trade negotiations. Third, the improved flow of information about markets of participating and other countries reduces risks and uncertainty in planning production levels. Finally, the group as a whole allows for the regional protection of 'infant industries' for which perhaps the national markets themselves may either be too small for efficient operation or unable to gain from economies of scale.

Counterbalancing these advantages, are a number of elements which mitigate against the formation of regional trading blocs. Higher prices may be paid for imports from regional partners – implying that production is less efficient and more expensive than elsewhere in the world. Regional integration may also lead to a retardation in the introduction of new technology because of less competitive conditions; and there may be a loss of potential gains from trade with the rest of the world due to the regional market orientation.

The position of ASEAN is that factor endowments between the various economies are nearly similar as are their respective comparative advantages in some commodities (Wong 1979). Benefits from integration may therefore be meagre. It is not surprising that the present level of economic integration, as manifested in trade flows among ASEAN nations, is still very low (Table 9.5). The member nations have a low absorptive capacity for primary commodities such as rubber, coconut and oil palm. Moreover, the

Table 9.5   ASEAN: intra-trade.

	1960	1970	1976	1980	1985
Value of intra-trade (million US$)	839	860	3 169	11 918	12 713
Exports to developing countries as a percentage of total ASEAN exports	32.8	31.7	30.3	35.6	35.2
Intra-trade of ASEAN as a percentage of total exports of group	21.7	14.7	13.9	17.8	17.9

Source: UNCTAD (1987, Table 1.13).

expanding export markets of developed countries are presently able
to take most of the commercial capacities of ASEAN primary
production. The reverse flow of imports to ASEAN countries in the
form of capital and intermediate goods mitigates against the policy
of attempting to keep trade intra-regional. Finally, much of the trade
with developed countries is reciprocal in nature, in response to the
tugs on the 'strings' attached to loans and foreign aid. Therefore,
the growth of intra-regional trade depends on the appropriate
restructuring of existing trade policies currently pursued by member
countries to minimize the disadvantages. A first step was taken in
the 1976 Bali Declaration of Accord which committed ASEAN
nations to preferential trade arrangements as a long-term goal. With
tariff concessions and formalized trade agreements, it can be
expected that intra-regional trade will grow further among ASEAN
nations.

Economic integration appears to be an obvious solution to the
problem of development for small nations. By combining markets,
and co-operating in the development process, there may be some
hope of improving their economic position. However, two major
constraints still need to be overcome. The first is political in that
national goals would need to be overtaken by supra-national goals
for economic gain. Many small nations may feel this to be too great
a sacrifice. Second, the problems of spatial dispersion of participant
countries may be a hindrance, especially in the distribution of the
benefits of integration. For ASEAN it seems that these constraints
have not yet posed serious problems. A relatively large population,
the size of its potential market, and its archipelagic location in
which all member countries are in nearly contiguous positions are
positive attributes. Most importantly, the use of typically Asian
devices has produced consensual agreements in which the
'traditional spirit of ASEAN friendship and solidarity' has been
preserved without sacrificing national objectives.

In examining the intra-ASEAN pattern of trade, it may be noted
that, for both exports and imports, the percentage shares are small,
being somewhat less than a fifth of total trade for all countries with
the exception of the Philippines. The Filipino component of trade
within ASEAN is by far the lowest among all member countries,
perhaps reflecting its strong historical links with the United States
where it enjoys preferential treatment in tariffs and trade subsidies.
It will clearly take some time before the flow of Filipino trade is
directed more towards the ASEAN region. There are also low levels
of trade with other Asian countries, excluding Japan. Trade with
Japan and the rest of the world is by far the largest, and this
pattern will continue until such time as ASEAN nations begin
manufacturing the capital goods and machinery that are presently
being imported.

The network of trade linkages between ASEAN nations is
dominated by one member nation – Singapore – whose primacy in
exports is matched by its lead position in imports from all other

ASEAN countries. Singapore can thus be described as the pivot in intra-ASEAN trade for both exports and imports. The concentration of trade with Singapore is heaviest for Malaysia and Indonesia but is more diffuse for the Philippines and Thailand, the latter two also recording the lowest trade values either for exports or imports to Singapore.

The next step seems to be the need for a regional trade policy for ASEAN, so that foreign demand can wisely be channelled. This may result in more efficient and profitable production especially in the agricultural sector. So far the evidence suggests that interdependence among developing nations can be nurtured and made to grow for the benefit of all participants. There appears to be less of a need for a conscious effort directed towards the strategy of import substitution than was the case for most countries immediately after independence. The strategy most required is one of export expansion, the logical extension to import substitution. Membership of the regionally integrated economies of ASEAN can produce this trade creation effect.

ASEAN provides a good lesson to the other developing countries in South East Asia, and especially to the coterminous and socialist Indochinese states in the beginnings of regional cooperation and the economic advantages of integration. Recent performances at the United Nations and other international fora also suggest that ASEAN is becoming a more important political force (Leifer 1989). Groupings like ASEAN may provide some solution to the acute balance of payments problems many developing nations face. Some of these problems stem from underdevelopment of resources which may require considerable external assistance to alleviate. This raises the question of international development assistance or 'aid'.

Development assistance

In the sixties the catch-word was 'trade not aid', and this came from newly independent countries whose march towards development saw many opportunities for trade. Some South East Asian countries were able to hop on to this bandwaggon of trade, whereas for others there could be little or no advantage because of the underdevelopment of their resources. Development and growth are the result of two or more interacting forces. On the one hand, the increase in primary and secondary production is the reflection of growth within any country. On the other, in order to 'cash in' on the increased productive capacity, trade acts as an external boost to economic growth. However, before these two interacting elements of increased production and growth in foreign trade may be set in motion, catalysts in the form of investments and development assistance are required. A number of countries in South East Asia are experiencing a so-called 'foreign exchange gap', which may imply that their productive capacities are either under-utilized or under-developed, at a stage when a fuller exploitation of these potentials is necessary. For these reasons, development projects initiated with foreign aid should have a significant impact on capacity utilization.

Aid refers to Official Development Assistance (ODA) grants, or loans with a grant element. A major source of financial aid is the World Bank, which has the objective of helping to raise standards of living in developing countries by channelling financial resources from developed countries to the developing world. World Bank loans generally have a grace period of five years, and are repayable over 20 years or less. One arm of the World Bank is the International Development Association (IDA) and its funds are called credits, to distinguish them from World Bank loans. Credits, made to governments only, have a 10-year grace period, 50-year maturities and no interest. An analysis of World Bank and IDA lending to countries in South East Asia reveals that the proportion of loans and credits is rather small relative to world-wide operations. World Bank loans to South East Asia amounted to US$21.2 billion, about 15 per cent of all loans, and there were US$2.0 billion in credits in 1987 (Table 9.6). The relatively small amount of credits is due to World Bank policy, since most credit is reserved for 'poorer' developing countries, and with terms that would bear less heavily on their balance of payments than would World Bank loans.

Most of the 391 World Bank loans granted to the region by June 1987 went to four countries, Indonesia (119), the Philippines (99), Thailand (86) and Malaysia (70). There were 14 loans to Singapore, making the domination of ASEAN countries virtually complete. This partiality of the World Bank towards the more successful economies is a reflection of World Bank confidence in them and its philosophy of supporting countries which are likely to meet repayments. Burma is the only other country to which World Bank loans have been granted, apart from ASEAN. The Indochinese states have yet to negotiate any loan.

Half of IDA credits have gone to Indonesia and almost a third to Burma. The Philippines, Thailand, Laos and Vietnam have also received IDA credits. Most of these have gone mainly to finance basic infrastructure such as transport and communication works, agriculture and population control.

The purposes of these loans and credits are given in Table 9.7, which provides data on East Asia and the Pacific, not South East Asia alone. This regional classification includes all ASEAN states, Fiji, Hong Kong, the Republic of Korea, Papua New Guinea and Taiwan. Excluded from it are Burma, which is placed under South Asia, and Laos, Kampuchea and Vietnam. The data defy further disaggregation and, while this regional classification by the World Bank complicates issues, it gives a rough idea of the relative emphases of project support. About one-fifth of all World Bank loans have been given for agricultural and rural development, and about a further one-fifth each to energy and to transportation

Table 9.6 South East Asia: World Bank and International Development Association (IDA) cumulative lending operations by borrower or guarantor, June 30 1987

Country	World Bank Loans Number	Amount (U.S. $ millions)	IDA Credits Number	Amount (U.S. $ millions)	Total Number	Amount (U.S. $ millions)
Burma	3	33.4	29	741.0	32	77.4
Kampuchea	–	–	–	–	–	–
Laos	–	–	7	82.9	7	82.9
Vietnam	–	–	1	60.0	1	60.0
Brunei	–	–	–	–	–	–
Indonesia	119	10 180.9	46	931.8	185	11 112.7
Malaysia	70	2 329.9	–	–	70	2 329.9
Philippines	99	4 808.7	3	122.2	102	4 930.9
Singapore	14	181.3	–	–	14	181.3
Thailand	86	3 692.6	6	125.1	92	3 817.7
South East Asia Total	391	21 226.8	92	206.3	483	23 289.8
World Total	2 818	140 286.8	1 699	43 307.7	4 517	183 594.5

Source: World Bank (1987b, Appendix 6.6, pp. 160–3).

Table 9.7   East Asia and Pacific:[1] World Bank and International
Development Association (IDA) cumulative lending by major purposes,
June 30, 1987 (in US$ millions)

Purpose[2]	World Bank Loans		IDA Credits	
	Value	Percentage	Value	Percentage
Agriculture and rural development	6 863.2	20.5	1 823.1	42.8
Development finance companies	2 979.0	8.9	205.5	5.4
Education	2 094.9	6.2	750.3	19.8
Energy	6 964.3	20.8	157.8	4.1
Industry	1 033.6	3.0	35.0	0.9
Non-project	2 479.3	7.4	–	–
Population, health and nutrition	309.9	0.9	236.3	6.2
Small scale enterprises	966.5	2.9	56.5	1.5
Technical assistance	23.0	–	75.7	2.0
Telecommunications	400.4	1.2	16.7	0.4
Tourism	25.0	–	16.0	0.4
Transportation	6 103.1	18.2	409.3	10.8
Urban development	1 993.4	6.0	100.0	2.6
Water supply and sewerage	1 133.4	3.3	106.4	2.8
Grand Total	33 369.0	100.0	3 788.4	100.0

Notes:
1. East Asia and Pacific includes: Fiji, Hong Kong, Indonesia, R.O. Korea, Malaysia,
   Papua New Guinea, Philippines, Singapore, Thailand, Taiwan.
2. Operations have been classified by the major purpose they finance. Many
   projects include activity in more than one sector or sub-sector.
Source: World Bank (1987b, Appendix 6.5, pp. 159–9)

(mainly land). Thus, these three headings account for a substantial
proportion of all Bank loans, and this proportion may be even
greater because some projects or activities are included in more than
one sector or sub-sector. The relationships between building basic
infrastructure, for example roads and highways, and agricultural and
rural development are rather obvious, one depending on the other
for successful implementation and outcome. With the 'oil crisis' of
the 1970s and rising fuel prices to 1986, proposed energy projects
found favour, because cheaper hydro-electricity may mean lower
production costs in industries. Support from the Bank has also been
forthcoming for development finance companies and for education,
the former mainly to enable governments to provide 'seed-funds' for
starting up industries and other economic projects. Loans for
education have been disbursed for the technical training of
personnel, agricultural extension workers and development advisers.
   The pattern of IDA credit distribution is similar to that of World
Bank loans with agriculture and rural development taking the lion's

share (about 43 per cent), and with education (20 per cent) and transportation (11 per cent) the other major beneficiaries. Significant credit has also been granted to energy projects and to population, health and nutrition. Together both World Bank loans and credits amount to about US$3.7 billion. The share taken by East Asia and the Pacific is about 10 per cent of the grand total.

It is clear that the grant of loans and credits is biased towards the 'basic' sectors of all developing economies, 'basic' referring here to those infrastructural facilities which are necessary for development, for example, roads and highways, irrigation canals and dams. 'Basic' also denotes those sectors of the economy which produce industrial raw materials, and facilitative items such as area development, rural banks, credit, research and extension work. In industry, the development of fertilizer and chemical plants has been given support, since these are crucial to rapid growth in the agricultural sector. Energy projects have also found favour because they can reduce costs of production in the factories, mills and mines. The granting of loans and credits, however, has yet to extend from its 'basic infrastructure' bias towards heavy industry and the production of capital goods. When that stage is reached it could be said that those countries may no longer need development assistance.

Trade and aid have been touted as two sides of the same coin. While this may be regarded as the very essence of commerce – aid to produce goods for trade – examples of such links are difficult to come by. Further, trade and aid are sometimes treated as one and the same by some donors. An example will serve to show this. For economic reasons, the Philippines copper industry plunged to its lowest level in the 1980s and closure of all major primary production of copper concentrates was threatened. The Philippines asked Japan for a concessional loan, in the form of an advance payment, to alleviate the ailing industry and enable the continuance of price support. However, the Japanese government refused this request, and instead suggested a commercial solution. The significance of this example is that Japan buys 95 per cent of its copper concentrate output from the Philippines. Also, there were other economic issues being negotiated by Japan, including lower tariffs on bananas, the removal of duty from Philippine-made cardboard, an air treaty for increased flights through Tokyo, increased official economic assistance and the financing of big projects like an integrated steel mill in northern Luzon. Japan is the country's second largest trade partner and investor after the United States. This example illustrates the differing perception of particular countries towards aid and trade and the fact that aid is not necessarily charity but rather one part of a complex and often commercial approach to the relationship of countries with each other.

Developing countries in South East Asia have had to combat other serious difficulties. First, increases in the cost of imports relative to export earnings means a reduced growth and in some cases a large deficit. To finance this, governments either borrow funds or seek aid. Second, as output and trade need to be restructured to meet new circumstances, there is a need for heavy investment and capital which these countries lack. The World Bank in particular gives so-called 'structural adjustment loans' (SALs) which are listed as non-projects in official publications. Thailand and the Philippines have benefited from such loans arising from their peculiar circumstances.

In Thailand, for example, structural problems have arisen from what is euphemistically referred to as a 'lack of far-sightedness in economic planning'. Thai economic planners concede that the import substitution policies of the 1960s have led to over-protection of industries because of high tariff walls and poorly executed fiscal policies such as tax collection; they have also been compounded by steadily rising energy costs. Thus, in the fifth *Five Year National Economic and Social Development Plan (1982–6)*, major efforts were directed to correct these structural problems. The experience of the Philippines of two decades of industrialization via 'import-dependent import substitution' has resulted in a badly distorted industrial structure in terms of output-mix (Wong 1979, p. 69). In the development plans, the government has taken measures to promote structural adjustment which include further encouragement to rapid growth of manufactured exports, increased industrial investment, and the de-regulation of the financial sector. These programmes are expected to reduce current account deficits. To finance their revised plans, both Thailand and the Philippines have sought structural adjustment loans from World Bank sources, to be repaid at the standard non-confessional World Bank rates.

'Aiding well-conceived and well-monitored programs makes a difference to the overall development effort' (World Bank 1981, p. 56). There can be no denying that the value of aid has been significant, for instance, in the family planning programmes in Indonesia, as well as in spheres of economic activity. However, there still remains the inherent danger that external factors, beyond the reach of local policy making, may sometimes intrude upon the social and political sovereignty of aid-receiving countries, and misinterpretations with unfortunate connotations may ensue. The pervasive influence of lending agency bureaucrats in the economic policy of developing countries is a much-vexed problem.

Future prospects

Only three countries in the region show reasonably consistent growth rates over the period since 1965: Thailand, Singapore and Malaysia (World Bank 1987a). Their rates of growth have been relatively high when matched against the background of inflation, periods of world-wide recession and the rising fuel prices of the period from the mid-1970s to the mid-1980s. The growth rates of other countries in South East Asia over the whole period since 1965 are not so remarkable. These rates of growth can be expected to continue under the present conditions of rigid economic structures that are slow to change. Growth can be expected to take place more rapidly when the optimum mix to maximize trade between primary production and industrial-manufacturing production has been found.

Within the regional context, trade-partner choice results from the relative needs of each country, and at present there is a greater volume of foreign trade flowing out of the South East Asian region than there is of internal trade. Colonial linkages and historical antecedents are continuing. The exchange of industrial and consumer goods for raw materials, agricultural products and mineral ores has been etched into the 'traditional' trading patterns. However, these patterns are changing, albeit slowly, towards something of a more 'inward-looking' structure. Singapore presents a good example, its volume of trade (exports and imports) within the South East Asian region being by far the largest. Most of the trade consists of transshipment cargo that uses Singapore as the intermediate location before shipment to a third country. Increasingly, however, some processing of the transshipped freight is taking place, not only to add value to the raw materials, but also to prepare them to suit a more fastidious industrial market. Singapore imports crude, low-grade Indonesian natural rubber, for example, and processes it into technically specified block rubber – Standard Singapore Rubber (SSR) – to the requirements of international consumers. In this process there is a value-added component, increasing industrial activity and employment. Similarly, refining activities for Indonesian and Malaysian tin ore and Indonesian crude oil and gas have become important components in Singapore's intra-regional trading pattern.

Through co-operative action there can be an increase in the efficient use of total regional resources. ASEAN has shown how regionalism could increase the collective bargaining power of member countries. Whilst ASEAN consensus is arrived at slowly and ponderously, because of the necessity for thorough consultations and frequent meetings, there can be no danger that decisions are made by domineering members. Even within ASEAN there are significant inequalities, but such differences in economic development need not pose overwhelming problems to regionally integrated economies. There could be policies aimed at inter-sectoral specialization, to capture benefits of comparative advantage in the agricultural and industrial sectors. It could be that regional integration may tune economies to each other and optimize their respective positive effects, with profits and losses shared equitably among the participant countries. ASEAN is still in its infancy and has a long road ahead before full economic integration can be achieved, yet even now it provides lessons to other countries in South East Asia.

South East Asia, like other developing regions of the world, has had to contend with inflation, the world recession and more expensive oil. In the 1990s and beyond, the liberalization of trade and positive moves towards co-operative efforts could pay off in higher productivity, growth and economic development. Whether these benefits eventuate will depend largely on the most valuable resource any country has – its people, who are both the means and the end of economic advance.

Further reading

Amin S (1976) *Unequal Development*, Harvester, London.

Balassa B (1961). *The Theory of Economic Integration*, Irwin, Homewood, Ill.

Brandt Commission (1980). Independent Commission on International Development Issues, *North-South: A Program for Survival*, MIT Press, Cambridge, Mass.

Broinowski A (ed.) (1982) *Understanding ASEAN*, St. Martin's Press, New York.

Cairncross A K (1962) *Factors in Economic Development*, Allen & Unwin, London.

Dicken P (1987) 'A tale of two NICs: Hong Kong and Singapore at the crossroads', *Geoforum* **18**, 151–64.

El-Agraa A M (1983) *The Theory of International Trade*, Croom Helm, London.

Frank A G (1969) *Capitalism and Underdevelopment in Latin America*, Monthly Review Press, New York.

Gilpin R (1987) *The Political Economy of International Relations*, Princeton University Press, Princeton, New Jersey.

Hamilton C (1987) 'Can the rest of Asia emulate the NICs?', *Third World Quarterly* **9**, 1225–56.

Harris N (1987) *The End of the Third World*, Penguin, Harmondsworth.

Kurian G (1987) *Encyclopedia of the Third World*, Facts on File, Oxford.

Leifer Michael (1989) *ASEAN and the Security of South East Asia*, Routledge, London.

Lewis W A (1980) 'The slowing down of the engine of growth', *American Economic Review* **70**, 555–64.

Linder S B (1986) *The Pacific Century: Economic and Political Consequences of Asian-Pacific Dynamism*, Stanford University Press, Stanford.

Milner C (1988) 'Trade strategies and economic development: theory and evidence', in Greenaway, D. (ed.) *Economic Development and International Trade*, Macmillan, London.

Nurske R (1961) *Patterns of Trade and Development*, Blackwell, Oxford.

Peet R (ed.) (1980) *An Introduction to Marxist Theories of Underdevelopment*, Department of Human Geography Publication HG/14, Australian National University, Canberra.

Prebisch R (1959) 'Commercial policy in the underdeveloped countries',: *America Economic Review* **49**, 251–73.

Riedel J (1988) 'Trade as an engine of growth: theory and evidence', in Greenaway, D. (ed.), *Economic Development and International Trade*, Macmillan, London.

Singer H (1988) 'The World Development Report 1987 on the blessings of outward orientation: a necessary correction, *Journal of Development Studies*, **24**, 2, 232–36.

Singh M S (1988) 'The changing role of the periphery in the international industrial arena', in Linge, G. J. R. (ed.), *Peripherality and Industrial Change*, Croom Helm, Beckenham.

Smith S and Toye J (1979) 'Three stories about trade and poor economies', in Smith, S. and Toye, J. (eds.), *Trade and Poor Economies*, Frank Cass, London.

South (1988) *South Survey: Indonesia*, October, 53–64

South (1989) 'Vietnam gets ready to ride the wave', May, 41.

Streeten P (1972) *The Frontiers of Development Studies*, Macmillan, London.

Todaro M (1985) *Economic Development in the Third World*, Longman, London.

United Nations (1988) *Economic and Social Survey of Asia and the Pacific 1987*, Bangkok.

UNCTAD (1987) *Handbook of International Trade and Development Statistics (Supplement)*, New York.

Wallerstein, I. (1974) *The Modern World System*, Academic Press, New York.

Wong J (1979) *ASEAN Economies in Perspective*, Macmillan, London.

World Bank (1981) *World Development Report 1981*, Oxford University Press, New York.

World Bank (1987a) *World Development Report 1987*, Oxford University Press, New York.

World Bank (1987b) *Annual Report 1987*, World Bank, Washington D.C.

Chapter 10

Population mobility

Paul Lightfoot

During a period of rapid social, economic and institutional change it is not surprising that rates of population movement tend to increase. Change creates new economic opportunities and destroys or devalues old ones. Modern communications make South East Asian people aware of new opportunities and dramatically increase their accessibility, whether it is at the agricultural frontier, in town or in the oil-rich countries of the Middle East. Thus almost every relevant survey or set of statistics in South East Asia indicates increasing propensities to move and predicts greater mobility in future. Until the mid–1970s, censuses and other large-scale surveys provided the only generally accepted statistical basis for statements about mobility; more recently a number of studies have demonstrated the inadequacies of these sources as measures of total mobility, and have drawn attention to far higher rates of mobility than were formerly acknowledged.

At the same time scholars have treated mobility more as an integral feature of the overall development and underdevelopment processes currently affecting the region, rather than as a demographic process of interest in itself. The inter-relationships between mobility and economic and social change have been explored more fully, with the result that many assumptions about the causes and consequences of mobility have come into question. Fundamental questions about these relationships remain to be resolved. Nevertheless, being aware of increasing rates of mobility and assuming various associated causes and effects, governments have increasingly regarded population mobility as within the aegis of planning. To a greater or lesser extent, and with respect to various categories of mobility, all South East Asian governments now regard population movement as both subject to control and an instrument of development policy. It is true, nevertheless, that actions likely to affect mobility are not usually concerted, and different developmental policy measures within the same country are often likely to have counteracting effects on rates and directions of movement. These contradictions reflect in part the problems which many South East Asian governments have in ordering developmental priorities, and in part the lack of any satisfactory theory linking mobility and development. But there has been progress, in that relationships between mobility and other aspects of development are now more often the subject of analysis rather than of assertion.

The very patchy nature of the evidence concerning all aspects of population mobility is a serious problem. Burma and the three countries of Indochina are extremely poorly served by both statistics and micro-studies which are relevant. This is unfortunate because the ideology of development in these four countries is very different in many respects from that of the remaining countries of the region; few generalizations are possible for them. Consequently much of what follows will relate to the non-socialist countries: Indonesia, Malaysia, the Philippines and Thailand. Singapore, being a city state, is of course a special case.

Levels and spatial patterns of mobility

National censuses provide far from perfect measures of population mobility. However, they have measured mobility in fairly consistent ways in each of the above four countries, and at more than one point of time. Thus within the limitations of their definitions of mobility, censuses provide a broad picture of spatial variations of mobility rates, and allow for comparisons over time.

It is not possible to make direct comparisons of mobility rates between these countries, as Pryor (1979) has pointed out. The 1970 or 1971 censuses, for example, yield information in all cases on the numbers living in a province or district different from that of their birth. But they record more recent moves in quite different ways. The Thai and Philippine censuses record differences in place of residence between the date of the census and a fixed date five years earlier. The Malaysian and Indonesian censuses define migrants as people who have previously lived in a place different from the place of enumeration, though without reference to any time period. In all cases, place of residence is recorded on a *de jure* rather than a *de facto* basis. In other words people were recorded as living at their 'normal' place of residence, even where they were living temporarily elsewhere at the time of enumeration. Again, interpretations of the term 'normal place of residence' vary between countries. Variations also occur in the spatial extent of the administrative areas which a person would have to leave or enter in order to count as a migrant. Indonesian provinces cover a mean area of 73 250 square kilometres, compared to 1 900 square kilometres for Malaysian districts. Therefore an Indonesian would normally have to move much further than a Malaysian to be recorded as a migrant. Finally, census organizations tabulate their migration data in different ways, so certain kinds of statistics presented in one census are not necessarily available in another, even though they could theoretically be computed from the data initially collected.

Thailand's 1960 and 1970 censuses suggest generally low levels of mobility, but provide unmistakable evidence of a marked increase in rates of movement. The proportion living outside their province of birth increased from 13 to 15 per cent between the. two censuses. The proportion classified as five-year migrants rose more markedly, from 3.8 to 6.9 per cent. Moreover, the number of migrants as a proportion of the population increased in every province, in some cases dramatically (Sternstein 1979). In both censuses most recorded moves occurred over relatively short distances. In 1970 39.5 per cent of the five-year migrants moved within the same province, and a further 21 per cent of migrants moved between contiguous provinces. And the predominant category of moves was that between two rural areas rather than between rural and urban areas. Seventy-two per cent of all recorded moves from 1965 to 1970 fell into this category. Many of these moves were to sparsely settled frontier areas such as Kampangphet and Phetchabun in the lower northern region, Nongkhai and Loei in the upper north east and Satun in the far south.

Only 12 per cent of all recorded moves between 1965 and 1970 were from rural to urban areas, and even this movement was counter-balanced by a large number of urban–rural moves, which contributed 6 per cent of all moves recorded. However, the number of rural-urban movers was still large in relation to the urban population, and migration contributes an estimated 50 per cent to the rapid urban growth which is occurring. The most striking characteristic of rural-urban migration in Thailand is its high degree of concentration on Bangkok. Most migrants originate from the Central Plain, within 150 kilometres of the capital, and those provinces also record the largest increases in rates of movement to Bangkok (Sternstein 1979). Large numbers also move to Bangkok from some north-eastern provinces. Migration from Nakhorn Ratchasima and Ubonratchathani to Bangkok made up two of the largest inter-provincial migration streams recorded in the 1970 census, while the rate of net out-movement to Bangkok was particularly high in Roi-et and Mahasarakham in both 1960 and 1970.

Evidence from peninsular Malaysia similarly suggests a modest increase in mobility rates. The proportion of the population enumerated outside their state of birth rose from 4.6 per cent in 1947 to 8.2 per cent in 1957 and 10.8 per cent in 1970. Saw (1980) has estimated that the number of inter-censal migrants who moved between states increased by 58 per cent from 1947–57 to 1957–70. However, as a proportion of the total population, inter-censal migrants increased only from 5.2 per cent in 1947–57 to 5.8 per cent in 1957–70; only 3.3 per cent of the population over the age of five years had moved across a civil division boundary during the period

1965–70; and since many migration streams have been reciprocated by moves in the opposite direction migration seems to have had little effect on the distribution of the population by state. Only 1.7 per cent of the 1970 population would have had to be moved in order to restore the population distribution to what it would have been had no migration occurred from 1957 to 1970 (Fernandez *et al.* 1976).

Certain states and areas within states have been considerably affected by migration on the evidence of successive censuses. The 1970 census showed that net migration was positive in only four states, most importantly Selangor and Pahang. Selangor contains Malaysia's main industrial centres of Kuala Lumpur, Klang and Petaling Jaya, and was the destination for 38 per cent of inter-censal migrants between 1957 and 1970. Pahang is a sparsely settled state where a number of large-scale land settlement schemes have been started, the number of inter-state lifetime migrants increased by 161 per cent from 1957 and 1970, and the estimated number of inter-censal migrants increased by 177 per cent, both of them more rapid rates of increase than for any other state. Malacca, Perak and Kelantan recorded the highest rates of out-migration in 1970.

As in the case of Thailand, most inter-state moves have been between rural areas rather than between rural and urban areas. Forty-nine per cent of all lifetime migrants were in rural areas in 1970, while 26 per cent were in one or other of Malaysia's eight largest cities with populations in excess of 75 000, and the remaining 25 per cent were in small-sized towns (Siddhu and Jones 1981). Also, local movement is considerably more common than long-distance movement. Almost 8 per cent of the 1970 population were classified as intra-state migrants, with high rates in Kelantan, Johore, Kedah and Perak. Rates of movement were particularly high in districts containing state capitals. And most inter-state movers moved to states adjacent to their state of origin: over half of Selangor's in-migrants came from neighbouring Perak and Negri Sembilan.

For Indonesia, census data suggest not only low mobility rates, but also very slow rates of increase in mobility. In 1930 3.2 per cent of the population were classed as lifetime inter-provincial migrants; by 1971 the proportion had risen only to 4.4 per cent. This aggregate change in the rate of movement disguises wide variations: the rate of movement from Inner to Outer Indonesia declined in relation to the population of Inner Indonesia, while the rate of movement in the opposite direction increased approximately fivefold. As in the case of Malaysia, migration as defined by the census has contributed only marginally to population redistribution in Indonesia. In spite of efforts over a long period to encourage movement from Java to the Outer Islands, Hugo and Temple (1975) estimate that the total net lifetime movement which has occurred represents little more than a single year's natural increase for the population of Java.

Again, however, the effects of migration have been concentrated. Two areas stand out as having a high incidence of in-migration: in Jakarta 41.1 per cent of the resident population in 1971 were in-migrants; and in Lampung, a province of extensive land settlement projects in southern Sumatra, the proportion was 36.8 per cent. In no other province were more than 17 per cent of the population classed as in-migrants. The incidence of out-migration is less concentrated with a maximum of 11.9 per cent of the population in Yogyakarta and West Sumatra and with rates of more than 10 per cent in three other provinces. Central Java and South Sumatra recorded the highest rates of net out-movement by lifetime migrants.

In the Philippines, the proportion of the 1970 population classed as lifetime inter-provincial migrants was 17.3 per cent, a marginal increase from the 1960 figure of 16 per cent. There was a similarly small increase in the proportion living in a region other than that of their birth, from 12.6 per cent in 1960 to 13.2 per cent in 1970. Between 1960 and 1970 4.7 per cent of the population moved within the same province; a further 2.0 per cent moved between provinces within the same region; and 7.1 per cent moved between regions (Perez 1978). Although it is difficult to make comparisons between countries, these figures suggest relatively high levels of mobility in the Philippines, with a particularly high proportion of long-distance moves.

Broadly, two trends can be discerned in both the 1960 and 1970 census data: moves to the Manila area, especially to the Province of Rizal; and moves from more crowded to less crowded rural areas, especially to the frontier areas of the southern island of Mindanao. Rizal in 1970 recorded by far the largest number of net five-year in-migrants, with a net gain of 338 000 over a five-year period, more than ten times the number gained by any other province. After Rizal the next six highest ranked provinces for net in-migrants were all in Mindanao. The main out-migration provinces were the city of Manila, which made a net loss of 189 023 migrants for 1965–70 as a result of suburbanization, and the central and western Visayas, Negros Occidental, Leyte, Samar and Cebu. The Philippines' largest rural–rural migration streams during the 1960s were from the Visayas to Mindanao, often related to governmentally sponsored land-settlement schemes.

Censuses provide a broad picture of mobility with respect to certain kinds of population movements, usually moves involving relatively long distances and relatively permanent changes of residence; but a very large number of moves do not meet the time and distance criteria of censuses, and therefore go unrecorded. A series of studies published since the middle 1970s indicate that temporary and short-distance moves predominate in the total picture of rural–urban movement in modern South East Asia (Goldstein 1978), as had already been demonstrated for other parts of the Third World. A great deal of work has been carried out in Indonesia, particularly by Hugo. His study of 14 villages in West Java showed that only one-third of all the moves which he observed would have met the time criterion of the national census; and of that third, the proportion within each village who had moved within the province of West Java, and therefore would not have qualified as migrants, ranged from 76 to 98 per cent (Hugo 1978). Contrary to the impression from the 1971 census of a sedentary population, three-quarters of all households in the 14 villages depend in part on income earned outside of their home villages, mostly in the two cities of Jakarta and Bandung.

Micro-studies elsewhere in South East Asia similarly suggest high and increasing levels of temporary movement. Hugo (1982) reviewed evidence from central Java and parts of Sumatra, Sulawesi and Kalimantan. The spatial and temporal patterns vary widely. In some areas modern patterns of circulation are based on long traditions of mobility, as among the Minangkabau of western Sumatra, Buginese traders of southern Sulawesi, and Banjarese people of Kalimantan. Elsewhere changing agricultural practices have created increasing numbers of itinerant harvesters who work the Javanese rice harvests, sago harvests of the eastern islands of Indonesia, and the coffee, pepper and spice harvests of Sumatra. Temporary movement to towns is a relatively recent phenomenon. Improvements in transportation services have made not only circular movement but also daily commuting feasible for large numbers of rural people. Mantra (1981) found that commuting was the predominant form of movement in the area of Yogyakarta, for example.

A number of studies suggest high and rising levels of circular movement in Peninsular Malaysia (Hadi 1981). Rice farmers in the Kelantan Plain work temporarily as agricultural labourers in the states of Kedah and Perlis for instance. According to Maude (1981) this pattern of temporary rural–rural movement had declined by the late 1970s, partly because of developments within Kelantan and partly because young people were increasingly moving instead to Singapore to work temporarily on construction projects. Sixteen per cent of the households studied by Maude had had one or more members go to Singapore within the previous twelve months. An alternative measure of the level of mobility and the continuing relationship between migrants and their homes is provided by Corner (1981), who found that 36 per cent of a sample of rural households in Kedah were receiving remittances from past migrants.

In the Philippines, Central Luzon has been greatly affected by temporary movement to Manila (Kuroda 1971). Anderson (1975) found that over 85 per cent of a sample of Central Luzon rural households had had a member work outside the *barrio* at some time. Moreover, the level of mobility had increased rapidly during the 1960s, with international movement increasing two and a half times and migration to other barrios almost doubling between 1962 and 1971. Studies in Thailand have produced similar findings, both from general, qualitative observations and the precise analysis of migration histories (Lauro 1979). Lauro found in his study village in Central Thailand that 36 per cent of villagers had ever been seasonal migrants, and that the rate of seasonal movement had tripled from the period 1962–6 to 1972–6. Other studies have testified to the increasing incorporation of the Central Plain into a Bangkok-centred social and economic system, population movements being simply one manifestation of the process. Whilst there is little evidence of movement from the Chiengmai area to Bangkok, circulation occurs between the northeastern region and Bangkok on a large scale. Lightfoot, Fuller and Kamnuansilpa (1983) found that 23 per cent of a sample of rural people in Roi-et province had spent some time in town during the previous three years. Movement of some kind had occurred from 77 per cent of all households; and 59 per cent of all households had had a member move to Bangkok, 700 kilometers distant.

Finally, temporary movement to foreign destinations has increased very rapidly since the middle 1970s. For Thailand and the Philippines, this category of movement has increased far more rapidly than any other. Filipinos have traditionally gone to Hawaii and the mainland United States to work temporarily in both skilled and unskilled jobs; but this pattern of movement has been largely superseded. A total of 324 233 Filipinos found employment overseas in 1982; 258 000 of these were in land-based jobs, of which 84 per cent were in the Middle East. Recorded cash remittances from overseas workers to the Philippines increased at an average annual rate of 35 per cent from 1978 to 1981, and reached an estimated US $ 869 million in 1981. Similarly 108 519 Thai workers went to the Middle East in 1982, four times the number for 1981. Overseas workers sent back US $ 521 million through the Thai banking system. There are also large numbers of Vietnamese working in the USSR; 10 000 are reported to be working in factories near Moscow, and there are unconfirmed reports that much larger numbers are involved.

Although knowledge of population movements is patchy, it is clear that there are two principal categories of movements in modern South East Asia: movement to frontier areas, some of which is planned by various governmental agencies; and movement to urban areas, including some foreign destinations. While national censuses probably measure the first category of movement fairly accurately, the temporary nature of much of the movement to cities makes it difficult to measure the scale of the process. However the evidence from a number of micro-studies strongly suggests that the rate of temporary rural–urban movement has been increasing. These studies also underline the fact that rural–urban movements initiate social and economic linkages and a continuous process of interaction between origin and destination areas. This aspect of population movement has been obscured by past reliance on the evidence of census data, and is a crucial consideration in any examination of the causes and consequences of movement.

Causes of population movements

Population movements result primarily from the uneven spatial distribution of economic opportunities and economic development, with net out-movement from areas of relative poverty and net in-movement to areas of prosperity or potential prosperity, as is already clear from the descriptive account above. A number of studies have more rigorously analysed these associations between movement and various measures of economic well-being using provincial or regional statistics. Sicat (1972) demonstrated a strong positive correlation between net population movement and output growth rates among regions in the Philippines. Cummings (1974) showed a clear relationship between migration and the changing fortunes of the tin and rubber industries in 14 southern Thai provinces. Population pressure on land resources has also been used as a proxy-variable for economic welfare, with predictable results particularly with respect to rural–rural migration streams. Cunningham (1958) explained the out-movement of Bataks from their West Sumatra homelands largely in terms of population growth resulting in lower incomes. In the Philippines, Simkins and Wernstedt (1971) in their classic study of movement to Mindanao and Ulack (1972), among others, have referred extensively to population density and the concept of 'carrying capacity' in explaining levels and directions of movement.

Studies like these provide an important over-view of patterns of movement. A number of both theoretical and empirical studies have focussed on behavioural aspects of population movement, testing hypotheses initially derived from aggregate data. From the viewpoint of individual migrants, population movements result from a perception that their aspirations are more likely to be realised elsewhere than at their present place of residence, after taking account of the economic and social costs and risks of moving. However, this apparently simple proposition conceals considerable ambiguity. 'Aspirations' include both economic and non-economic goals, though in some cases economic needs become so pressing as to amount to a desperate search for survival. Even so, the risks involved in leaving the safety-net of village society may dissuade from moving those who would appear to have least to lose and most to gain. After moving, the migrant's initial perceptions of the place of destination may turn out to have been based on false information, making his move difficult to understand for the outside observer. Thus the calculation of costs and benefits associated with moving varies widely from person to person, depending on economic resources and expectations, social obligations and the information available about opportunities elsewhere.

Nevertheless there is a large literature which attempts to generalize about the decision-making process among migrants and prospective migrants. Analysts seem to have little difficulty in coming to terms with migrants's decisions with respect to rural–rural movement, which follows broadly predictable patterns leading to relatively sparsely populated areas. Rural–urban movement has attracted a great deal more attention, mainly because it is hard to understand why the process continues and accelerates in the face of alarming levels of poverty and apparent unemployment in the main destination cities.

This is the problem tackled in an important paper by Todaro (1969). He argued that because of unemployment, prospective rural–urban migrants would expect to wait in effect in a 'job queue', perhaps for several months, before securing employment. However, the difference between expected earnings in rural and urban areas had become so great that a long wait in the job-queue would soon be justified economically once employment was finally found. The longer the prospective migrant's planning 'horizon', especially if it extended to his children's future, the more likely he was to move. Todaro's formulation has been the most widely accepted behavioural model of rural–urban migration for the last decade, with respect to South East Asia as well as elsewhere. The model implies certain conclusions for policy-makers, most importantly that planning to accommodate migrants in cities is in one sense counter-productive because each newly created urban job is likely to attract many more than one new migrant. Consequently, controlling rural–urban movement depends on increasing income opportunities in the countryside to reduce the rural–urban wage differential.

The Todaro model contains some important truths which in general terms apply to both rural-urban and rural–rural movement. There is no doubt that the great majority of movement to both rural and urban destinations occurs primarily for 'economic' reasons, and an arduous period of forest clearance, education lasting several years, or a spell in the 'job-queue' may be seen as a necessary and reasonable price for long-term security. However the Todaro model also ignores some features of the migration-decision process which have important implications for his conclusions with respect to planning.

The nature of the economic stimulus varies greatly from household to household. In some cases, particularly in West Java, many poor farmers are literally unable to survive until the next harvest without either going more deeply into debt or earning some non-farm income (Hugo 1982). Typically they move, or have their children move, on a seasonal basis either to town or as itinerant harvest workers; they may become permanent migrants if the land- and income-concentration processes in the countryside remove any remaining prospect that they could earn a living wage in their home villages. An important weakness of the Todaro model is that it does not provide a satisfactory framework for the analysis of rural–urban movement which is either forced on the migrants by their inability to meet their economic needs by staying on their farms, or deliberately planned as a short-term, circular move.

It would be wrong to characterize all migration as a strategy for survival. Lipton (1980) has pointed to the 'bi-polar' nature of migrant selectivity, with the highest probabilities of movement occurring among households at each end of the spectrum of economic well-being within rural communities. Survey data from Roi-et in North East Thailand reveal migration rates marginally higher among households who consider their farms 'adequate' to meet their basic economic needs than among those who consider their farms 'inadequate' (Lightfoot, Fuller and Kamnuansilpa 1983). Indeed, the poorest are often dissuaded from moving by the financial outlay involved and the risks of failing to find a satisfactory source of income at the destination, whether urban or rural. On the other hand, households who can afford this risk and are able to invest in their own and their children's future are most likely to encourage either temporary or permanent movement, spreading the available labour over a number income-earning opportunities on the farm and elsewhere, either as a 'risk-minimization' strategy or as a relatively reliable means of increasing the household income.

A number of South East Asian studies have stressed the importance of social networks in affecting the probability of moving and the direction and consequences of movement. Simkins and Wernstedt (1971) noted that many of the rural–rural migrants from the Visayas to Mindanao moved at the suggestion of close relatives. Bruner (1961) drew attention to the continuous nature of interaction between rural Bataks and urban outliers of their community in the Sumatran city of Medan: at the urban end of the migration system the Batak community deliberately re-established many features of their traditional rural culture, and served as a bridgehead for newly arriving Bataks from the countryside. Similarly, in the poor rural province of Roi-et, North East Thailand, virtually all of a sample of 15–39 year olds knew and could provide personal details of at least one person living in a town (Lightfoot and Fuller 1983). In these and other cases, the existence of social networks linking rural and urban areas has been a major stimulus for continued movement. These networks introduce into villages a generally attractive image of the city which heightens villagers' perceptions of rural–urban differences (Lightfoot, Fuller and Kamnuansilpa, 1981). Social networks present prospective migrants with the choice of moving into urban communities with which they are already somewhat familiar, and which offer housing and companionship. They often also offer employment, either directly or indirectly, and in either the formal or the informal sector of the urban economy. Consequently, unemployment rates among recent migrants are often lower than among native city-dwellers.

The importance of social networks, rural–urban communication channels and other non-economic considerations in encouraging migration (Savasdisara 1984) suggests that Todaro's behavioural assumptions are overly simplistic, as is his conclusion that rural development will reduce the rate of rural–urban movement. Within South East Asia, a number of rural development programmes have been introduced with the reduction of out-migration as either a primary or secondary goal. Examples are Indonesia's *Kabupaten* programme and Thailand's Rural Job Creation Project, as well as numerous irrigation and land-development projects. There have been, as yet, no comprehensive analyses of the effects of these programmes on migration, but knowledge of migrants' decision-making suggests that even if these programmes stimulate an improvement in local living standards they will have little effect on rates of rural–urban movement. Rural development almost invariably entails greater awareness of opportunities elsewhere. Most categories of rural development are more likely to increase rural–urban movement than reduce it. The problem is that while rural development projects and programmes often raise living standards locally, economic growth continues at an equal or faster rate in the cities, maintaining or increasing the gap between rural and urban living standards.

This underlines the inadequacy of analysing the causes of movement only in terms of the migrant's decision-making process (Parnwell, Webster and Wongsekiarttirat 1986). From another point of view, migration is an inevitable response to the 'uneven impact of development under capitalism' (Forbes, 1981). Douglass (1982), for example, sees rural–urban movement as one manifestation of the 'incorporative drive' whereby spatially concentrated economic growth, particularly in primate cities, draws increasingly and inexorably on the resources of poorer, peripheral regions, including their labour resources.

Within such a structural analysis, the causes of movement most worthy of investigation lie not in the decisions of individual migrants but in the prevailing ideology of development in most of South East Asia, an ideology which creates the inequalities to which individuals respond in predictable ways. Structural analyses offer a refreshing alternative to the study of population mobility, avoiding the 'tug-of-war' methodology of those who agonize fruitlessly over whether 'push' or 'pull' factors are more important in causing migration. The importance of an analysis which rises above the level of the individual migrants is even more apparent with respect to the consequences of population movements.

Consequences of population movements

According to neo-classical theory, the transfer of labour spatially and between different sectors of the economy results in benefits for those who move, and for the places of both origin and destination. Migrants move in the expectation of improved material standards of living, and if these expectations were not realized then the migration process would soon cease. Destination areas benefit from additional supplies of labour, and areas of origin benefit from both reduced pressures on resources and the cash which rural–urban migrants in particular almost invariably send 'home'. Thus population movements are represented in neo-classical theory as equilibriating processes which encourage greater aggregate production and allow the benefits of development to be shared more equally between regions and economic sectors.

Evidence from empirical studies suggests that in practice these benefits do not always occur. Migration is not at all an easy option for the individuals who move, and economic success is far from certain. Many Javanese farmers, for example, have tried unsuccessfully to come to terms with the different ecological conditions of Sumatra, resulting in a recreation of the poor and overcrowded conditions typical of their Javanese home villages. Moreover, their intrusion has often been deeply resented by neighbouring non-migrant communities in southern Sumatra and elsewhere. In Cebu in the Philippines, the destruction of the environment has followed uncontrolled in-movement by farmers out for quick profits. The poverty of migrants in cities has frequently been commented upon. According to Lopez and Hollnsteiner (1976) with reference to Manila. 'Although migrants could cope with the difficulties of living in an urban environment, the adjustments for the majority meant a bare existence rather than a satisfying life.' One national survey of migrants in the Philippines indicated that as few as 53 per cent thought their economic conditions better after they had moved; the social costs of moving seemed to be even higher, with only 45 per cent of the sample prepared to say that moving had improved the social conditions of their lives (Filipinas Foundation 1975). Moreover, urban planners have tended to see in-migrants as contributors to problems of overcrowding and unemployment, rather than as assets for urban growth and development.

Although the casualties of the migration process cannot be disregarded, the weight of the evidence shows that most of those who move improve their material living standards as a result. Simkins and Wernstedt (1971) found that migrants from the Visayas to Mindanao increased the size of their holdings by an average of 62 per cent, and also improved their productivity and income as a result of moving. Rural-urban migrants have also been economically successful on the whole, in spite of the fact that many appear to be living in a state of poverty. A series of studies of migrants to Bangkok has documented the process whereby urban lifestyles are adopted and employment secured: 56 per cent of males and 75 per cent of females found work within seven days of arriving in the city, and 84 per cent of both males and females had jobs within 30 days (Chamratrithirong 1979). Similarly, Papanek (1975) found that in Jakarta even those engaged in the lowest level jobs earned more than in their home villages.

Almost all studies of rural-urban movement report that migrants send substantial amounts of cash from town to village. Textor (1961) found that northeasterners working temporarily in Bangkok as tri-shaw drivers sent or took more than most rural households could expect to earn in cash from a year of farming. A more recent study of Bangkok migrants revealed average remittances equal in value to 48 per cent of the mean income of rural northeastern Thailand (Lightfoot and Fuller 1983). There have been similar results from work in the Philippines and Malaysia (Corner 1981, Maude 1981). Many rural households in Sulawesi have come to depend on cash income earned seasonally by driving tri-shaws in the regional capital, Ujung Pandang. Moreover, as noted earlier, significant amounts of cash are remitted by international migrants, particularly Thais and Filipinos working in the Middle East. In short, the evidence with respect to remittances strongly indicates that 'from a short-term perspective certain economic benefits usually accrue to the individual movers, their families, and to some extent their villages' (Hugo 1982).

In spite of these inflows of cash, the net effect of migration on communities of origin is difficult to assess. First, out-migration reduces the amount of labour available for work within the village. This can occur even where there is a seasonal pattern of circular movement, with departures and returns ostensibly planned to coincide with variations in the demand for farm labour. In Roi-et, northeastern Thailand, most circular movers extended their stay in town over at least one farming season as well as one or more non-farming seasons, with the result that some 10 per cent of the adult population are absent from the village even at the time of the harvest, the season of greatest demand for labour. These somewhat unpredictable absences have an effect beyond the household of the migrants concerned, because they disrupt village labour-exchange networks whereby groups of households have traditionally co-operated at times of peak labour demand (Lightfoot Fuller and Kamnuansilpa 1983). These effects might be further compounded by the selective nature of migration, migrants generally being both young and better educated than non-migrants. However it must be said that while the fact of age and educational selectivity is reasonably well documented, its social and economic effects have not been adequately analysed.

Turning again to neo-classical theory, marginal productivity of labour in rural areas of less developed countries has been widely assumed to be zero. If this is so, the losses of labour which result from migration will not affect aggregate production, and will increase the productivity and presumably the consumption levels of the remaining population. However, the marginal productivity of rural labour has been largely a matter of speculation; the few relevant analyses in South East Asia and elsewhere suggest that farmers do not have long periods of idleness; and labour shortages can act as a constraint on increasing production. Farmers in northeastern Thailand commonly cite lack of labour as both the main disadvantage of temporary and permanent out-movement by members of their households, and the main reason for not planting dry-season crops. The counter attraction of urban jobs and amenities is probably the most important reason for the gross under-utilization of dry-season irrigation facilities in the northeastern region: only about 15 per cent of the land potentially irrigable by pumping stations is actually used.

A final point concerns the use of cash remitted from towns and cities. With such large amounts involved, particularly in the case of international migrants, there appears to be considerable scope for investment either in the farm or in non-farm rural enterprise. In fact studies consistently demonstrate that remittances are used overwhelmingly for either immediate consumption needs, luxury consumer items or improvements to the structure of the house. Each of these uses has at best a very limited impact on the productive capacity of local resources and produces very restricted local multiplier effects in creating employment and demand for local materials. Therefore while there have been remarkable transformations in the surface appearance of some villages, especially those involved in temporary movement to the Middle East, there have been relatively few changes in their productive methods or levels. Indeed, there is a real danger that the availability of cash from urban work will reduce any incentives for farmers to modernize their farms.

As in the case of the causes of movement, it is important in analysing its consequences to look beyond the perspective of the individual migrant and his or her family. Although those most directly concerned appear to benefit from migration, often considerably, it is less clear that the communities of origin as a whole gain from the process. In particular it is far from clear that the process of circulation which predominates among rural–urban movers, or urban–rural remittances, represent a desirable long–term development strategy for the villages concerned. The urban economy stands to gain from this process, since employers are able to secure labourers more or less as and when they want, with minimal responsibility either individually or collectively for raising or educating their workers, or caring for them when their labour is not needed. The village continues in the last resort to provide a safety-net of social security at no cost to the urban economy. On the other hand, aside from the losses of rural labour, the prospects for development of production systems within the villages of origin are compromised by their continuing social obligation to periodically returning migrants and by the fact that infusions of cash reduce the incentive to improve their production methods and levels.

Planned movement

Government action has far-reaching implications for all categories of population movements. Governments oversee the macro-conditions and create the communications infrastructure which stimulate and channel population movements. In this sense most movements can be described as 'planned'. However, governments have also become progressively more involved in attempts directly to influence the levels, patterns and consequences of movement, and it is in this sense that the term 'planned movement' is normally used.

In South East Asia, a number of categories of governmental intervention with respect to population movement can be identified, including the resettlement of political refugees; the resettlement of rural populations in strategic villages for security reasons; and the resettlement of evacuees from reservoir basins. Each of these categories has had a considerable local effect since the 1950s. Rural settlement patterns in peninsular Malaysia and southern Vietnam were temporarily transformed during the Malayan Emergency and Second Indochina War, respectively, as a means of denying logistical support to guerilla fighters. Large numbers of refugees, predominantly Roman Catholics, left northern Vietnam for the south and for Thailand in 1954. During the war in Laos, almost one-third of the population of the country were refugees at the peak period of 1971, requiring elaborate resettlement measures; and since 1975 large numbers of Vietnamese, Laotians and Kampucheans have left their respective countries either for permanent resettlement in the United States or Europe, or temporarily, crossing and recrossing borders in their tens of thousands to avoid the latest outbreak of fighting or repression. Reservoir projects have similarly required the evacuation and planned resettlement of thousands of people at a time, often including ethnic minorities (Lightfoot 1978, Anti-Slavery Society, 1983).

Governmental intervention with respect to these categories of movements has been localized and often carried out on an emergency basis, primarily attempting to minimize the inevitable traumas resulting from enforced relocation. Governments have in other cases intervened both more consistently and with greater developmental effect, first to stimulate and control the settlement of frontier areas; and second to control the growth of urban populations.

Indonesia, Malaysia, the Philippines, Thailand and Vietnam have specific policies and programmes with respect to settlement in frontier areas, encompassing large numbers of people. Under Indonesia's *Transmigrasi* programme, designed to transfer people from Java to the Outer Islands, 800 000 people had moved between 1905 and 1978, and at an accelerating rate during the 1970s and into the 1980s; 350 000 people had moved to various types of settlement schemes in Malaysia since the start of the land settlement programme in 1957 and 1976, including 200 000 to Federal Land Authority (FELDA) schemes; 400 000 people moved to organized settlements in the Philippines from 1913 to 1975; and 1 200 000 people had moved to Thai settlements from 1935 to 1979. Since 1976 Vietnam has adopted an ambitious land settlement and population movement policy, with 1 500 000 people already moved by 1980 and plans to move a further 3 or 4 million by 1985 and 10 million by the year 2000. Most of this movement will be from the densely settled northern delta to the highlands and delta of southern Vietnam.

Each of these settlement programmes has essentially the same objectives, though in a different order of priority. The redistribution of people from crowed to frontier areas has been an important goal in Indonesia, the Philippines and Vietnam. In Thailand the main objective is to control frontier settlement, allocating land on as rational and equitable a basis as possible. Ostensibly the intention since 1975 has been to allocate land to the landless as part of a land reform programme, though in practice little has been done to recruit settlers from the provinces near Bangkok most affected by the concentration of land ownership. In Malaysia on the other hand, a much greater emphasis is given to using the land settlement process to create a commercially oriented, dynamic agricultural sub-sector which will contribute significantly to export earnings. As a result Malaysian settlements, particularly those initiated by FELDA, are highly capitalized and geared towards the production of particular crops, mostly rubber and oil-palm, using modern production and processing technology and planned marketing. FELDA schemes have been seen as rural 'growth poles', involving up to 30 000 people in parts of Pahang where settlements have been located in close proximity to each other (Hackenberg 1980). A somewhat similar philosophy has been adopted for the *Transmigrasi* programme since the middle 1970s, with substantial funding from the World Bank. Settlements in the Philippines, Thailand and Vietnam depend to a greater extend on 'self-help' on the part of the settlers, who are normally responsible for clearing and managing their own plots, with government providing roads, infrastructure and loans in both cash and kind.

Land settlement schemes have received a great deal of attention in various national plans, and represent a cornerstone of rural development policy in Malaysia, Indonesia and Vietnam. Even in Thailand and the Philippines considerable amounts have been invested in governmentally sponsored settlement areas. Is the investment justified? Uncontrolled frontier settlement has its dangers, including the tendency for both the fragmentation of holdings and the concentration of ownership to occur among spontaneous settlers. However, the philosophy of planned settlements has been criticized as 'ecological imperialism' (Palmer 1974). There is evidence too that privately funded migrants to the frontier resettle themselves more cheaply than government funded migrants, and are able to generate higher incomes after resettlement. This difference has been demonstrated among both reservoir evacuees (Lightfoot 1978) and Filipino settlers on the island of Palawan (James 1983). Moreover the numbers of spontaneous, or privately funded, migrants are generally much greater than the numbers of governmentally sponsored settlers. For example an estimated 1 200 000 people moved privately from Java to the Outer

Islands during the same period as 800 000 people were helped by the *Transmigrasi programme* (MacAndrews 1982). Therefore while Malaysian FELDA schemes have been economically successful, public investment below the level necessary to create an integrated, efficient and highly commercial agriculture appears to be more difficult to justify.

Government policies towards urban growth have in many respects been contradictory. Macro-economic policies have on the whole promoted urban-based sectors of the economy in both the socialist and non-socialist countries. Yet at the same time various steps have been taken to attempt to control the resulting rapid increase in the urban population. Simmons (1979) has identified four strategies for slowing metropolitan city growth: reversing the flow of rural–urban movements; imposing administrative and legal controls on movement to prevent people from leaving the countryside or entering certain cities; diverting prospective urban migrants to the rural frontier; and redirecting rural–urban migrants towards intermediate-sized towns instead of towards the metropolitan centres on which most rural–urban migration tends to be concentrated.

Within South East Asia there has been nothing directly equivalent to the *hsia fang* ('sending down') of Maoist China, where city-dwellers at all socio-economic levels were required to return to the countryside either temporarily or permanently. Urban Vietnamese are exhorted to return to the countryside to help with agricultural tasks at certain seasons, very much as happens in the non-socialist countries of the region. And the Khmer Rouge government summarily evacuated all Kampuchean cities immediately after taking power in 1975. With these exceptions, planned urban–rural movements have involved specific groups of people identified either by living in illegal squatter housing or by working in occupations considered undesirable or redundant. For example after pedal-driven tri-shaws were banned from Bangkok in 1960 a land settlement area was prepared 100 kilometers north of the city to accommodate the drivers affected. Clearances of squatter housing have been less humane and less successful. Sixteen thousand people were transfered from Manila's Tondo foreshore in 1965 to a settlement outside the city at Sapang Palay; with no adequate source of livelihood most of those resettled returned to Manila immediately.

Legally enforced controls on movement have been used in modern times in Indochina and in Indonesia, where the mayor of Jakarta instituted a 'closed city' policy in 1970. In-migrants who had not secured employment within six months were required to leave the city and return to their homes; and people identified as recent migrants were occasionally rounded up and summarily transported out of the city. These actions seem to have had little effect on the growth of Jakarta. Preventing or redirecting rural–urban movement through rural development has similarly had no more than local effects, most notably in the case of highly capitalized development projects such as Malaysia's Muda irrigation scheme and the FELDA land settlements discussed above.

Considerable attention has been given to the promotion of intermediate-sized cities both as centres of economic growth and migration destinations alternative to the major metropolitan centres. Each of the major countries of the region has a decentralization policy focussed on growth poles, industrial estates or freeports. With respect to migration management, this approach has the advantage of acknowledging the likelihood of continuing large-scale rural–urban movement in association with economic development, while at the same time seeking to retain local labour resources within their regions of origin, to be utilized for the benefit of those regions. The spread effects of a decentralized urbanization strategy would reach further into the rural hinterland, while the shorter travel distances involved would enable migrants to maintain contact more completely with their homes, and therefore reduce the locally disruptive consequences of outward movement. There would also be social benefits, since there is evidence that migrants to local towns adjust more successfully to their new surroundings than migrants to metropolitan cities. The position has been summarized by Hackenberg (1980).

'If rural development programmes create urban centers in formerly rural places and stimulate migration to those short-range destinations, the resulting urban migration tends to redress regional imbalances (and) redistribute income . . . .. Rural to primate city migration has the opposite consequences.'

Whether decentalization policies will affect spatial patterns of population movement in desirable ways remains to be seen. Some growth poles have grown in population quite spectacularly, for example the population of Khon Kaen in northeastern Thailand more than quadrupled from 1960 to 1980. Yet this growth seems not to have affected local patterns of migration as expected, since the province of Khon Kaen beyond the city limits experienced the largest net out-movement of all of Thailand's 71 provinces between 1965 and 1970 (Sternstein 1979). Moreover, Thailand's Labour Department consistently reports an excess of job vacancies over applications for northeastern towns, the opposite of the situation in Bangkok; yet there is a heavy net movement out of the North East to the capital. These anomalies could be explained by imperfections in the availability of information about jobs, a hypothesis which has been strongly supported experimentally in this region (Lightfoot and Fuller 1983). It would seem therefore that a sound information dissemination programme should be one component of a general strategy to decentralize both economic growth and spatial patterns of rural–urban movement.

Conclusion

Until recently the literature on population movements has focussed upon migrants as consumers, each responding to the promise of greater wealth and each responsible for some degree of consumptive pressure on local resources in his place of origin and destination. In this sense it is reasonable to see population flows as equilibriating processes, reducing pressures in the place of origin where fewer resources are available and increasing pressures in the place of destination where resources are more abundant. It is important also to analyse population movements as initiating flows of resources, including the movement of labour power and the reciprocal transfer of cash to the area of origin. Both these resources have considerable developmental potential, though precisely what are the developmental consequences of these resource flows is not clear. It is unclear also what are the effects of governmental intervention on migration, though there is good deal of evidence that intervention in most cases has done little to either control population movements or maximize their developmental consequences. Properly understanding the causes and consequences of population movements depends on an analysis which incorporates the wider social and economic setting within which migrants and their families operate. Similarly, migration management must be closely coordinated with other development policies, particularly those pertaining to the decentralization of economic growth. There is plenty of scope for policy-oriented empirical studies to clarify the murky inter relationships between migration, economic processes and planning.

Further reading

Anderson J N (1975) 'Social strategies in population change: village data from central Luzon', in Kantner, J F and McCaffrey, I. (eds.), *Population and Development in South East Asia*, Lexington Books, London.

Anti-Slavery Society (1983) *The Philippines: Authoritarian Government, Multinationals and Ancestral Lands*, Anti-Slavery Society, Indigenous Peoples and Development Series, No. 1, London.

Bruner E M (1961) 'Urbanization and ethnic identity in North Sumatra', *American Anthropologist* **63**, 508–21.

Chamratrithirong A (1979) *Recent Migrants in the Bangkok Metropolis*, Institute for Population and Social Research, Mahidon University, Bangkok.

Corner L (1981) 'Linkages, reciprocity and remittances: the impact of rural out-migration on Malaysian rice villages', in Jones, G. and Richter, H V (eds.), *Population Mobility and Development in South East Asia and the Pacific*, Australian National University, Development Studies Centre Monograph No. 27, pp. 117–36.

Cummings F H (1974) *Migration and Regional Development: Implications* *for Planning in Indonesia, Thailand and the Philippines*, University Microfilms, Ann Arbor.

Cunningham C E (1958) *The Postwar Migration of the Toba Batak to East Sumatra*, Yale University Cultural Report Series, New Haven.

Douglass M (1982) 'The Incorporative Drive: Central Plains migrants in the Bangkok metropolis', Mimeo, Institute of Social Studies, The Hague.

Fernandez D Z, Hawley A H and Pridaza S (1976) *The Population of Malaysia*, Department of Statistics Research Paper No. 10, Kuala Lumpur.

Filipinas Foundation (1975) *Understanding the Filipino Migrant, an In-Depth Study of the Motivational Factors behind Internal Migration*, Filipinas Foundation Inc., Makati, Rizal.

Forbes D (1981) 'Mobility and uneven development in Indonesia: a critique of explanations of migration and circular migration', in Jones, G. and Richter, H V (eds.), *Population Mobility and Development in South East Asia and the Pacific*, Australian National University, Development Studies Centre Monograph No. 27, pp. 51–70.

Goldstein S (1978) *Circulation in the Context of Total Mobility in South East Asia*, Papers of the East-West Population Institute No. 53, Honolulu.

Hackenberg R A (1980) 'New patterns of urbanization in South East Asia: an assessment', *Population and Development Review* 6, 391–420.

Hadi A S (1981) ' "Permanent' or "circular" migration in Peninsular Malaysia: the experience of three villages in Negri Sembilan', *Akademika* **19**, 39–57.

Hugo G (1978) *Population Mobility in West Java*, Gadjah Mada University Press, Yogyakarta.

Hugo G (1982) 'Circular migration in Indonesia', *Population and Development Review* **8**, 59–83.

Hugo G and **Temple G** (1975) 'Inter provincial migration in Indonesia', Mimeo, Department of Demography, Australian National University, Canberra.

James W E (1983) 'Settler selection and land settlement alternatives: new evidence from the Philippines', *Economic Development and Cultural Change* **31**, 571–86.

Kuroda E (1971) 'Atomization in the social relations in a peasant community of Rizal Province, the Philippines: a case study of sitio Pulong Kumanoy, Lagundi, Morong, Rizal', *Developing Economies* **1**, 29–314.

Lauro D (1979) 'The demography of a Thai village: methodological considerations and substantial conclusions from field study in a Central Plains community', Ph.D. thesis, Department of Demography, Australian National University, Canberra.

Lightfoot P (1978) 'The costs of resettling reservoir evacuees in northeast Thailand', *Journal of Tropical Geography* **47**, 63–74.

Lightfoot P and **Fuller T D** (1983) 'Circular movement and development planning in northeast Thailand', *Geoforum* **14**, 277–87.

Lightfoot P, Fuller T D and **Kamnuansilpa P** (1981) 'Impact and image of the city in the northeast Thai countryside', *Cultures et Development* **13**, 97–122.

Lightfoot P, Fuller T D and **Kamnuansilpa P** (1983) 'Circulation and interpersonal networks linking rural and urban areas: the case of Roi-et, Northeastern Thailand,' Papers of the East-West Population Institute No. 84, Honolulu.

Lipton M (1980) 'Migration from rural areas of poor countries: the impact on rural productivity and income distribution', *World Development* **8**, 1–24.

Lopez M E and **Hollnsteiner M R** (1976) 'People on the move: migrant adaptations to Manila residence', in Bulatao, R A (ed.), *Philippine Population Research*, Population Center Foundation, Makati, Rizal, pp. 227–50.

MacAndrews C (1982) 'Land settlement policies in South East Asia' in Jones, G W and Richter, H V (eds.), *Population Resettlement Programs in South East Asia*, Australian National University Development Studies Centre Monograph No. 30, pp. 9–23.

Mantra I B (1981) *Population Movement in Central Java*, Gadjah Mada University Press, Yogyakarta.

Maude A (1981) 'Population mobility and rural households in North Kelantan, Malaysia', in Jones, G and Richter, H V (eds.), *Population Mobility and Development in South East Asia and the Pacific*, Australian National University Development Studies Centre Monograph No. 27, pp. 93–113.

Palmer G (1974) 'The ecology of resettlement schemes', *Human Organization* **33**, 239–50.

Papanek G (1975) 'The poor of Jakarta', *Economic Development and Cultural Change* **24**, 1–27.

Parnwell M, Webster C and **Wongsekiarttirat W,** (1986) 'Development in North East Thailand: Case Studies in Migration, Irrigation and Rural Credit', Centre for South East Asian Studies, University of Hull, Occasional Papers No. 12, Hull.

Perez K J (1978) 'Internal migration', in United Nations, *Population of the Philippines*, Economic and Social Commission for Asia and the Pacific, Country Monograph Series No. 5, Bangkok, pp. 44–77.

Pryor R J (1979) 'South East Asia: migration and development', in Pryor, R J (ed.), *Migration and Development in South East Asia*, Oxford University Press, Kuala Lumpur, pp. 3–16.

Savasdisara T (1984) 'The non-economic factors in rural-urban migration in Thailand', *Singapore Journal of Tropical Geography* 5, 175–83.

Saw S H (1980) 'Estimation of inter-state migration in Peninsular Malaysia, 1947–1970', Institute of Southeast Asian Studies, Research Notes and Discussion Paper No. 24, Singapore.

Sicat G P (1972) *Economic Policy and Philippine Development*, University of the Philippines Press, Quezon City.

Siddhu M S and **Jones G W** (1981) *Population Dynamics in a Plural Society: Peninsular Malaysia*, UMBC Publications, Kuala Lumpur.

Simkins, P D and **Wernstedt F** (1971) *Philippine Migration: The Settlement of the Digos-Padada Valley, Davao Province*, Yale University Press, New Haven.

Simmons A B (1979) 'Slowing metropolitan city growth in Asia: policies, programs and results', *Population and Development Review* 5, 87–104.

Sternstein L (1979) 'Thailand: changing patterns of population movement, 1960–1970', in Pryor, R J (ed.), *Migration and Development in South East Asia*, Oxford University Press, Oxford, pp. 19–29.

Textor R B (1961) *From Peasant to Pedicab Driver*, Yale University, South East Asian Studies Cultural Report Series, No. 9, New Haven.

Todaro M P (1969) 'A model of labour migration and urban unemployment in less developed countries', *American Economic Review* **59**, 138–48.

Ulack R (1972) *The impact of industrialization upon the migration and demographic characteristics of Iligan City, Mindanao*, Ph.D. thesis, Pennsylvania State University.

CHAPTER 11

Urbanization

Denis Dwyer

South East Asia forms part of a developing world that has now
assumed dominance in terms of global urbanization. As Fig. 11.1
shows, during the mid-1970s for the first time a greater proportion
of the world's urban population was located in the developing
countries than in the industrialized countries; and towards the end
of the present century more than two-thirds of the world's urban
population will be located within the developing countries. This
most significant change in global urban population distribution
carries important consequences for the developing countries in terms
of their socio-economic development. In the modern era, cities have

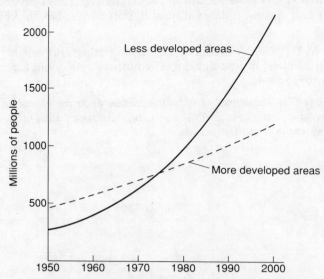

Fig. 11.1 World urban population, 1950–2000. Note: urban population is
that living in settlements of 20 000 people or more. (Source:
Dwyer 1979, based on UN statistics)

278

always been centres of change. They have been viewed in such roles in several aspects important to both regional and national development, and geography had held a significant place in most concepts since a critical factor has been the reduction in distance, or the elimination of the friction of space, implied in urban concentration.

In the economic sense, it is under such conditions of urban space that the external economies associated with business enterprises can most easily be made available to neighbours. For this reason, and because of other less particularly economic characteristics outlined below, the city has been historically the principal centre of change and growth in industry and commerce (Dwyer 1972). The industrial revolution of today's 'developed' countries was essentially urban-based, although change in agriculture was also vital to its accomplishment. The city has also been, and is, the mainspring of regional development. The economic history of today's industrialized nations shows their economic growth to have been led in spatial terms by rapidly developing regions associated with thriving urban centres, which at some stages of growth have stood in marked contrast to other lagging regions in those nations. At certain periods, counter-balancing forces, sometimes impelled by direct government policy, have acted in favour of equalization and the more even spread of growth. But today marked regional inequalities remain a feature of national economies even among the economically advanced nations. Not surprisingly, the creation of new urban centres in backward regions, or the strengthening of existing ones, has become an important feature of regional planning both in the industrially advanced countries and in South East Asia and other parts of the developing world. In the developing countries – and virtually every country in South East Asia typifies this – the desired industrial take-off on a national scale is being sought almost wholly through city-oriented policies.

A second, and possibly more important, way in which cities have been acknowledged as centres of change is in their role as catalysts, intellectually, socially and politically. Flows of information and ideas, for example, achieve their highest acceleration under conditions of maximum accessibility, that is, within urban centres. Cities have therefore been designed, consciously or unconsciously, not only to provide better physical access to goods and services but also as communication centres and storehouses of information. The number of face-to-face contacts possible in a given time at urban population densities is of a markedly higher order than that possible at rural population densities. Because of their function as collecting places for the most advanced information, and also because of the coming together within their boundaries of people of diverse origins (for example, the addition by immigration of the overseas Chinese and Indians to South East Asia), cities have functional as vital social melting-pots in the development of states. Cities have repeatedly bred movements designed to establish new forms of order as national ways of life. In national plans in South East Asia or elsewhere, therefore, it is not surprising that urban life is very commonly seen as the basis for modernity and economic progress.

Even so, generalizations drawn from past urban experience – which usually in the literature of the social sciences means the urban experience of the industrially developed countries – should not automatically be accepted for considering contemporary situations in developing regions such as South East Asia (Dwyer 1979); and there are several factors relative to urbanization in South East Asia that need to be considered in this context. It is doubtful, for example, whether urban hierarchies similar to those in the industrially advanced countries are in the process of evolution in the region. Relatively high proportions of urban population in several South East Asian countries tend to be focussed upon a few urban places and, in general, urban growth seems to be concentrating on the larger urban places rather than producing the broader-based urban structure more characteristic of the urban situation of maturity in the industrially advanced countries.

Second, as several aspects of the present chapter will demonstrate, there is considerable doubt as to whether towns and cities are playing the same developmental role within South East Asia as has occurred in the West. It is clear, for example, that the overall growth of towns and cities demographically is not being matched by the same levels of growth in manufacturing industry. Urbanization in South East Asia in this sense is much more a demographic phenomenon than an economically induced one, except in the sense that it is undoubtedly the lack of economic development in the countryside that provides a major motivating force for a high proportion of migrants to the towns and cities. Overall, the urban situation in South East Asia is such that it is necessary to question much of the application of concepts and models developed to explain the Western experience of urbanization to the urbanization process in the region. A great deal more academic work, particularly in the field, remains to be carried out before a full range of relevant concepts and models is evolved (Dwyer 1977, 1979).

The historical background

The above paragraphs are not meant to imply that urbanization in South East Asia is a shallow-rooted phenomenon of recent origin. What is being emphasized is that in all probability urban places are emerging there, as elsewhere in the developing countries, that are of fundamentally different types from those which emerged from industrialization in the West. Today's urban growth in South East Asia is both strongly rooted in the socio-economic circumstances of the region and firmly related to its historical past. Reid (1980, p. 235) has emphasized that there were trading cities and trade-based kingdoms in South East Asia as far into the past as the records at present allow us to go. The period from the late fourteenth to seventeenth centuries was one of exceptionally rapid growth of trade and with it of a network of indigenous cities. It was the Crusades that brought a new demand for Eastern luxuries in the West. Much of this expanding East-West trade was directed initially along land routes through central Asia. But by the end of the fourteenth century, the Red Sea route was flourishing while the overland 'Mongol' route had completely collapsed. Chinese trade was becoming more and more important, and with it imperial missions were sent to South East Asia, for example by the Emperor Yung Lo (1402–24). With the growth of trade came an influx of Chinese traders who even at this early date played a significant role in the rise of such cities as Melaka and Ayuthya. It was in the ports of South East Asia that goods from China were sold to Muslims from the west who would take them further along the trade route to Europe. A supplementary trade also developed in cash crops grown within the region: by 1416, for example, pepper was being cultivated in northern Sumatra for export and by the time of the arrival of the Portuguese in South East Asia in 1511 its production was rivalling that of the older Malabar source in India (Reid, 1980, p. 236). Cloves, nutmeg and other spices soon after added considerably to the region's export trade.

It is probable that prior to the fourteenth century surge in trade the ports of the region were very small. Indian seamen and merchants had, however, entered South East Asia well before the birth of Christ, and with them Indian cultural characteristics were imported (Leinbach and Ulack 1983, p. 373). Indian concepts of royalty, particularly of the god-king, of politics and of religion characterized early indigenous kingdoms in South East Asia and the first true towns and cities of the first century AD. According to early Chinese accounts, Vyadhapura had emerged about this time as the capital of the earliest important kingdom of the region, Funan, located in the Mekong area of what is now Vietnam and Kampuchea. For five centuries, until the decline of Funan, Vyadhapura was a leading city of the region but travellers also remarked upon such towns or cities as Pandurang and Indrapura, capitals of Champa located on the coast of central Vietnam; Srikshetra, the capital of the Pyu kingdom located in lower Burma; and the capitals of ancient Dvaravati in the lower Cao Phraya valley of Thailand (Keyes 1977, p. 261). In contrast to the trading ports of the region, these were sacred, or temple, cities. Ultimately, they were the built expressions of the religious power upon with which the rulers of the state depended for much of their legitimacy. They were viewed as the meeting point of heaven, earth and hell, and only secondarily as capitals of kingdoms. Auspicious locations for such cities were constantly sought: hence their locations were regularly changed, and their urban plans strictly reflected geomantic principles. Massive religious buildings were dominant in such cities.

Armies of peasants were conscripted both to build the temples and to provide the necessary food for the urban populations, but there was also a trade function and internal economic activity and therefore such agglomerations were not merely cult centres.

The classic temple cities reached the apex of their development in the twelfth and thirteenth centuries, a peak typified by the achievement of the Khmer dynasty in building the inland city of Angkor (Coedes 1963) as the capital of a kingdom which included as its territory present day Kampuchea and parts of Laos, Thailand and Vietnam. Angkor was established in the ninth century and eventually abandoned in the fifteenth because of invasions by the Thais. By the twelfth century, its population was several hundred thousand, in a city dominated by huge temple complexes which survive as perhaps the region's most important archaeological heritage to the present day. Its size pre-dated the urban expansion of the fourteenth century in the port cities and towns. Reid (1980 p. 239) claims that, unsatisfactory as the statistical evidence may be, it suggests that between the fourteenth and the sixteenth centuries urban populations of 50 000 to 100 000 were being reached by a significant number of South East Asian cities. For Melaka in the early sixteenth century, for instance, a hundred junks were required annually for importing rice. This would mean an annual import of about 6000 tonnes, enough to feed about 50 000 people. though there were possibly even more within the city since undoubtedly some food would also have come overland or in smaller craft (Reid 1980, p. 237). This was at a time when few European cities exceeded a population of 40 000. The South East Asian urban experience of the time was even more remarkable in view of the general lack of population within the region. In relation to its total population, Reid claims (p. 289), South East Asia in this period must have been relatively highly urbanized.

A change within the region amounting to a complete urban revolution began to occur with the European intrusions, which date from the Portuguese capture of Melaka in 1511. For a lengthy initial period, the urban effects of the growth of colonialism within the region were slight, principally because the early Europeans sought trading points rather than territorial gains. They captured a few coastal cities such as Melaka and Ambon, the chief port of the Spice Islands in eastern Indonesia, but their traders entered most coastal cities under treaties with the local rulers. In the Philippines, however, the Spaniards established widespread territorial control, and many settlements, from the mid-sixteenth century. Manila and Cebu were therefore very early colonial cities. Java, with what became its dominant city of Batavia (now Jakarta), was also an exception because of its widespread penetration by the Dutch during the seventeenth century.

From these beginnings arose within the region a series of dominant colonial cities at points from which the surge in the development of the resources of South East Asia which characterized the colonial period could be directed (Dwyer 1968). The only exception among today's great cities of the region was Bangkok, which was created as a royal capital in 1782. The growth of the other great cities was directed by the respective colonial powers at points from which large tributary areas could be commanded, and largely wherever access to the sea coincided with this requirement. In almost every case, indigenous settlements upon the site were small before foreign enterpreneurs and colonial government officials created the cities as essential funnels, service points and administrative centres. The points for urban development chosen during the colonial period were those at which exportable products could be most conveniently assembled; the same access worked equally well in reverse for imports and, of course, for the needs of colonial administration. Major traditional lines of movement – the rivers of mainland South East Asia and the sea lanes in the island realm – thus commonly focussed upon the site choices of the colonial powers. Modern transport services were introduced to reinforce these traditional lines of movement. The new cities became the seats of colonial government. Hence they rapidly developed into the largest settlements in their respective countries.

Thus Manila, founded in 1571 by the Spaniards as an administrative, military and missionary centre, grew as its hinterland in Luzon developed but, more particularly, as the sea connections between China and the Philippines, already in existence in pre-Spanish times, were intensified during the early colonial period and expanded to include trade between China and Acapulco in New Mexico (and ultimately Spain) by means of the famous Manila galleons (Schurz 1939). Later, a further basis for growth was provided for the city, especially when it came under American rule at the end of the nineteenth century, with the expansion of export crop production, principally sugar, tobacco and Manila hemp, which characterized the period after 1820 and for which Manila served both as the organizational headquarters and the main port of shipment. The other great colonial cities had broadly similar patterns of development. Nearer the other end of the time-scale for the colonial period in South East Asia, Singapore's rapid growth after its foundation by Raffles in 1819 was based both upon its position at the point of overlap in world ocean lanes with the inner seas of the Indonesian archipelago, parts of the islands of which, notably Java, had reached a high level of export crop production under the Dutch, and also upon the internal development of Malaya, largely by British and Chinese interests, for tin and rubber production. Rangoon grew upon the general internal development which resulted from the incorporation of Burma into the British colonial system, and especially upon the export trade in rice of the Irrawaddy Delta, which was created during the colonial period. Saigon grew in sympathy with the French development of Indo-China, while Batavia was essentially the headquarters from which the Dutch ruled their possessions in the East Indies and from

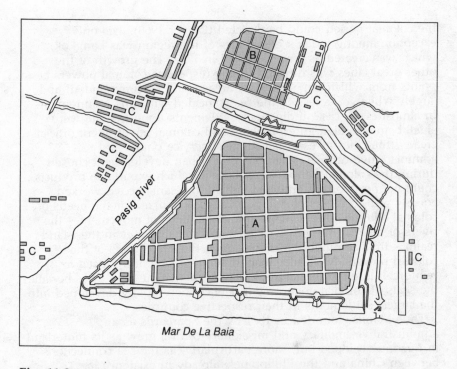

Fig. 11.2 Manila in 1671, after the map of Ignacio Muñoz
A = Intramuros, the Spanish walled city.
B = The Parian, the Chinese quarter.
C = Areas of indigenous settlement and building forms.
(Source: Author's interpretation)

which radiated their official shipping line, collecting the commerce of the Outer Islands to that port.

Although, for example, seventeenth-century maps of Manila show areas of recognizably indigenous building form (Fig. 11.2), it is primarily as centres of international trade that the growth of South East Asian cities during the colonial period must be viewed, that is, as international much more than local creations. In the economic sense, such cities were fundamentally outposts of the economies of the colonial powers rather than intrinsic products of the internal evolution of the countries in which they were located. They were thus founded and developed on a very different basis than that of the cities of the Industrial Revolution in the West. The great colonial cities of South East Asia were international too in their population composition, for they quickly became very cosmopolitan. Large numbers of Chinese flooded into South East Asia to participate in the economic opportunities created there by the colonial powers; Indians came too, especially into Burma. The impact of immigration was felt most in proportionate terms in Malaya, where the highest level of economic development in the region was achieved only through significant inputs of immigrant labour. Kuala Lumpur was founded as a Chinese trading post in 1859. Twenty years after its foundation its population was only 2600, but by 1911, largely because of the development of tin mining in the vicinity by Chinese interests, this had increased to 47 000 (Lee 1976, p. 42). The population growth of Kuala Lumpur, like the development of eastern Malaya generally, was largely the result of large-scale immigration of Chinese and, to a lesser extent, Indians. As late as 1970, Kuala Lumpur remained essentially a Chinese settlement but more recently the in-migration of Malays has changed its ethnic composition considerably. Significant proportions of immigrants settled in the other colonial cities of the region with the result that all contain sizeable non-indigenous minorities and this is now a fact of considerable significance not only economically but also politically. Singapore, of course, remains in essence a Chinese city state, a fact which in the post-colonial period has at times led to considerable tensions with its near neighbours Malaysia and Indonesia.

Within today's great cities of South East Asia considerable dichotomy in physical form emerged during the colonial period, and such dichotomy, still existing as a major element within contemporary townscapes, remains significant as an indicator of their socio-economic character. A basic contrast developed, and has been perpetuated during the ensuing period of national independence, between what can be identified as their Western and non-Western parts. The Western parts of the South East Asian city comprise the Central Business District, even though such areas in South East Asia are usually not so specialized in their functions or as sparsely inhabited by permanent residents as would be the case in Western Europe or the United States. The Western parts also comprise high-class residential suburbs, which are now tending to become even more completely demarcated from the remainder of the city than they were even during the colonial period, port facilities and, during the period of national independence and national economic planning, areas reserved for modern factory industry as industrial estates. It is in the non-Western areas of the city, however, that the majority of the population lives and still, to a very large extent, circulates. In almost every South East Asian city there is a Chinatown, often composed of shophouses; indeed, most of the smaller towns of Malaysia are still very largely Chinese in terms of their built environment. More surprising, perhaps, is the existence of very large urban areas of indigenous building forms, the *kampong* areas as they are called in Malaysia and Indonesia. Such areas interpenetrate cities and even come close to their central cores. They are characterized by the use of local constructional materials and by building forms similar to rural types. In marked contrast to the cities' Western parts, they are also characterized by serious deficiencies of physical facilities, usually even the absence of sanitation and safe water supply, by gross overcrowding (as, indeed, are the Chinatowns), and by a lack both of modern employment opportunities and even the most rudimentary community facilities.

Such dichotomy in physical form has been a feature of the colonial cities of South East Asia from their very foundation. From its earliest days, Manila exhibited a marked contrast between the walled Spanish quarter of Intramuros, the Parian, where the immigrant Chinese were required to live by the Spanish colonial authorities, and surrounding areas of indigenous rural-type building (Fig. 11.2). Similar circumstances prevailed in the other colonial cities, and as late as the 1820s Singapore also was being laid out on a plan which provided for separation of its districts on racial lines, a separation which in time tended to become confirmed in building types (Fig. 11.3). Today, prestige public buildings continue to be built as in colonial times; the opulent new housing of the tiny minority of the indigenous rich is the modern equivalent of the government quarter provided for colonial civil servants; in central business districts buildings of ever-increasing height are added to the existing colonial commercial symbols; but at the same time the indigenous parts of the cities continue to grow, perhaps even more, and the most disturbing expression of the post-colonial period within the great cities has become the arrival on a massive scale of the squatter hut. Clearly, the built environment alone of the colonial and post-colonial city in South East Asia is indicative of a socio-economic development at both national and local scales which differs very significantly from that of the West.

Demographic factors

Whilst it must be stressed that many of the circumstances relating to urbanization in the pre-colonial period – particularly Reid's speculation about the possibility of a high level of urbanization – must remain conjectural because of data deficiencies, the same unfortunately is also true of many aspects of the modern period. The quality of statistics in the various national censuses of recent years is of a questionable nature (Bidani 1985, p. 15). Registration of births and deaths, though improving, remains far from complete and data on population movement, including movement to and from towns and cities, is often fragmentary and extremely difficult to interpret. Writing of Thailand, Lightfoot et al. (1981, p. 98) have claimed that the basic data, although extensively used in various publications both officially and by academics, are extremely limited 'even in such apparently well-documented areas as rates of rural-urban movement'. The decennial census records for Thailand give the numbers of people living in a different location on the day of the census compared to where they were five years previously. Thus this vital source reveals nothing about short-term migration,

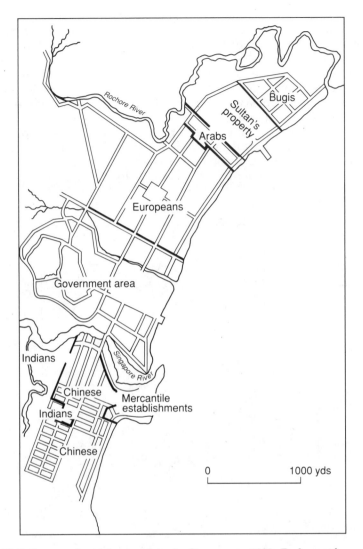

Fig. 11.3 Proposed racial groupings in Singapore, 1828. Redrawn from a plan by Lt P. Jackson, showing the recommendations of the Town Committee. (Source: Hodder 1953)

including the seasonal moves to towns and cities which are known from field studies to be an important element of migration in Thailand (Textor 1961). In Thailand, there is also a residential registration system which in theory records all changes of place of residence (Sternstein 1974): it is thought, however, that less than 10 per cent of rural–urban migrants register their moves in this manner, and practically none of those involved in temporary, seasonal movement.

These kinds of difficulties apart, analysis of the urbanization process in South East Asia is also hampered in some cases by the sheer lack of data even of the most basic kind. Vietnam, Kampuchea and Laos have had either none or only one census so far. Between 1930 and 1980, Indonesia conducted four censuses. The Philippines and Thailand have conducted censuses at irregular intervals. All in all, it is difficult to measure even the total urban population of South East Asia with any precision.

An associated problem concerns the definition of urban population within the region. Census definition of urban areas varies from country to country and it sometimes changes from one census to another in the same country. In Malaysia, for example, the

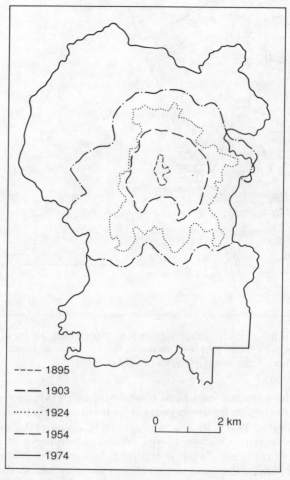

Fig. 11.4 The municipal boundaries of Kuala Lumpur, 1895–1974. (Source: Hill 1986)

Table 11.1 South East Asia: urban population

	Average annual growth rate			As a percentage of total population	Percentage of urban population in largest city	
	1965–73	1973–83	1980–5	1985	1960	1980
Indonesia	4.1	4.8	2.3	25	20	23
Thailand	4.8	3.6	3.2	18	65	69
Malaysia	3.3	3.5	4.0	38	19	27
Singapore	1.8	1.3	1.2	100	100	100
Burma	4.0	3.9	2.8	24	23	23
Philippines	4.0	3.8	3.2	39	27	30
Lao PDR	4.6	5.7	5.6	15	69	48
Vietnam	5.5	2.4	3.4	20	32	21

Note:
Kampuchea: no data; Brunei; no data.
Source: World Bank (1985, 1987).

definition of urban areas changed from that of all gazetted areas of over 1000 population in the 1957 census to all gazetted areas with population of over 10 000 in the 1970 census. Thailand has moved from a definition of places with 2000 or more households to considering all municipalities as urban in the 1960 census without specifying any minimum number of households. As Bidani (1985, p. 16) points out, the Philippines has a relatively complicated definition which takes density and other urban characteristics into consideration. In the Philippines, and also Indonesia, urban definition is thus an administrative or political decision rather than a purely statistical one. Brunei, Burma, and the Communist countries of Vietnam, Kampuchea and Laos either do not have a formal definition for urban status or have not made it known.

A city or town can grow in population in one of three ways or any combination of them. It may expand its municipal boundaries, thus taking more people into its statistical area; grow through the natural increase of its population; or receive incoming migrants. Frequently, towns and cities are underbounded, that is, urban development occurs well outside their administrative boundaries. Boundaries are frequently extended to meet this problem, for example the administrative area of Kuala Lumpur (Hill 1986, p. 24) has been extended from 21 square kilometres in 1903 to 247 square kilometres in 1974 (Fig. 11.4). Change in statistical areas is thus another factor which greatly complicates assessment and analysis. Perhaps the most contentious issue in terms of the causes of urban growth in contemporary South East Asia is, however, the extent to which rural-urban migration is contributing to it.

As Table 11.1 shows, the region has a relatively low level of urbanization. Apart from the city states of Singapore and Brunei, even by the year 2000 only the Philippines, and possibly Malaysia, will have passed the 50 per cent mark in terms of the relationship of its urban population to its total population. In 1985, the Philippines had only 39 per cent of its population urban but this was well ahead of such countries as Indonesia (25 per cent), Burma (24 per cent), and Thailand with a very low 18 per cent. There is no data of this nature available at present in respect of Kampuchea but the other two Communist countries of the region, Vietnam and Laos were recording generally low levels of urbanization also, at 20 per cent and 15 per cent respectively (World Bank 1987, p. 266). Overall in 1985, the level of urbanization for the region (excluding Singapore, Brunei and Kampuchea) was 25 per cent.

For most countries in the region, until well into the next century their greatest absolute population growth will occur in rural not urban areas given both the present relatively low level of urbanization and a possible second factor, relatively slow urban growth. It is thought, for example, that between 1950 and 1970 the proportion of urban to total population in South East Asia rose from 13.6 to 20.1 per cent, a twenty year change believed to be the smallest among the developing regions of the world, excluding South Central Asia, Oceania, and East Africa (Bidani 1985, p. 29). However, this situation may be changing, since the data for the decade 1970–80 indicate an acceleration in the pace of urbanization (Ogawa 1985, p. 92). These newer circumstances are causing attention to be focussed once again on the components of urban growth in the region.

These matters have been examined in some detail in respect of the ASEAN countries of South East Asia in a recent paper by Ogawa (1985). He concludes that in Indonesia 54.8 per cent of urban growth in the 1970s was due to internal migration and reclassification of urban areas (unfortunately it is not possible to separate the two), while in peninsular Malaysia the contribution of these components was 50.8 per cent. The figures for the Philippines and Thailand were 39.7 per cent and 61.1 per cent respectively. As a whole, his results lead Ogawa to question previously widely held views that the rate of urbanization in the ASEAN countries at least is low primarily because of the insignificant role of internal migration. He suggests that these four countries have now advanced into the first stages of high socio-economic mobility and that this is the reason for the increased pace of urbanization characterizing the decade 1970–80, an acceleration which it is probable will also characterize decades to come. The increased contribution of internal migration and reclassification to urban population growth is also attributable, Ogawa points out, to recent sharp declines of fertility in urban areas. 'This suggests,' Ogawa states (1985, p. 92), 'that the mobility transition is presently underway in parallel to the fertility transition in ASEAN society.'

It is far from clear, of course, how far these conclusions might apply to Burma and to the Communist states of Kampuchea,

Vietnam and Laos. For the Communist states at least, the situation is rather that, as in China, population mobility is deliberately restricted by bureacratic control within the established totalitarian system.

Besides the level and pace of urbanization, there are other demographic factors that significantly affect both the role of the towns and cities of South East Asia as centres of change and their practical problems of physical planning. As Pryor (1979) has observed' these largely concern the characteristics of migration and include the fact that all major migrational streams need to be assessed rather than that of rural–urban migration alone. In the case of peninsular Malaysia, it is clear that as elsewhere in the region there are substantial movements of people within both the rural and urban areas. There is a great deal of short distance migration within peninsular Malaysia and, in fact, rural to rural migration was the most important component of internal migration recorded in 1970 (Nor Laily Aziz, quoted by Lai and Tan 1985, p. 268). Almost 40 per cent of total migration was of this nature whereas only 15 per cent comprised rural-urban migration. One-third of migration was from one urban area to another and, in addition, there was a considerable return flow from urban to rural areas (Table 11.2).

Table 11.2 Internal migration by rural–urban destination, Peninsular Malaysia, 1970

Rural–rural	Rural–urban	Urban–urban	Urban–rural	Total
949 200	368 100	799 700	295 900	2 412 900
(39.3%)	(15.3%)	(33.1%)	(12.3%)	

Source: Nor Laily Aziz, quoted by Lai and Tan (1985, p. 268).

Clearly, the future demographic prospects for the cities need to be considered within this wider context of migrational movement and where there are prospects for land settlement in rural areas, as in Malaysia, potentialities for the deflection of a substantial part of migration streams away from the towns and cities can be significant. Whether there will be an expanding agricultural frontier within the region after the turn of the century, however, appears doubtful.

Other considerations of importance concern the permanence of migration and the age, sex and family circumstances of the migrants. Recent research has emphasized the importance of distinguishing between relatively permanent changes of location, properly termed 'migration', from daily, periodic, seasonal or even fairly long-term 'circulation'. The classic work of Textor (1961) analysed the circumstances of seasonal migration to Bangkok by rural dwellers from the north east of Thailand. Such circumstances of circulation may have their own particular urban employment and housing needs which are significantly different from those of true migrants. In addition, they very rarely involve whole families since usually the family remains behind in the rural area to await the return of the migrant. Such migrants are predominantly male and thus conform to the general historical pattern of rural–urban migration within the region in terms of sex. Migration to towns and cities has been male oriented but the proportion of females within migrant streams is increasing, and in Thailand and the Philippines the sex ratio of the urban population was recorded in the 1970 censuses for the first time in favour of females. The reduced mortality of females compared to males in urban areas has raised their relative numbers but the number of females in the migrant stream to urban areas in these two countries is also a significant factor (Bidani 1985, p. 26). The migrants of both sexes are predominantly young adults, 15 to 34 years of age. This again shows up in the aggregate figures for the urban areas; in the Philippines, for example, whereas 51.4 per cent of the national population was recorded in the age group 15–64 years in the 1970 population census, the comparable figure for metropolitan Manila was 58.1 per cent. The implications for urban employment prospects of this aspect of migration structure within the region are, of course, considerable. In all cases, it must be remembered that natural increase within towns and cities is in itself creating major additional difficulties in terms of strain upon urban infrastructures and growing problems of employment.

Given the growth of such problems, which will be outlined below, the question of motives for migration from the rural areas becomes both an interesting and an important one. In 1981, Lightfoot *et al.* published a study which addressed this question through questionnaire surveys in six villages in the Atsamat district, Roi-Et province, an area chosen because of its history of a high rate of out-migration, especially to Bangkok. Whilst acknowledging other work in the literature which has emphasized motives of moving to find work or to earn more money, Lightfoot *et al.* stressed the need for the analysis of broader aspects of the question such as why peasant migrants desire more money to the point of taking jobs which most of them do not like and at the expense of at least temporarily giving up their positions in rural society. Such migrants overwhelmingly come from families which do not have long histories of needing or using cash. What leads them, therefore, to wish to abandon the subsistence ethic? In Thailand, migration to urban areas is predominantly focussed upon Bangkok, which has more than forty times the population of the second largest city of the country, Chiengmai. Yet there is evidence, the authors, state, that many of the migrants to Bangkok are leaving agricultural areas where there is a shortage of unskilled labour, rather than being pushed from the land by rural over-population, and moving to a major city which, paradoxically, has both a surplus of unskilled labour and a substantial unemployment problem.

The 1970 census indicates that net out-migration to Bangkok from Roi-Et province over the previous five years was 1.4 per cent of the population. But in recent years the rate of movement between the study villages and the capital has increased. Bus services have been introduced direct from the district to Bangkok with no intermediate stops. In these circumstances it was not surprising that the younger respondents to the survey indicated a relatively high level of knowledge about the capital. Sixty-seven per cent of them knew the name and some personal details of at least one person in Bangkok. In the villages, there are thus well-established personal channels of communication through which information about the city is constantly imported.

In general, a considerable majority of villagers saw their home village as the most desirable place in which to live. The nearby cities of the North East were not considered attractive in comparison with Bangkok, a finding which has considerable implications for regional planning. Only a few saw their prospects of making a better living as being improved in the city in comparison with their village. A very high percentage also saw their village as offering a less risky life than the city, in spite of the fact that in this part of Thailand poor harvests are relatively frequent. It was largely for non-economic reasons that Bangkok was attractive to the largest number of respondents. A search for status within the village community, ultimately as a returned migrant, was clearly significant in terms of reasons for moving to the city. The North Eastern cities were held in very low regard in terms of this criterion. Bangkok was also attractive because of the entertaining lifestyle it was believed could be found there.

The general impression from the survey was that people in the villages found Bangkok attractive at least for short periods of residence but that they tended to be drawn to it by its non-economic aspects as well as its economic opportunities. Returned migrants were thought to have improved in their appearance and personalities, as well as in their economic status, because of living in the city. These were the positive features of migration which provided the mirror image to the drudgery and relative poverty of life in the villages.

The Lightfoot study underlines what Bidani (1985, p. 25) characterizes as a magnetic force, namely the presence of relatives, friends and fellow villagers in urban centres pulling new migrants from the countryside who, in turn, influence another set of new migrants. This form of chain migration is common everywhere in South East Asia. It is impelled both by the attractive force of the cities, particularly the opportunity for economic gain, and also by the various push factors from a countryside that in general is lacking both in progressive economic development and the extension of amenities at a reasonable pace. Mounting population and poor prospects are expelling people from the countryside in many parts of the region, and this is one reason why migrants continually add to already overcrowded urban labour markets. However, the operation of push factors in the rural areas cannot be seen within the context of poverty only, for there is much evidence that only a minority of rural migrants originates from the poorest group of agricultural labourers. Migration tends to be more positively correlated with an increased economic status and better education in the countryside, and a fundamental paradox inherent in rural development is that the extension of education increases the propensity to migrate. The factors involved in migration to urban areas throughout South East Asia are therefore exceedingly complex and so far research has not got very far beyond the stage of the individual empirical study. One area which clearly deserves further research investigation is what seems to be the ever-increasing capacity of the region's towns and cities to absorb migrants. This is reflected in the continuing strength of migration streams; the findings of surveys which show that very real economic gains continue to be made in the towns and cities by migrants from the countryside; and levels of urban living for the poor which appear to continue to be broadly tolerable.

Problems of absorption

Employment

There was very little modern manufacturing industry within South East Asia during the colonial period, except for the rudimentary processing of natural products. Small-scale production often existed to meet local demands but the importance of handicrafts in general declined due to the competition of mass-produced, imported goods. Essentially, South East Asian cities were cities of colonial administrators and workers in service industries. Rangoon, for example, had no less than 62 per cent of its labour force engaged in tertiary occupations in 1931 (Keyes 1977, p. 275).

Not surprisingly, during the subsequent period of national independence a great deal of emphasis has been placed in economic planning within the region upon industrial development, and in the absence of robust policies for regional development the tendency has been for industrial growth to concentrate in the largest urban places. In Malaysia, the manufacturing sector in 1957 consisted largely of the processing of agricultural products and accounted for less than 8 per cent of the gross domestic product (Lai and Tan, 1985, p. 273); by 1980, the share of the manufacturing sector had risen to 20 per cent. This kind of relatively rapid manufacturing development has been paralleled in various degrees in recent years in the other countries of South East Asia within the capitalist system. Frequently, as in Malaysia, import substitution policies have been implemented since the early 1960s: tariff barriers have been erected; tax incentives have been given to 'pioneer' industries; and factory sites have been allocated on favourable financial terms in industrial estates. In Malaysia, a Pioneer Industries Ordinance was brought into effect in 1958, allowing tax holidays ranging from two to five years for qualifying establishments. Similar legislation followed, including the Industrial Incentives Act of 1968 which allowed tax relief status to non-manufacturing establishments and extended the period of tax holiday to a maximum of ten years. As in the other capitalist-oriented countries of the region, Malaysia has actively wooed foreign industrial development as well as having tried to stimulate indigenous entrepreneurship. However, even after vigorous government action, industrial employment throughout the region has failed to keep pace with the growth of urban populations, and even Malaysia is now facing a large and widening gap between the numbers of people entering the job market and opportunities for employment (Fig. 11.5).

In these circumstances, all the towns and cities of South East Asia face extremely serious problems of absorbing their growing numbers of people in employment. Basically, there are two reasons for what has become an intractable situation in this respect. In the first place, the number of people pressing upon the economic resources of the towns and cities is both very large and growing substantially. Second, the technology used in industrial development is almost wholly imported from the West. There has been little indigenous technological development and South East Asia, in common with the other developing countries as a group, both lags far behind the West and is experiencing a continuously widening gap. The West's experience of industrialization and urbanization over at least the last one hundred years has broadly been one of shortage of urban labour. The result has been a technological development that has been economical of labour through the substitution of capital and the deployment of ever more sophisticated machinery. The point has now been reached at which whole factory floors can be controlled by one or two operatives using computers but, in one sense, this form of development is exactly the opposite to what is required in South East Asia's towns and cities. Such towns and cities are characterized by large numbers of very poor people desperate for work. At the same time, developmental capital is exceedingly scarce. What is needed, then, in terms of employment is a labour-absorptive technology rather than a labour-saving one.

Faced with this problem of labour absorption, South East Asian countries have tended to respond in three ways. First, production for export has been emphasized more recently in order to enlarge the possibilities for industrialization beyond those of the domestic market. Thus about a quarter of Malaysia's total manufacturing output is currently exported. Second, new means have been sought to make industrialization within the region still more attractive. In particular, attention has been actively directed towards multinational corporations seeking relatively cheap labour and, in parallel, free trade zones have been developed through which products can pass

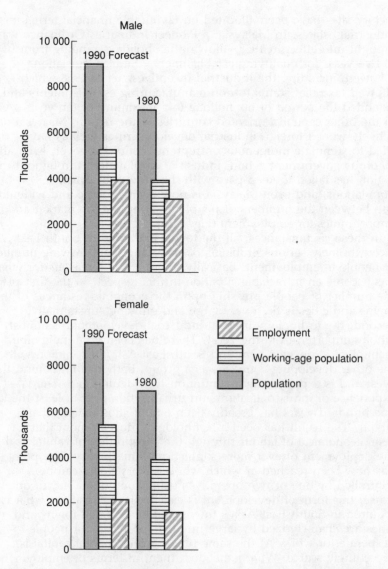

Fig. 11.5 Population and employment in Malaysia, 1980 and 1990. (Source: Government of Malaysia (1986)

for processing *en route* from one branch of a multinational corporation to another. This type of development has, for example, been very significant in the development of Penang since 1970. Among the industries attracted to the region, electronics is by far the most important, though garments and textiles have also become significant. By 1976, the number of electronics companies in

Malaysia had increased to 138 and they were employing 47 000 workers, predominantly female (Armstrong and McGee 1985, p. 207). Many of these factories were located in the Penang area.

The third way in which capitalist-oriented South East Asia countries have reacted to the growing problems of labour absorption within their urban areas has been to some extent to reconsider the whole question of protecting their industries from foreign competition. It has become much more widely recognized than previously that there are only minimal incentives to efficiency within highly protected national markets of such small dimensions that possibly only two or three manufacturing plants exist within each industrial sector. Further, there is a serious contradiction between such featherbedding and the exceptionally high degree of industrial efficiency that is required for the successful penetration of export markets. In Thailand, the Five Year National Economic and Social Development Plan, which commenced in October 1986, was designed to emphasize the increase and diversification of exports (Sricharatchanya 1986, p. 56). For the first time in Thai history, in 1985 rice lost its position as the biggest export item, to textile products. Such industrial progress had, however, by no means eliminated the employment problem; indeed, during the new plan four million additional workers were anticipated, only half of which could be absorbed within agriculture. At the start of the previous national development plan, in 1982, the government had already selected ten major protected industries for a restructuring process, the broad thrust being to reduce and unify tariffs on competing imported products in order to force the local industries to become more competitive. The results were poor, because of local industrial opposition and also because of lack of political will to implement the programme. It is now hoped to carry the programme further during the current plan but it remains to be seen whether it will be successful and, if so, whether negative effects upon industrial employment overall can be avoided.

Although there are sizeable and well-established streams of return migration from the towns and cities of the region to the countryside (Table 11.2), they are composed to a high degree of migrants returning on retirement and those involved in circulatory movements returning to the countryside on a temporary or seasonal basis. There is no evidence of migrants proper returning to the countryside in sizeable numbers as a result of a changed evaluation of the relative merits of life in the urban areas as opposed to the countryside, and this suggests that even for the poor life in the towns and cities remains broadly tolerable, including access to employment opportunities.

It is clear that in all of the urban places the urban poor are, in fact, employed in a wide variety of occupations outside of formal employment structures. It is because of this circumstance – the growing importance of what has become known as the 'informal' sector in urban employment – that the proportion of economically active population recorded in service occupations continues to be very large in South East Asian cities. This fact by no means demands the same interpretation as it would do in industrially developed countries. Indeed, it tends to convey a false picture of the prosperity of the cities because in the West a full tertiary sector is usually associated with a high level of economic development. In South East Asia the very reverse tends to be the case. Many more persons are supported in the cities than the formal economic base warrants through what is essentially a shared poverty system. The proposition here is quite simple and realistic in the demographic circumstances. It is that if a proportion of the urban population is to find the means of economic subsistence within the urban environment, a much larger number of jobs has to be created than is reconcilable with western concepts of labour absorption and productivity. Hence street sellers and clean-your-car boys abound; relatively large numbers of domestic servants are kept; and there is a multitude of office employees, particularly in government service. There are many middlemen in commerce also, each taking a small share of the profit on the movement of goods. This type of city employment may be an urban interpretation of an ethic originally created in response to purely agricultural demands, that is, an extension of the pattern of shared poverty in the countryside, where complicated renting and sub-renting patterns have developed which allow a great number of people each to claim a small proportion of the agricultural output from a single piece of land (Dwyer 1979, p. 94).

In recent years these employment circumstances, in the developing countries generally as well as in South East Asia, have been intensively studied by social scientists. In a very early contribution to this field, Geertz (1956, p. 143) drew attention to the social pressure upon entrepreneurs in the prevailing demographic circumstances to inflate employment opportunities as far as possible. Writing of a town in Java, he stated:

'One of my informants set up a cigarette factory in a shed behind his house. He began with two workers – girls – rolling cigarettes by hand in corn sheaves provided by the workers themselves. The factory grew to employ a workforce of twenty girls, the number not being determined by economic considerations but by the entrepreneur's and the girls' notions of the 'correct' number which should be employed given the work involved. The result was an extremely uneconomically operated factory. Unable to accumulate enough capital to provide sufficient tobacco to keep twenty girls working even six hours a day at full capacity, the entrepreneur merely apportioned out regulated quantities of the available tobacco to each girl each day, and the girls worked at a very slow speed, producing only 1000 cigarettes in a working day where they might easily have produced 1500–2000. The outcome was typical: twenty workers and an entrepreneur made a semi-adequate living and no one made a good one, with the added consideration in this case that this economically

298

Table 11.3 Characteristics of the two circuits of the urban economy in developing countries

	Upper circuit	**Lower circuit**
Technology	capital-intensive	labour-intensive
Organization	bureaucratic	primitive
Capital	abundant	limited
Labour	limited	abundant
Regular wages	prevalent	exceptionally
Inventories	large quantities and/ or high quality	small quantities poor quality
Prices	generally fixed	negotiable between buyer and seller (haggling)
Credit	from banks, institutional	personal, non-institutional
Profit margin	small per unit; but with large turnover, considerable in aggregate (exception = luxuries)	large per unit; but small turnover
Relations with customers	impersonal and/or on paper	direct, personalized
Fixed costs	substantial	negligible
Advertisement	necessary	none
Re-use of goods	none (waste)	frequent
Overhead capital	essential	not essential
Government aid	extensive	none or almost none
Direct dependence on foreign countries	great; externally orientated	small or none

Source: Santos (1979, p. 22).

inefficient operation reduced even further the opportunities to amass enough capital to increase output and hire more workers. As a matter of fact, the business failed after a while, and the Javanese entrepreneur fled his Chinese creditors.'

In more general terms, Santos (1979, p. 22) has conceptualized these circumstances as part of a model of what he terms the upper and the lower circuits of the urban economies of the developing countries (Table 11.3). He sees the upper circuit as comprising relatively large, formally organized and capital-intensive operations. Within this circuit, there is a tendency to limit employment. There is regular wage labour and prices are generally fixed. This is the employment sector which governments recognize and are prepared to assist in a variety of ways in their search for national development. In contrast, the lower circuit of such urban economies is highly labour intensive; is limited organizationally, frequently to individual families; is very restricted in its access to capital; does not pay regular wages; does not produce highly standardized goods by the methods of mass production; is heavily involved in providing services of various kinds as well as in small-scale production; and is usually not recognized by governments either in terms of providing official aid or even of physical accommodation through urban planning. Whilst the upper circuit may even employ some foreigners and is certainly closely linked to trans-national capital and the operations of multinational companies, the lower circuit is essentially a local creation. This is not to say that the two circuits are not linked, for linkages do occur at many points. Sometimes, for example, parts assembled in factories within the upper circuit are made in small workshops that belong more to the lower circuit; and the price of labour in the lower circuit is undoubtedly strongly influential in keeping down payments to labour in the upper circuit.

Although detailed estimates are lacking for almost all the towns and cities within the region, it appears that large proportions of the urban labour force – at least a third and possibly much more – are already engaged in informal sector activities, and that these proportions are growing steadily as urban labour forces continue to swell. A great deal of research work still remains to be accomplished in terms of analysing the multitude of small-scale economic activities, many of them informal in their organization, that characterizes South East Asian towns and cities. Significant work has already been accomplished in some respects, particularly in terms of teasing out the interconnections within individual cities in respect of the production, marketing and distribution of certain commodities (see, for example, Ebery and Forbes 1985), and in examining hawking as an important activity (McGee and Yeung 1977). Detailed field studies of hawking, for example, have revealed distinctive features in various cities. Sharp differences in length of residence of hawkers in Kuala Lumpur, Manila, and Jakarta are clear from the studies (Table 11.4) and also age-sex composition is quite different in each case (McGee and Yeung 1977, p. 98). These and other significant differences revealed by this research work indicate the

Table 11.4 Percentage of hawkers by length of residence in city

Length of residence in city	Kuala Lumpur	Manila	Jakarta
3 years and below	12	11	31
4–10 years	14	12	31
11 years plus	27	39	22
Native-born	46	34	13
Living outside	1	4	3

Source: McGee and Yeung (1977, p. 99).

dangers of generalization both about the informal sector and about urban conditions generally in the region (Dwyer 1985). The fact that such work has been a product for less than two decades also reveals how flimsy has been much of the factual basis for urban planning, particularly since such planning has tended to be very largely directed towards accommodating more formally organized economic activities within the urban fabric. As for the future, it appears that in South East Asia, as elsewhere in the developing countries, towns and cities with an employment mix that is very different from those in the West are in process of evolution. Much the same can also be said about urban residential characteristics in the region.

Infrastructure and housing

A basic fact concerning the infrastructure of South East Asian cities is that to a substantial degree what would be considered in the West as the two basic elements for urban life, a safe water supply and hygienic means of sewage disposal, are absent. In Jakarta, for instance, 40 per cent of households depend upon water vendors for their supply, and at prices five times higher than the charges for piped water (Yeung 1985, p. 50). Jakarta has no water-borne sewerage system. Only 15 per cent of its residents have access to the city's water supply. In Manila, a sewerage system which was constructed in 1909 for a population of not more than 440 000 is now the only such facility for a population that has grown to over six millions. What Carrol (1966, p. 585) wrote some years ago about the city remains true today.

'In the past twelve months Manila has had a water shortage in which some 70 per cent of the metropolitan area was without regular service . . . it has had a garbage crisis . . . there has been a school crisis, though a minor one this year . . . electric services went through a bad period some months ago . . . mail is in a constant state of crisis and in general it seems better to give up trying to use the telephone . . . police and fire protection are unreliable . . . the constantly increasing burden of traffic and the condition of the roads discourages one from venturing beyond walking distance. In other words social organization in the Manila area has not been able to maintain these services in the face of population increase and normal wear and tear on the facilities; and at times it appears we are returning to a *barrio* [village] type of existence.'

Unfortunately, such problems abound throughout the region, particularly in the smaller towns. As with employment, the extension of urban infrastructural services has not been able to keep pace with population growth. Islands of modern development in the physical sense seem in the process of becoming more and more divorced from the physical fabric of major parts of the cities within which the poor live and circulate. This has been typified most of all in recent years by the appearance of exclusive residential areas surrounded by walls and guarded at their entrances by security personnel (Dwyer 1977). The mass of the built residential environment reflects the poverty of the overwhelming majority of the cities' inhabitants. It is characterized by high residential density interdigitated with small-scale industrial and commercial activities in which official land-use zoning plays little part. Many land-use assemblages continue to be dictated by walking distance, or at least the range of non-mechanized transport. Most of these areas are seriously, or even completely, deficient in basic urban services.

All of the towns and cities of the region are facing critical housing problems. Residential building, both public and private, has been completely outstripped by population increase. One result has been the emergence of huge problems of slum living in areas within which housing has become sub-standard through age, neglect or the excessive sub-division of rooms. The inner parts of the cities, in particular, suffer from this increasing phenomenon whilst the rich live well insulated in peripheral locations. Even the extreme sub-division of existing housing units has not been able to solve the accommodation problem. A resultant phenomenon within the residential sector has therefore been the extensive growth of squatter areas (Fig. 11.6). As with so many other aspects of South East Asian urban development, reliable estimates of squatter populations are very hard to find, but Kuala Lumpur, for example, is thought to have a squatter population of over 250 000 or approximately 25 per cent of its total population. Excluding the Communist countries, the other large cities of the region are in at least as serious a position in this respect. The smaller towns are similarly affected, though to varying degrees.

The response of governments to the rapidly increasing housing problem has been weak, with the exception of Singapore. In that city state what has now become a massive housing programme was introduced in 1960 on the basis of constructing high-rise flats. Singapore has succeeded in rehousing more than half its total population, an achievement which is rivalled only by that of Hong Kong. Housing in Singapore has been designed to serve not only socio-economic but also political ends, and in the process of the programme 'a largely foot loose immigrant community of diverse ethnic stocks has been transformed into a united and increasingly bourgeois society with stabilizing roots' (Yeung 1985, p. 46). The programme has also made direct contributions to economic growth and employment generation in Singapore, a positive aspect of housing construction which the other countries of the region have been slow to recognize. Outside of Singapore, the view of the United Nations that housing' . . . is not something to be dealt with after development; it is a part of it' (quoted by Yeung 1985, p. 48) has been largely ignored. There has been a good deal of the bulldozer approach not only to the renewal of inner urban areas rather than their conservation and improvement but also to squatter

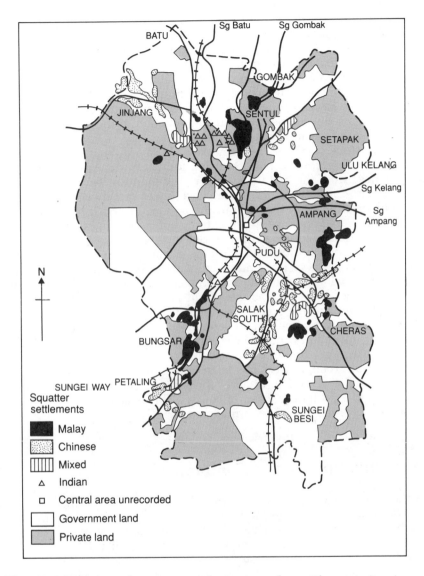

Fig. 11.6 Location of squatter settlements and resettlement sites in
Kuala Lumpur and the Federal Territory 1976. (Source:
Johnstone 1983)

areas with the aim of eliminating local problems and sometimes
even relocating squatters well outside the city (Dwyer 1975).

In Bangkok, public housing meets only about 6 per cent of total
needs, yet housing activity in Thailand is, in fact, concentrated upon

303

the capital. This is the usual situation within the region. In Malaysia two-thirds of conventional urban dwellings are built by the private sector and more than half of them are concentrated within the Kuala Lumpur conurbation. The problem is that the dwellings built are too expensive for over 90 per cent of the population (Johnstone 1983, p. 249). As elsewhere, the construction of truly low-cost housing is largely in the hands of the squatters. The same situation applies in Malaysia's smaller towns, the housing situation in them being even more difficult than in Kuala Lumpur.

As Johnstone (1983, p. 254) remarks of Malaysia. 'scarce public funds are being used to build housing for those who are more able to purchase from the private market rather than for low income groups who can only afford to purchase lower cost dwellings'. This is a universal problem throughout the region. In part it has a political explanation since, to an extent, public housing is used as a form of patronage to reward privileged insiders. More generally, though, the countries of the region have failed to get to grips with the actual situation of large numbers of the urban poor by adopting unconventional approaches to the problem. Schemes that have been advocated by bodies such as the World Bank to provide the urban poor with sites for the erection of their own housing, together with the provision of minimal urban services to such settlements, have as yet hardly moved beyond the experimental stage in South East Asia. Equally, the upgrading of existing squatter settlements through the provision of urban services has been both slow and generally inadequate (Dwyer 1975). In all of the squatter settlements serious problems of land tenure exist, of course. This is another area in which little progress has been made so far.

Urbanization under Communism

If the above paragraphs indicate that the urban record of the capitalist-aligned states in the region is generally poor, that of the Communist states can only be considered substantially worse. In the case of Kampuchea, the Communist victory in April 1975 was accompanied by the immediate forced evacuation of the capital, Phnom Penh, at a time when, because of war, the city had swollen to a population in excess of three millions. Within little over a week, Phnom Penh was almost totally devoid of civilian life. Even hospitals were emptied, and the seriously ill simply left to die by the roadside. The dreadful fate of the inhabitants of Phnom Penh subsequently became part of that which befell the population of Kampuchea in general. In addition to those who were sent to forced labour in new agricultural areas in remote districts, or placed in concentration camps, some hundreds of thousands were murdered. Since the Vietnamese invasion of Kampuchea and the fall of the then-existing Communist regime in January 1979, some life appears to have returned to Phnom Penh and its population may now be approaching 500 000.

In Vietnam itself, the Communist regime in the North adopted a policy of freezing urban growth as far as possible after 1954. The newly independent North Vietnam was predominantly an agricultural country, only 7 per cent of its population being urban and its largest city, Hanoi, having a population of about 400 000. As part of the peace negotiations, a period of months was allowed in which the population of North and South Vietnam could choose which zone to live in. As many as 900 000 people left the North at that time, including almost all of the urban entrepreneurial class. From the Communist point of view, the cities of the North were therefore already cleansed of possible dangerous elements, and there does not seem to have been any significant policy of de-urbanization pursued thereafter. Totalitarian control has ensured that population migration to the cities has been closely supervised. Such migration, together with natural increase, had resulted in an estimated population for Hanoi in excess of 2.5 millions by 1984, although approximately one million of this population resulted from the extension of the administrative boundaries of the city (Thrift and Forbes 1986, p. 89).

In contrast to the North, after the reunification of Vietnam under Communism in 1975, a plan of moving people out of the southern cities was quickly formulated. This was part of a more general plan of population relocation for the country as a whole (Thrift and Forbes 1986, p. 130). Substantial movement within the North itself was initiated, mainly away from the urban areas towards new agricultural zones in the hinterlands. There was also a programme to move large numbers of people from the northern rural provinces to the South, not only for agricultural colonization but also for strategic and political purposes. The programme for the South itself consisted largely of de-urbanization, and over the period 1976–80 perhaps one million people were moved from the cities and towns into the countryside (Thrift and Forbes 1986, p. 132). Subsequently, even larger population movements have been targeted with the aim in part of holding constant the size of cities in the South. Progress appears to have been uneven and the programme is probably running well behind schedule.

Another aspect of the attitude of the regime towards Ho Chi Mihn City and the other urban places of the South has been distrust of the urban middle class and in particular of the ethnic Chinese who are very important within it. The traditional anti-Chinese sentiments of the Vietnamese have become accentuated during the period since the reunification of the country. In part this stems from the economic position of the Chinese in the towns and cities of the South and the result has been an important Chinese element within the 700 000 refugees who had departed from Vietnam by the end of 1982, including the tragic boat people. Given its dismal economic performance, and the extremely low level of infrastructural

305

investment that appears to be taking place in its towns and cities, it is likely that urban life in Vietnam today is exceedingly harsh.

Further reading

Armstrong W and **McGee T G** (1985) *Theatres of Accumulation: Studies in Asian and Latin American Urbanization*, Methuen, London.

Bidani N D (1985) 'Demographic characteristics of the urban population in Southeast Asia', in Krausse, G. (ed.), *Urban Society in South East Asia*, Asian Research Service, Hong Kong.

Carrol J J (1966) 'Philippine social organization and national development', *Philippine Studies* **14**, 573–90.

Coedes G (1963) *Ankor: An Introduction*, Oxford University Press, Hong Kong.

Dwyer D J (1968) 'The city in the developing world and the example of South East Asia', *Geography* **35**, 353–69.

Dwyer D J (1972) (ed.). *The City As A Centre of Change in Asia*, Hong Kong University Press, Hong Kong.

Dwyer D J (1975) *People and Housing in Third World Cities*, Longman, London.

Dwyer D J (1977) 'Economic development: development for whom?', *Geography* **62**, 325–34.

Dwyer D J (1979) 'Urban geography and the urban future', Geography **64**, 86–95.

Dwyer D J (1985) 'Urban growth, form and density: The transactional city and its counterparts in the Third World', in Hills P., (ed.), *State Policy, Urbanization and the Development Process*, Centre for Urban Studies, Hong Kong, pp. 86–99.

Ebery M G and **Forbes K** (1985) 'The "informal sector" in the Indonesian city: a review and case study', in Krausse, G. H. (ed.), *Urban Society in Southeast Asia*, Asian Research Service, Hong Kong, pp. 153–70.

Geertz C (1956) 'Religious belief and economic behaviour in a central Javanese town: some preliminary considerations', *Economic Development and Cultural Change*, **4**, 134–58.

Government of Malaysia (1986) *Fifth Malaysia Plan*, Kuala Lumpur.

Hill R D (1986) 'Land use change on the urban fringe', *Nature and Resources* **22**, 24–33.

Hodder B H (1953) 'Racial Groupings in Singapore', *Malayan Journal of Tropical Geography* **1**, 25–36

Johnstone M (1983) 'Housing policy and the urban poor in peninsular Malaysia', *Third World Planning Review* **5**, 249–71.

Keyes C F (1977) *The Golden Peninsula*, Macmillan, New York.

Lai Y W and **Tan S E** (1985) 'Internal migration and economic development in Malaysia', in Hauser, P. M. Suits, D. S. and Ogawa, N. (eds.), *Urbanization and Migration in ASEAN Development*, University of Hawaii Press, Honolulu, pp. 263–89.

Leinbach T R and **Ulack R** (1983) 'Cities of Southeast Asia', in Brunn, S. D. and Williams J. F. *Cities of the World*, Harper and Row, New York.

Lee B T (1976) 'Patterns of urban residential segregation: the case of Kuala Lumpur', *Journal of Tropical Geography* **43**, 41–8.

Lightfoot P, Fuller T and **Kamnuansilpa P** (1981) 'Impact and image of the city in the Northeast Thai Countryside', *Cultures et Developement* **13**, 97–122.

McGee T G and **Yeung Y M** (1977) *Hawkers in Southeast Asian Cities: Planning for the Bazaar Economy*, International Development Research Centre, Ottawa.

Ogawa N (1985) 'Urbanization and internal migration in selected ASEAN countries: trends and prospects', in Hauser, P. M., Suits, D. S. and Ogawa, N. (eds.), *Urbanization and Migration in ASEAN Development*, University of Hawaii Press, Honolulu, pp. 83–107.

Pryor R (1979) 'Migration patterns and housing needs in South-East Asia', in Murison, H. S. and Lea, J. P., *Housing in Third World Countries*, Macmillan, London, pp. 3–12.

Reid A (1980) 'The structure of cities in Southeast Asia, fifteenth to seventeenth centuries', *Journal of South East Asian Studies* **9**, 235–50.

Santos M (1979) *The Shared Space*, Methuen, London.

Schurz W L (1939) *The Manila Galleon*, Dutton, New York.

Sricharatchanya P (1986) 'A shift to the city', *Far Eastern Economic Review* 24 July.

Sternstein L (1974) 'Seasonality of migration to Bangkok', *Pacific Viewpoint*, **15**, 75–80.

Textor R (1961) *From Peasant to Pedicab Driver*, Yale University: South East Asian Studies Cultural Report Series, 9.

Thrift N and **Forbes D** (1986) *The Price of War: Urbanization in Vietnam 1954–85*, Croom Helm, London.

World Bank (1985) *World Development Report 1985*, Oxford University Press, New York.

World Bank (1987) *World Development Report 1987*, Oxford University Press, New York.

Yeung Y M (1985) 'The housing problem in urbanizing Southeast Asia', in Krausse, G. H. (ed.), *Urban Society in South East Asia*, Asian Research Service. Hong Kong. pp. 43–66.

Index

Africa, 38, 50, 59
Agglomeration, 15
   see also Core
Agrarian land law, 36
   see also Agriculture
Agriculture, 2, 17, 39, 41, 55, 132, 134, 145,
   156, 181, 199, 234, 248, 297
   aid for, 148
   clearance for 10, 99, 100–2, 116–17,
     150
   commercial, 30, 31
   development of, 27, 32, 35
   fuels from, 150, 153–4
   growth in, 168
   milling, 201
   Sawah, 170, 173–8, 181
   systems of, 170–6
   and Fisheries *See* Chap 7
   fishing, 18, 186–90
   see also Aquaculture, Conservation,
     Cultivation, Deforestation, Forestry
Agrochemicals, use of 102–3
Aid, 9, 65, 67, 247–52
   bilateral, 50
   multilateral, 50
   foreign, 6, 65, 246, 248
   military, 65
Alternative energy *see under* Energy
America *see* United States of America
Angkor, 282
Anglo-French agreement, 41
Annamite (Vietnamese), 20
Anti-Chinese activity, 12
Anti-colonialism, 27
Aquaculture, 88, 93, 96, 189, 190
   see also Agriculture, Cultivation
Arabs, 22
Asean, 6, 8, 9, 10, 194, 195, 198, 199, 206,
   208, 209, 211, 218, 221, 226, 232, 236,
   237, 244–7, 253, 290

Asia, 59
   Asia Eastern, 10
   Asia Minor, 2
   Asia monsoon, 1
Asian Development Bank, 195, 211
Australasia, 2
Australia, 2, 225, 233, 242

Bali, 105, 174
Balkan bridge, 2
Bandung, 12
Bangkok, 18, 19, 41, 42, 60, 61, 106, 145,
   146, 148, 218, 244, 258, 262, 267, 271,
   273, 283, 291, 292, 303
Banjarese, 261
Barrio, 261
Basic needs theory, 53, 66–7, 74
   see also Development theory
Batak people, 263, 265
Beijing, 8
Biomass *see* Fuels
Birth control, 13
Birth rate, 13
   see also Fertility Population
Boat People, 305
   see also Refugees
Borneo, 37, 81, 118, 171
   climate, 84–5
Britain *see* United Kingdom
British East India Company, 204
Brunei, 6, 10, 11, 44, 231, 240–1
   agriculture, 132
   economy, 231, 232
   energy exports, 141, 142
   fertility levels, 129
   fuel supply, 151, 157, 158
   industry, 201, 116, 138; recession in
     141
   population, 114, 289
   trade, 234, 236, 237, 242, 244

urbanization, 289, 290
Buddhism, 18
Buginese, 261
Burma, 3, 4, 5–6, 18, 31–4, 38, 41, 42, 44, 79, 82, 232
    agriculture, 96, 171, 174, 176
    aid, 248
    climate, 84, 87
    colonial development, 283, 284
    conflict, 125
    debt, 220
    decolonization, 124, 125
    economic policies, 195, 208, 231, 240–1
    energy consumption, 141
    history, 281
    immigration, 123
    industry, 194, 195, 199, 201, 226
    mortality rates, 126, 127
    population, 110, 111, 116, 118, 121, 135, 257, 289
    trade, 242
    vegetation, 94

Cambodia see Kampuchea
Cam Ranh Bay, 8
Capital
    accumulation, 15
    external, 27, 30
    intensive sector, 35, 299
    investment, 44, 220
    domestic, 65
Capitalism, 54, 69, 70, 71, 72, 186
Capitalist mode of production, 69, 70–1, 73
Central business district, 285, 286
Centrality, 15
    see also Core
Ceylon, 30
Chiengmai, 292
China, 1, 2, 4, 8, 9, 10, 17, 18, 20, 21, 22, 23, 24, 28, 32, 50, 52, 53, 65, 66, 111, 121, 225, 283, 291
    migration from, 122–3, 281
    population, 135
Chinese, 3, 11, 12, 42, 182, 187, 189
Cholon, 39
Cloves see Spices
Cochin-China, 130
Coconuts, 27, 37, 179, 184, 185, 186, 236, 245
    see also Plantation economy
Coffee see Cultivation
Colonialism, 1, 17, 15–46, 48, 117, 118, 124, 126, 193, 282–3
    British, 32
    exploitation, 39
    spontaneous, 42
    effects of, 122, 181
    internal, 17, 134
    indirect, 50
    Japanese, 9–10

see also Decolonization, Neo-colonialism
Commercial Development, 33, 34, 193
    see also Industrialization
Communism, 3, 5, 68
Conflict, 3, 41, 44, 52, 58, 185, 188, 189, 234
    racial, 11, 33, 52, 305
    class, 68; analysis 72
    internal, 125
Confrontation, 3 see also Conflict
Conservation, 33, 95, 96, 106–7
    see also De-forestation, Forestry, Pollution
Consumerism, 55, 199
Core, 16, 18–43
    capital, 74
    community, 15
    convergence, 16
    development, 18–19
    dispersal, 46
    economies, 193–4
    European impact on, 23–4
    status, 15
    intercore-warfare, 20
Cotton see Cultivation
Craft industry, 21, 55, 132, 237, 294
    see also Informal sector
Cuba, 66
Cultivation, 27–8, 34, 35, 36, 37, 39, 42, 87, 126, 170, 171, 173, 180, 181, 182
    shifting, 17, 21, 26, 31, 36, 96, 107, 170–3
    High Yielding Variety (HYV) crops, 174, 176, 181
    see also Agriculture, Conservation, De-forestation, Forestry, Plantation economy, Pollution
Culture system (Corvee), 4, 15, 35, 36, 41, 118–22, 281;
    change, 17
    see also Integration

DARG xii
Debt, 40, 169, 173, 204, 264
Decolonization, 43–6, 48, 50, 124–5, 181–2
Deforestation, 92, 93, 95, 96, 99–100, 102, 104, 150, 154–6, 162, 172–3
    see also Agriculture, Conservation, Cultivation, Forestry, Pollution
Dependency, 27, 38, 40, 45, 52
    ratio, 132
    theory, 63–6, 72, 225
    see also Development theory
De-urbanization, 132, 305
    see also Development, Industrialization, Resettlement, Urbanization
Development, 52, 225, 226–9
    capitalist, 51
    concepts of see Development theory

geography input, 52, 56, 65, 68
global, 52
goals, 3
  of infrastructure, 24, 27
  planning, 51–54
  problems, *xii*, 44, 48
  structures, 16–17, 145
  uneven, 266
  see also Underdevelopment,
    Industrialization, Urbanization
Development theory, 15–16, 51, 59–66,
  68–9, 68–74 and Chapter 3
  linear, 52, 54–8, 63, 65, 66, 74
  marginalization, 53, 59, 60, 125
  Marxist, 68
  neo-classical, 52
  neo-Marxist, 51, 52, 68, 69–72, 73
  World Systems Theory, 63, 65, 66, 69,
    72–4
Division of labour, 71, 204, 227
  see also NIDL
Djambi (Malaysia), 21
Double-cropping, 40
  see also Agriculture, Cultivation
Dualism, 17, 33, 36, 37, 53, 59–63, 71
Dutch, 28, 282, 283
Dutch East India Company, 23, 34

East Indies, 123
East Timor, 44
Economic centrality, 15, 246
Economic decline, 4, 134, 146
  see also World economic recession
Economic enterprise, 63
  see also Cores
Economic growth, 4, 16, 27, 30, 36, 55, 58,
  72, 122–3, 140, 142, 194, 195, 221, 225,
  227, 247, 252–3, 273, 279, 280
  see also Development,
    Industrialization, Urbanization
Economic imperialism, 17
  see also Colonialism
Economic policies, 278–9
  see also Economic centrality, Economic
    growth
Education, 134–5
Employment, 27, 292, 294–301
  unemployment, 57, 292
  see also Under-employment
Energy: resources, 93, 163, 164 and
  Chapter 6
  aid for, 248
  alternative, 96–7, 106, 161–4
  co-efficient, 140
  consumption, 142–8
  crisis, 140
  domestic use, 142, 144, 146
  human, 142–3
  logging, 98, 150, 106
  see also Fuels

English East India Company, 23, 28
Environmental planning, 102–6, 197
Environmental resources, Chapter 4
  see also Agriculture, Cultivation,
    Conservation, Energy
Equality, 68
Ethnicity, 3, 5, 118, 122–9
  see also Conflict, Culture systems
Europe, 2, 28, 57, 141, 193, 195, 198, 217,
  220, 285
Exploitation, 59, 64, 65, 71, 204
  see also Colonialism
Export processing zone (EPZ), 214, 217,
  218, 219
Export substitution, 104, 140, 206, 209,
  211, 219, 230, 232, 237, 238, 243
  diversity of 236–40
  see also Trade

Family planning, 131, 136, 252
  see also Fertility, Population
Famine, 168
Federal Land Development Agency
  (FELDA), 182, 183, 270, 271, 272
Fertility, 87, 129–31
  decline in, 130–1, 136–7, 290
  government programme, 131
  see also Family planning, Population
Feudalism, 26
Fiji, 248
Finance, 42
Fisheries *see* Agriculture
Food production, 40, 146, 147, 199
  see also Agriculture, Cultivation,
    Plantation, Economy
Forestry, 33, 93–102, 156–7, 211–13, 237
  clearance, 29, 153
  re-aforestation, 106
  see also Agriculture, Conservation,
    De-forestation, Trade
Formal sector, 60, 61, 62, 216
  see also Economic growth,
    Industrialization, Informal sector,
    Self-help
France, 23, 39, 242
Free market economies, 231
Free trade, 54, 227
  Zones, 295
  see also Trade
Fuel, 141, 142, 144–5, 147, 148
  non commercial, 140, 142, 146
  biomass, 144–5, 149–157, 164
  collection of, 150
  crisis, 151, 154
  fossil, 157–61, 164
  gasohol, 157
  sugar cane, 157
  LPS, 148
  water, 92–3
  wood, 172

Geopolitics, 6
Global division of labour, 74 *see also* NIDL
Global economy, 220
Global production, 74
    GNP, 4–5, 56, 58, 140, 146, 168, 186,
    231
Green revolution, 170, 173, 174, 176, 186
    see also Conservation, Pollution
Guam, 7

Handicrafts *see* Craft industry, Industrial
    development, Informal sector
Hanoi, 6, 39, 130, 232, 305
Hawking, 61, 300
    see also Informal sector
Health, 14, 40, 126, 127, 128, 129, 169, 185
Holland, 23, 38
    see also Dutch
Hong Kong, 27, 65, 195, 206, 217, 218, 230,
    242, 248, 302
Housing, 301–4
    see also Squatters
Human resources, Chapter 5
    see also Energy resources

IMF, 186
Immigration, 3, 11, 12, 29, 33, 35, 40, 42,
    48, 122–3, 126, 170, 261, 263, 269, 279
Imperialism *see* Colonialism
    ecological, 271
Import substitution, 42, 54, 56, 71, 106,
    209, 210–11, 228, 230, 232, 252, 294
India, 1, 2, 11, 18, 21, 22, 28, 29, 33, 68,
    111, 121, 123–4, 135, 211
Indo-China, 3, 4, 8, 96, 231, 283
    French, 38–40, 130
Indonesia, 1, 3, 4, 6, 9, 12, 17, 34–8, 48, 60,
    62, 63, 78, 82, 94, 96, 99, 106, 132, 134,
    182, 226
    agriculture, 169, 172, 177, 179, 184, 187
    aid, 248, 252
    debt, 220
    economy, 34, 231–2, 234, 236, 240,
        242, 244, 247
    employment, 196–8
    energy resources, 141, 142, 145, 146,
        147, 149, 151, 154, 157, 163, 164
    health, 169
    industrialization, 194, 195, 198, 199,
        201, 208–9, 218
    investment, 206
    population, 115, 116, 118, 126, 129–30,
        131, 135, 136, 257, 259, 260, 270,
        271, 272, 285, 288, 290
Industrial development, 37, 43, 50–1, 52,
    54, 56, 63, 67, 68, 193, 194, 195–202,
    202–16, 219–21, 228
    decline, 41–42
Inequality, 16, 24, 26, 28, 31, 36, 43, 50, 55,
    56, 63, 66, 67, 176, 184, 229, 253

Infant mortality, 6
    see also Health
Informal sector, 37, 61–2, 132, 179, 216,
    297, 300
    see also Self-help
Infrastructure, 30, 32, 35, 39–40, 41, 43, 45,
    55, 61, 140, 248, 251, 271, 292, 301
Instability, political, 6, 8–9, 11–12, 14, 42,
    50, 54, 124–6, 178
Integration: Regional, 6–10, 17, 44, 52, 56,
    57, 58, 74, 221, 226, 230–1, 244–7, 253,
    299
    see also Culture system
International division of labour
    see NIDL
International trade *see* Trade
Investment, 24, 27, 30, 33, 36, 45, 55, 74,
    134, 204, 230, 238
    Japanese, 9
    Foreign, 51, 205–6, 232, 237
Involution, 176–7
Iran, 53
    see also Ethnicity, Immigration, Islam
Irrigation, 32, 35, 42, 67, 101, 117, 118,
    173–4, 190
Islam, 10–12, 18, 22, 28, 53, 185, 281

Jakarta, 12, 146, 260, 261, 282, 300, 301
Japan, 8, 9, 48, 187, 195, 206, 214, 217, 225,
    232, 242, 243, 246, 251
Java, 2, 12, 13, 22, 23, 28, 34, 35, 36, 37,
    38, 80, 82, 84–5, 92, 94, 101, 298
    agriculture, 169, 174, 177, 181, 189
    energy resources, 147, 149, 153, 154,
        156, 160, 163
    population, 115, 118, 126, 130, 169,
        259, 270, 271, 272, 282

Kampongs, 285–6
    see also Malaysia
Kampuchea, 4, 6, 8, 9, 20, 38, 40, 41, 84
    agriculture, 168, 169, 179, 189
    education, 135
    economy, 208, 226, 231, 232, 242
    energy resources, 141, 142, 148, 151
    industrialization, 194, 195, 281, 290,
        304
    population, 110, 114, 132, 135, 233,
        288, 289, 290, 304
Khmer Rouge, 9, 121
Klong Toey, 60–1
Korea, 50, 218, 248
    south, 65, 66, 135, 195, 206, 230
Kuala Lumpur, 29, 31, 106, 218, 259, 285,
    289, 300, 302, 304

Labour, 73, 134–5, 204, 300
    cheap, 61, 71, 211
    child, 132
    elite, 62

female, 71, 206, 214, 219
forced, 304
intensive industries, 35, 56, 300
law, 211, 214
urban/rural, 131–4
see also NIDL
Laos, 4, 40, 41, 111, 114, 132, 135, 141,
142, 148, 151, 169, 172, 194, 195, 208,
226, 231, 232, 234, 242, 248, 270, 288,
289, 290, 291
Latin America, 50, 59, 63, 211
Linear theory *see* Development theory
Logging *see* Energy resources

Malacca, 2, 22, 28, 29, 30, 34, 92, 282
Malacca straits, 18, 21, 22, 28, 34, 78, 173,
179, 188, 189, 204
Malaysia, 3, 4, 6, 8–11, 14, 24, 28–31, 41,
52, 56, 62, 64–5, 70, 79, 81–2, 84–5, 92,
94, 99–103, 124, 182, 283, 285, 304
agriculture, 168–9, 174, 176, 179, 184,
186, 189
aid, 248
economy, 125, 231, 252, 294–5
education, 135
energy resources, 141–2, 144, 146,
148, 153, 157–8
industrialization, 194–8, 201, 206,
208–9, 214, 218, 226, 229, 232, 290
pollution, 104–6
population, 111, 116–8, 124, 126–9,
131, 135, 257–8, 268, 270–1, 288–9,
290–1
trade, 234, 236, 242, 244, 247
Manila, 24, 27, 58, 146, 148, 218, 260, 261,
267, 272, 282, 283, 286, 292, 300, 301
Manila core, 24–8, 31
Manufacturing, 24, 40, 41, 42, 50, 55, 56,
71, 125, 144, 146–7, 172, 193–4, 198,
200–1, 203, 205–6, 209–10, 211, 214,
216–17, 219, 225, 236, 237, 238, 295–7
and Chapter 8
Marginalization theory, *see* Development
theory
Marxist theory *see* Development theory
Mexico, 24
Middle-East, 50, 256, 262
Migration, 2, 15–18, 27, 37, 50–2, 59, 60–2,
96, 103, 118–24, 130, 132, 170, 176,
158–9, 263–5, 268–9, 271–4, 280–1, 284,
286, 291–4, 297, 305
in–migration, 29, 51–2, 260, 290–1
out–migration, 259, 260, 266, 268, 273
rural–rural, 263–4
rural–urban, 103, 262, 264, 267, 272–3,
285–6, 289, 292
urban–rural, 258–9
short–distance, 260–2
temporary, 261, 264
data on, 257–8

government intervention in, 269,
273–4
Mining, 24, 30, 41, 42, 43, 82, 104, 141,
144, 146, 147, 157, 160–161, 181, 189,
201, 232, 237, 263, 283
Missionary activity, 18, 38
see also Colonialism
Morbility *see* Immigration, Migration,
Population
Modernization, 16, 36, 41, 43, 52, 54, 64,
132
Molucca islands, 22, 28
Multi-national corporations, 58, 64, 71, 74,
134, 203, 204–5, 206, 208, 210, 216, 217,
218, 220–1, 230, 233, 238, 295–6
see also Investment

Nationalism, 12, 39, 40, 44, 54, 124–5, 134,
181
Neo-classical theory *see* Development
theory
Neo-colonialism, 45, 72
Neo-marxist theory *see* Development
theory
NIC, 65–6, 141, 195, 198, 218, 220, 226,
221, 230
NIDL, 71, 72, 193, 194, 202, 204, 217
North Vietnam, 305

ODA, 248, 252
Oil, 10, 12, 37, 44, 48, 82, 105, 141, 144–7,
157, 160, 189, 210, 232, 237, 250
palm, 2, 37, 95–6, 104–5, 179, 180–5,
237, 245, 271
seed, 32
see also Cultivation, Manufacturing,
Mining
Oil palm *see* Oil

Pakistan, 135
Papua New Guinea (PNG), 17, 37, 248
Pedicabs *see* Trishaws
Penang, 28, 29, 30, 31, 70, 84, 105, 214,
218, 296
Periphery, 15–18, 33–4, 37–8, 40–1, 43, 46,
52, 63–4, 68, 71, 73–4, 193, 225, 229, 281
see also Core, Semi-periphery
Petroleum *see* Oil
Philippines, 3, 4, 6, 10, 11, 12, 24, 26–7,
50, 78, 81–2, 85, 96, 102–3, 170, 282, 283,
292
agriculture, 154, 169, 173, 176, 179,
181, 185–8
aid, 248, 252
debt, 186, 220, 251
economy, 231, 234, 236, 242, 244, 246
education, 135
energy resources, 141–2, 144, 146–8,
150–1, 156, 161–3

industrialization, 194–5, 196–9, 201,
206, 208–9, 214, 218, 226, 232, 290
population, 110, 122, 126–7, 129, 132,
134–5, 257, 263, 289
Plantation economy, 36–7, 43, 170, 178–86,
204
Political change, 1, 15, 46, 48
Political control, 1
Political core, 21
Political divisions, 50
Pollution, 93, 102, 103–5, 107
Population, 1–3, 13, 110–16, 125, 260,
294–304, 305, Chapter 10
absorption, 209, *see also* Chapter 7
density, 32, 257–62, 278
growth, 4, 12–13, 20, 26, 28, 37, 40,
42, 43, 55, 103, 111–15, 118, 122–3,
125–8, 132, 133, 136, 141, 171, 173,
237, 263, 282, 289, 290–1, 295
movement, 52, 115–18, 256–9, 261–6,
266–9, 269–73, 286
statistics, 286–7
see also Immigration, Migration
Portugal, 23, 28
Poverty, 28, 37, 40, 59, 60, 168–70, 264,
267, 294, 297
Primary production, 6, 50, 68, 193, 236,
237, 243

Raffles, Sir Stamford, 28, 29, 204, 283
Rangoon, 5, 32, 33, 283
Refugees, 8, 12, 270, 305
Remittances, 261, 262, 266, 268, 269
Republic of the Phillippines *see* Philippines
Resettlement, 22, 23, 151, 183, 259, 270–1
Rice, 26, 32, 39, 42, 101–2, 116, 201, 237,
241–2, 283, 297
wet, 26, 27, 35, 117, 132, 173, 177–8
see also Agriculture, Cultivation,
Rubber
Riots *see* Conflict, Instability
Rubber, 2, 30–31, 35, 37, 39, 40, 41, 42, 50,
104, 157, 172, 179, 180, 183, 184, 201,
237, 245, 263, 271, 283
see also Agriculture, Cultivation,
Plantation economy, Rice, Sugar,
Tobacco
Rural development, 248, 266
Russia *see* USSR

Sabah, 105
Saigon, 38, 39, 283
Sector theory *see* Dualism
Self-help, 61–2, 63, 271, 304
Semi-periphery, 225
Singapore, 1–3, 4, 6, 10–12, 28–29, 31, 44,
64, 65, 84, 105–6, 134, 135, 211, 283, 289,
302
agriculture, 179, 186, 189

aid, 248
economy, 146, 230–3, 237–8, 240, 244,
246–7, 252, 253, 285
energy resources, 140–2, 146
industrialization, 194–5, 198–201, 206,
208–9, 214, 218–19, 226
population, 111, 114, 125, 129, 131,
132, 135, 257, 261
Soil erosion, 86–7, 88–91, 95, 102, 107, 173
Spain, 24, 26, 283
Spices, 22, 24, 28, 30, 35, 179, 281
see also Agriculture, Cultivation
Squatter settlements, 42, 59, 60–2, 96, 103,
173, 272, 286, 302
Straits settlements, 29–30, 31, 122
Sugar, 27, 35, 36, 42, 179, 186, 236, 201,
236, 242, 283
see also Agriculture, Cultivation,
Plantation economy, Rice, Rubber,
Tobacco
Sumatra, 36, 37, 82, 84–5, 92, 94, 96, 153,
163, 171, 172, 173, 178, 260, 261, 267, 281

Taiwan, 10, 65, 66, 195, 206, 218, 230, 248
Thailand, 1, 3, 4, 6, 8, 10–11, 40–42, 50, 52,
60, 81, 84, 85, 87, 92, 96, 99, 101, 105, 303
agriculture, 103, 168–9, 172–4, 176,
181, 184–5, 187
aid, 220, 248, 252
economy, 226, 229, 231–2, 234, 236–7,
242, 244, 247, 252, 297
energy resources 141–2, 144–6, 148,
150–1, 153, 154, 156–7, 161–2
industrialization 67, 193–5, 196–8, 199,
210, 206, 208, 209, 218
population, 116, 118, 121, 123, 125,
127, 129, 131, 134–5, 257–259, 261,
262, 268, 271, 273, 281, 286, 290–2
see also Klong Toey
Third World, 10, 48, 53, 54, 58–60, 63–4,
65, 67–9, 70, 71, 73, 142, 164, 168, 184,
189, 208, 219–20, 244
Timber *see* Forestry
Tobacco, 24, 26, 27, 35, 37, 201, 283
see also Agriculture, Cultivation,
Plantation economy, Rice, Rubber,
Sugar
Trade, 2, 21, 23, 24, 27, 29, 33, 39, 41, 42,
54, 225, 226–9, 231–44, 247, 253, 280–2,
283 and Chapter 9
restrictions, 210, 230, 297
policies for, 229–31, 241
Transfer of value, 71
Transmigration, 117, 172, 178
see also Immigration, Migration,
Population
Transnational corporations *see*
Multinational corportations
Trishaws, 62, 70, 71, 268, 272

Underdevelopment, 16, 17, 38, 54, 56, 59, 62, 63
  see also Development
Unemployment, 58, 134
  see also Underemployment, 64, 185, 219
United Kingdom, 38, 198, 233
United States of America, 6, 9, 27, 57, 65, 73, 163, 185, 193, 195, 198, 214, 217, 220, 225, 233, 242, 251, 262, 285
Urbanization, 45, 55, 278, 279, 280–6, 286–94, 301, 302, 304–6, Chapter 11
USSR, 6, 8, 9, 225, 242, 262

Value added, 198, 237, 253
Vegetation, 93, 94–5
Vietnam, 2, 3, 4, 5, 6, 7, 8, 9, 10, 40, 48, 50, 262, 305–6
  agriculture, 169, 172, 176, 179

aid, 248
economy, 195, 208, 231, 232, 234, 236, 242
energy resources, 141–2, 144, 148, 153, 154, 157, 161
population, 110, 111, 114, 116, 121–2, 132, 135, 270–1, 281, 288, 289, 290, 291

World Bank, 63, 67, 145, 163, 183, 186, 230, 231, 232, 234, 248–51, 252, 271, 290, 304
World capitalist economy, 17, 63, 73, 74
World economical system, 4, 36, 50, 51, 73, 208, 220
World recession, 4, 12, 31, 73, 130, 203–4, 238, 252, 253
World systems theory
  see Development theory